HOW THE NEW WORLD BECAME OLD

PRINCETON MODERN KNOWLEDGE

Michael D. Gordin, Princeton University, Series Editor

For a list of titles in the series, go to https://press.princeton.edu/series/princeton-modern-knowledge

How the New World Became Old

THE DEEP TIME REVOLUTION IN AMERICA

CAROLINE WINTERER

PRINCETON UNIVERSITY PRESS
PRINCETON & OXFORD

Copyright © 2024 by Princeton University Press

Princeton University Press is committed to the protection of copyright and the intellectual property our authors entrust to us. Copyright promotes the progress and integrity of knowledge created by humans. Thank you for supporting free speech and the global exchange of ideas by purchasing an authorized edition of this book. If you wish to reproduce or distribute any part of it in any form, please obtain permission.

Requests for permission to reproduce material from this work should be sent to permissions@press.princeton.edu

Published by Princeton University Press
41 William Street, Princeton, New Jersey 08540
99 Banbury Road, Oxford OX2 6JX

press.princeton.edu

All Rights Reserved

ISBN 978-0-691-19967-2
ISBN (e-book) 978-0-691-26545-2

British Library Cataloging-in-Publication Data is available

Editorial: Eric Crahan and Rebecca Binnie
Production Editorial: Jill Harris
Jacket Design: Chris Ferrante
Production: Danielle Amatucci
Publicity: Alyssa Sanford and Carmen Jimenez
Copyeditor: Molan Goldstein

Jacket image: *Panorama from Point Sublime* by Clarence E. Dutton, illustration, 1882. Courtesy of the Library of Congress, Geography and Map Division.

This book has been composed in Arno

Printed in the United States of America

10 9 8 7 6 5 4 3 2 1

CONTENTS

Abbreviations vii

Introduction	1
1 Why the New World Was New	14
2 Beginnings	30
3 Fossil Futures	61
4 The Oldest South	92
5 Mammals, the First Americans	121
6 Glacial Progress	145
7 The Dinosaurs Go to College	180
8 The Caveman within Us	219
9 Pterodactyls in Eden	259
Epilogue	271

Acknowledgments 275
Notes 279
Bibliography 297
Illustration Credits 331
Index 341

ABBREVIATIONS

ANSP   Academy of Natural Sciences of Philadelphia (since 2011, Academy of Natural Sciences of Drexel University)
EDCP   Edward Drinker Cope Papers
EOWHP   Edward and Orra White Hitchcock Papers
LAC   Louis Agassiz Correspondence and Other Papers
LSGM   Letterbook of Samuel George Morton
PAE   Papers of Amos Eaton
PBS   Papers of Benjamin Silliman
PBSB   Papers of Benjamin Smith Barton
PGWF   Papers of George William Featherstonhaugh
PJT   Papers of John Torrey
PLFW   Papers of Lester Frank Ward
PRN   Papers of Rush Nutt
PSGM   Papers of Samuel George Morton
PSH   Papers of Samuel Hildreth
PTAC   Papers of Timothy Abbott Conrad

HOW THE NEW WORLD BECAME OLD

Introduction

AS WE PEDAL our mountain bikes through the scorching Utah afternoon, we finally come to what we're looking for. Dinosaur tracks. Each one is about the size of one of my bike's wheels, a rock dimple big enough for me to sit in. Which I do, plopping myself down to pose for a picture. I'm sitting in a dinosaur footprint, the exact spot where millions of years ago a giant sauropod wandered along on a mysterious errand. Its feet sank into the mud on just the right day, at just the right time, such that the prints endured for eons. Other tracks, of hiking boots and bicycle tires, tell us we are not alone. We've all come to look for America, the ancient America that lies beneath our feet, the remains of a lost world that came long before ours.

For many citizens of the United States, the antiquity of the land is a core feature of their national identity. National parks are cathedrals of primordial American nature: Yellowstone, the Grand Canyon, Dinosaur National Monument. Among the oldest exposed rocks on the planet are those of the so-called Canadian Shield, which cover much of eastern Canada and stretch into the northernmost regions of the United States. They are over four billion years old. And you can walk right on them, your soles touching the basement of time. People come to the New World to see the oldest world of all.

This book tells the story of how and why this view of ancient America came to be. It shows how in the span of a mere century many Americans changed their minds about the age of the Earth and the continent they inhabited. In the late eighteenth century, many people thought our planet was only about six thousand years old, a number derived by calculating the years and generations relayed in the Bible. They also believed that the Americas were the last lands to emerge from the Earth-drowning flood described in Genesis. Barely a century later, a new possibility seemed far more plausible: not just that Earth was billions of years old, but that some of the oldest lands on the planet could be seen in the United States. This idea seemed obvious to many Americans by the early twentieth century.

Today we call this idea *deep time*. Deep time refers to the billions of years over which the planet was born and life upon it appeared and evolved. The term was coined in 1981 by the writer John McPhee.[1] It exploits two meanings of *deep*. The first one is literal. Old rocks usually lie buried far underground, so it makes sense to call the time in which they were formed *deep*.

But the idea of such an enormous amount of time is also profound. It challenges our minds to the point of paralysis. It seems to burrow into the very nature of things. With the term *deep time*, we feel the same delicious thrill of other twentieth-century projects that hint at a fundamental reality only partially glimpsed: deep space, deep cover, deep structure, deep state. Although people in the nineteenth century did not use the term *deep time*, they did invent a lot of other terms that expressed the profound and unknowable quality of this huge chunk of time. They talked about the immensity of time, the night of time, a past eternity, and the dark abyss of time.

The lengthening chronology was revolutionary because it reshaped many aspects of American life. John Adams once said that the American Revolution was not the war waged by soldiers on the battlefield, but the fundamental change in the minds and hearts of the American people. The same can be said of the deep time revolution. It was not only the fieldwork conducted by geologists and paleontologists wresting dinosaur femurs from cliffsides. It was a transformation in the outlook and experience of ordinary Americans, for whom deep time became a lived reality embedded in their daily lives. The deep time revolution was largely complete by the early twentieth century. We live in its shadow today.

Yet curiously, we have no history of this profound transformation, when the earth literally shifted underneath Americans' feet. We have many excellent histories of the sciences most involved in the transformation (geology, paleontology, and archaeology), of the mania for dinosaurs that gripped the nation, and of the challenges posed by Charles Darwin's theory of evolution by natural selection. I have benefited greatly from those works. But here I am telling a different story, a story about how Americans began to think differently about one of the most essential categories that structures human reality: time. The deep time revolution is one whose contours we have sensed while tracking other stories. But here the focus is on deep time itself, with those other momentous changes cast as supporting characters. The payoff, I hope, is a new history of the United States, from the nation's birth in 1776 to the turn of the twentieth century, when Americans first called themselves "modern." They were modern in part because of the deep time revolution unveiled here.

When we talk about deep time, we often refer to *absolute* or *actual* years—that is, time measured by counting Earth's revolutions around the sun. We say that an asteroid slammed into Earth around sixty-six million years ago, wiping out the non-avian dinosaurs. But reliable measures of actual years did not appear until the turn of the twentieth century, with the invention of radiometric dating. This technique compares the abundance of naturally occurring radioactive isotopes in a mineral sample to their decay products, which form at a known and constant rate. In 1907, the Yale chemist Bertram Boltwood published "actual ages" for rocks in the United States of 1.9 billion years.[2]

The events in this book unfold mostly before the era of radiometric dating, when only *relative* times could be hazarded. Relative dating was based on the so-called law of superposition, first announced in the seventeenth century by Danish cleric and philosopher Nicolaus Steno, which stated that older rocks usually lie under newer rocks. Whether by tides, winds, floods, or volcanoes, rocks generally settled into place from bottom to top. A rock became meaningful in time only in relationship to the rocks lying above and below. "The relative position of the strata we are respectively examining," wrote one American geologist to another in 1833, was "invaluable to us both." Despite having only relative dates, people in the nineteenth century often proposed actual dates, whether as legitimate guesses, extravagant daydreams, or provocations to enemies. These fanciful dates ranged from thousands to millions of years. In other words, time got a lot longer for people in the nineteenth century even though they did not know how long that time "actually" was.[3]

This is our first clue that deep time is interesting for its meanings rather than for the total year count. The same can be said for time in general. Unlike space, time is maddeningly elusive. We somehow know it is there, but we need material objects to make it real to our senses. We experience time in terms of space: the hands of a clock ticking forward, the sun rising and setting, children growing into adults. This is why many scholars agree that it is more interesting and rewarding to see time not as a brute fact of nature but as an artifact created by human beings to communicate about this world of mundane projects and the cosmic world of the divine. If we see time as a social experience, we can suspend judgment about whether time objectively exists "out there" and instead study the many different forms that the experience of time has taken in the numerous societies that left their chronometric fingerprints behind. What seems obvious and natural in one society seems strange and wrong in another.[4]

The people who deepened time in nineteenth-century America were not simply seeing or discovering part of the natural order that others had missed. They were looking at what had always been there—mountains, prairies, river valleys, lakes—and imagining something fresh. The United States played a starring role in this drama of discovery and imagination. The rocks that had seemed to announce the newness of the New World were soon declaring its vast antiquity. By 1849, one South Carolina geologist could tell a Charleston audience "to assign millions, rather than thousands of years, as the age of the earth." Soon after the Civil War, another naturalist ventured a more radical conclusion: "Thus again we discover that the 'New World' is in reality the oldest." That rocks tell time was a new way for Americans to talk to one other, to the rest of the world, and to God.[5]

This book follows those conversations as they emerged and developed during the nineteenth century. Deep time, first, was a way to forge a national identity. Americans used rock strata and fossils to claim a place on an international stage dominated by great European powers. They argued that while their nation was young, their continent was very ancient indeed. As they claimed temporal equivalence with Europe, they soon moved to asserting temporal priority to promote agendas of all kinds, from the economic exploitation of natural resources to chattel slavery, Native American removal and genocide, and social movements from feminism to eugenics. The deep time revolution in the United States did not roll out all at once. Instead, Americans dug little holes of time here and there, in New England valleys, New Jersey marl pits, Alabama cotton fields, parched Dakota gullies, and the severe granite flanks of Yosemite. They unearthed a world unsuspected, a lost era of armored fish, coal-producing forests, plodding dinosaurs, and snarling saber-toothed cats. They slowly sewed these pieces together into a tapestry of national glory, a continental antiquity surfacing as the new United States.

This task was largely complete by the early twentieth century, when the United States enlisted dinosaur diplomacy to announce the nation's imperial aspirations. Like a triumphant Roman emperor hauling an Egyptian obelisk back to Rome because he could, Americans shipped gargantuan dinosaur casts to Europe to signal that they had finally arrived on the international stage. It was with especially pleasurable schadenfreude that Americans presented Dippy the Diplodocus to King Edward VII of England. Americans still dream with dinosaurs, from the docile ruminants in Rudolph Zallinger's murals to the genetically engineered monsters of *Jurassic Park* (fig. I.1 and plate 1). By the early twentieth century, the New World/Old World distinction no longer referred to a geological—let alone a theological—reality. With Americans deploying radiometric dating to show that they too lived atop primordial rocks, no longer could Europeans allege that their continent was literally older. The major difference between the United States and Europe was now seen to be cultural, as one French observer noted between the two world wars. It was gleaming cars versus collapsing castles, the

FIGURE I.1. Gentle ruminants populate a segment of Rudolph Zallinger's mural, *The Age of Reptiles* (1947), at Yale University. Created before the asteroid extinction theory of the 1980s, the mural suggests gradual extinction through such agents as volcanos steaming ominously behind the oblivious T-Rex.

efficient and materialistic New World contrasted with the genteel charms of the Old.[6]

Who were these first people to imagine this more ancient America? The pioneers were paleontologists and geologists, born during the Revolutionary era and eager to professionalize the gentlemanly hobby of fossil collecting. Filling private cabinets and new scientific academies with fossilized oysters, stony fern fronds, and fearsome reptile teeth, they mixed science with piety and nationalist ambition. Some had trained for the Protestant ministry before turning their attention to fossils. Many were medical doctors, using their anatomical knowledge to identify and reconstruct the remains of long-extinct animals. In Britain, some of them are still remembered in the names of diseases they described, such as Hodgkin's lymphoma, named for Thomas Hodgkin. Among Hodgkin's leading correspondents in the United States was the physician Samuel George Morton of Philadelphia, notorious today for his enormous collection of human skulls that he measured to insist on the racial superiority of white people. Here, however, we will recover the lost Morton: the internationally influential paleontologist who helped to establish the Cretaceous as a transatlantic geological formation shared by Europe and North America. A few fossil collectors were women. Plants and shells were deemed appropriately feminine

scientific pastimes, and women were counseled to avoid mentioning that flowers were sexual organs analogous to those of animals. Most difficult to recover are the voices and experiences of the many Native and enslaved peoples who formed part of the deepening of time in North America, but I have tried to incorporate them into the story told here. Living within their own chronological schemas, American Indians were subjected to temporal imperialism over the course of the nineteenth century, by which the new language of deep time was used to exclude or belittle other ways of measuring time. Planters deployed slaves to gather and haul fossils, while at the same erecting imagined racial hierarchies that consigned black people to the lowest, most primitive rungs of human social development. Deep time, in short, was a language of exclusion as well as of inclusion. Educated white naturalists deemed some peoples and not others capable of imagining and living in deep time.[7]

Organizers of knowledge rather than abstract theorizers, nineteenth-century American fossil hunters tucked their finds into mahogany cabinets and small wooden boxes. This was God's work. Believing that the order of nature reflected the harmony of the divine, they penned tiny, precise labels for each specimen, some of which survive today as yellowed fragments in America's older natural history museums. They nestled their precious samples in hay-lined wooden crates and dispatched them to colleagues in the United States and abroad. The grateful response was seldom long in appearing: "Your present of Rocks has arrived." For these were not just rocks. They were "hidden treasures of the Physiological & fossil world." The gnashing dinosaur jaws and curling ammonites decorating the collectors' many letters remind us of the sheer excitement and joy pulsing just underneath the sober scientific prose. From Europe, Mexico, India, and the Far East, fossils and rocks also poured into the United States. All these specimens, accumulated slowly but surely, formed a mighty citadel. The United States was not new, Americans announced from the ramparts, but rather old—as old as Europe, and perhaps older still.[8]

Deep time also bolstered state and corporate power. From theory in the eighteenth century, deep time became infrastructure in the nineteenth. Coal, gold, iron: the industrial age set its table with the fruits of the Earth. Americans monetized rocks and minerals on a grand scale, transforming them into commodities that industries could reliably locate and exploit for profit. The chemical revolution of the late eighteenth century, which prioritized observation and standardization in lieu of idiosyncratic system building, allowed for a much easier assaying of rocks and the separating out of precious metals. Naturalists coined the term *natural resource* to describe the rocks and minerals that powered the Industrial Revolution. American cotton and sugar cane planters, miners, canal diggers, and railroad engineers soon joined in, taking note of stratigraphy and fossils. Whether to dig, drill, or farm above or below a particular stratum could involve a major investment of money and time. By 1860, approximately thirty state geological surveys had been founded, which were joined in 1879 by the United States Geological Survey. These agencies hired a professionalizing cadre of geologists and paleontologists to probe North America's rocky substratum.[9]

Americans did not work alone. Collaboration with European naturalists was essential to the formulation of the idea of deep time. Plunging under the Atlantic Ocean and resurfacing on the opposite shore, rock formations shared by Europe and the United States were transformed during the nineteenth century into an economically legible antiquity. *Synchronous* was the word Americans began to use to describe the ribbons of same-age rock that encased the planet. Locally inspired names for these synchronous bands (such as the Devonian, named for Devon, England) became a shared international code that facilitated resource extraction across borders. These names also flagged the scientific credentials of the modern nation where they were first recognized. Geologists contributed many names, from the Permian (for Perm, Russia) to the Jurassic (for the Jura Mountains on the France-Switzerland border) to the Mississippian and Pennsylvanian. Deep time underlay the international latticework of capitalist institutions.[10]

The twentieth century dawned with deep time embedded further still in American life. Colleges, universities, natural history museums, magazines, popular books, and silent films spread the idea of deep time to the public. The discovery of prehistoric human remains in Europe in the 1850s energized Americans further still. Human antiquity was a wrenching realization for the many Americans who had long believed that deep time extended only to Earth and its plant and animal inhabitants. Humanity itself, they believed, was cordoned off as God's final creation, lovingly placed in the Garden of Eden about six thousand years ago. But the *caveman*—a word coined in the 1850s to refer to prehistoric humans—suggested that humanity, too, had now joined the rest of creation in deep time, with humans evolving from ape-like ancestors thousands and perhaps millions of years before.

Americans turned to the caveman as an anchor for a nation adrift in the swirling currents of modernity, cut loose from traditional beliefs and ways. What could the caveman teach modern Americans about their society and culture, the dreams and darkest yearnings that lurked beneath their conscious perception? By the 1920s the caveman had become the blank screen onto which a diverse group of influential American intellectuals projected their aspirations: the sociologist (and paleobotanist) Lester Frank Ward; the feminist Charlotte Perkins Gilman; the museum director Henry Fairfield Osborn; Charles Knight, the first major American painter of prehistoric subjects; and the psychoanalyst Beatrice Hinkle, among the first American popularizers of Swiss psychiatrist Carl Jung. These Americans, most of them born after the Civil War when the national compass sought new directions, were the first to imagine that humanity's remote antiquity could point to a better future for the United States. Filmmakers, etiquette advisers, and cartoonists carried their message to the masses in cities, towns, and farms, where it penetrated unevenly and with varying results. The point is that prehistory and modernity were not separate developments but simultaneous inventions, one the companion of the other. The gleaming future promised by modern progress was rooted in deep time's immensely long backstory, as evolution by natural selection replaced Genesis as the controlling narrative of life on Earth. Modernity needed prehistory just as surely as prehistory needed modernity.

Finally, deep time was an inward journey. A book about deep time is not obviously a book about God. And yet the American Protestants who populate this book often described deep time in the same terms they used to discuss the divine. First and foremost, they were preoccupied by the fundamental unknowability of deep time. Mid-nineteenth-century Americans described the long chronology just as they described the unknowability of the divine. Deep time was a concept that could not be conceived, a span of time so great that the finite and material human brain failed to grasp it. The naturalist David Dale Owen observed that the early Earth was "a period so remote as to defy human conception," while his contemporary Jacob Green deemed the time span between then and now "incalculable."[11]

As a new concept that could not be conceived, deep time presented an exciting cognitive problem different from the long-standing Christian idea of eternity. Christian theologians had long argued that God existed outside of space and time: the divine eternity was both unseen and unchanging. This was so foreign to human beings' lived experience of a visible and changing world that eternity became one of the attributes that made the divine deliciously mysterious and incomprehensible. Visible, material, and changeable, deep time was therefore the opposite of the Christian eternity. Yet it was still so immense as to be as incomprehensible and awesome as the deity. In his history of the American Civil War, published in 1867, the naturalist-historian John William Draper felt compelled to begin at the very beginning: the pre-Cambrian origins of the North American continent (a decision that may explain why his history necessitated three volumes). Draper saw "something majestic and solemn" in the "vast lapses of time" that "our finite faculties vainly try to grasp." One gets the distinct sense from Draper, Owen,

Green, and others that American naturalists enjoyed trying to grasp the ungraspable antiquity of rocks and fossils. They had embarked on a new quest, a journey whose goal was more than empirical knowledge. It was a journey into the soul, what the poet Emily Dickinson, who had studied geology at the Amherst Academy, called the "subterranean freight."[12]

To make deep time real to the senses, Americans crafted a virtual deep time of portable objects: tiny plaster models of trilobites, wooden blocks representing coal strata, linen posters painted red with exploding volcanoes, magic lantern shows featuring primordial landscapes, and magazines and books teeming with scowling sauropods. With this proxy world, they hoped to reveal to the senses what was inconceivable to the mind. They often referred to the *sensorium*: the entire sensory apparatus that included the five senses and the brain's reception and interpretation of sensory stimuli. They imagined that not just humans but also the tiny creatures of long ago possessed this sensorium, a gift from the loving Creator. Piercing the oceans of the infant planet, the light of heaven had reached the "sensorium" of the trilobites drifting below, they wrote. How "marvellous" this was: literally a thing of supernatural wonder. In turn Americans activated their own sensorium to make that ancient world palpably real to their eyes and hands. By midcentury they had crafted a new, portable proxy world small and light enough to be mailed, hoisted aloft on the public lecture circuit, displayed at the front of a college auditorium, or squirreled away in a fossil cabinet. These little things pulsed with meanings that were scientific, educational, ornamental, and spiritual all at once. Casts, posters, books, and magazines further expanded the number of participants in the deep time revolution. While published naturalists often took the credit, these objects were in fact created by artisans and craftspeople whose names appear in tiny script at the edges of book engravings—if they appear at all.[13]

As the mood ripened for communion with the past, the mobile deep time objects were joined by large, immobile objects that created a total experiential world. You went to them; they did not come to you. Murals, moving panoramas, dioramas, painting sequences—and of course the enormous dinosaur skeletons in museum lobbies—these appeared in American cities, towns, and classrooms beginning in the

1850s. Monumental and grand in its sheer size, this immobile deep time gallery enfolded the American public in wondrous lost worlds lying just beyond the here and now. Like a stained-glass church window, the brontosaurus in the lobby pointed to another realm tantalizingly out of reach to all but the human imagination.

Deep time also absorbed another aspect of divinity: omnipotence. A vast canvas of time became the stage for the greatest changes in the history of the planet. Time's agency swelled, with all kinds of natural processes unfolding across its broad lap. The most important was climate, an old agent that was now assigned a starring role in the history of life. Although Aristotle and other ancients had assigned agency to climate, the modern idea of climate change was born in the nineteenth century, when oscillations in planetary temperature over eons were filled with godly power and nationalist purpose. Americans gazed at their continent and saw God's handwriting in the scraped rocks and valleys of the last glacial era. They concluded that God had breathed long global winters and springs, creating and extinguishing life according to his hidden plan.

Slowly but surely, deep time became a new way to talk about God. The long chronology superimposed onto a new, naturalistic timescale the older categories of the divine: infinity, unknowability, wonder, and meaning. The shift from the short Genesis chronology to the billions of years of deep time transcended sects and creeds. Like Manifest Destiny, the mid-nineteenth-century idea that Americans were ordained by God to conquer the North American continent, deep time appealed to many denominations of Christians. Deep time in fact bolstered Manifest Destiny by supplying the opening chapter to the glorious narrative of American progress. In the beginning, God created North America. Then a new nation, consecrated by nature and nature's God, expanded westward across a continent whose antiquity was slowly unveiled and sewn into a golden national future. Swollen with the importance of their vocation, American geologists and paleontologists styled themselves explorers of "time-worlds," their vocation "sublime." The new language of deep time gave Americans a vocabulary with which to frame their nation's place in the cosmic order. Even as it reduced them to specks in a planetary history spanning billions of years, deep time amplified Americans' perceived role in the cosmic order by exposing the sheer immensity of the scale on which their actions unfolded. By the twentieth century, deep time had become a moral outlook. Far from diminishing Americans' moral responsibility in a planetary history so old it defied conception, deep time inserted Americans into breathtaking temporal vistas filled with purpose and shimmering with meaning.[14]

But deep time has also brought Americans—and many others—deep anxiety. Our modern preoccupation with our role in changing Earth's climate—epitomized by the new term, *Anthropocene*—makes us uneasy in part because we are assuming the role that the first theorizers of deep time had assigned to God: control of the future. Nineteenth-century Americans saw primordial life forms culminating in their own Age of Man, a future that God had seen at the outset of everything. Their Age of Man was distinguished from earlier ages by what they thought were the unique moral capacities of human beings. Today, we have seamlessly applied this view to our own era, engaging in the moral imperialism pioneered by nineteenth-century Americans. We project our ethical vision into a distant planetary future that we have convinced ourselves is our own responsibility. One geologist has recently called this conjunction of deep time and morality "timefulness." The Anthropocene makes us anxious about salvation, reconceived in natural terms. That our brains cannot conceive of this awesome past and future time span only adds to the sense that deep time caresses divinity itself.[15]

How did we get here? The answer lies in the deep time revolution, the century between roughly 1800 and 1900 during which Americans invented a primordial antiquity for their continent, their nation, and their innermost selves. If not the traditional God, then some creative force, the invisible behind the visible, walked for eons upon eons with the beings into which he had breathed life, lavishing his children with his creative energies, to their endless wonderment. Or so it seemed to a growing number of Americans, whose story is told in the pages that follow.

1

Why the New World Was New

THE NEW WORLD was new for two reasons. The first reason emerged in the wake of Christopher Columbus's arrival in the Caribbean. Europeans were awestruck by the shattering novelty of a whole other half of the Earth, full of peoples, plants, and animals. The Florentine voyager Amerigo Vespucci announced this *mundus novus* to his European patron in letters published in 1503. The landmasses were much bigger than originally imagined, entitling them to be not just new territories, but a whole new world. The Americas were new because they were new to Europeans.[1]

Our story takes off from the second meaning of new. This meaning was added in the sixteenth and seventeenth centuries as the Americas became entrained in European debates about how to reconcile the Genesis story of Creation with the realization that other peoples had set their beginnings thousands of years earlier. Jesuit missionaries and other travelers had returned to Europe with texts from China, the Americas, Egypt, and the Middle East. European scholars studying these texts became aware that other chronologies existed for everything from the origin of the Earth to the rise of the civilizations on it. The Chaldeans, the Egyptians, and the Chinese possessed histories that set their beginnings earlier than the biblical account. The stakes were high. In the broader context of these histories, the Bible risked losing its place as the unique, universal history of the world. The Bible might instead be seen as the particular history of the Hebrew people, and the Genesis flood as local rather than global. Some alarmed observers saw an atheistic plot to promote the eternity of the physical world in lieu of the Genesis account of a world with a definite beginning, created by God at a particular time. Concerned European scholars in the seventeenth century defended the Bible's sacred history as more than one model among many others. They insisted that the Bible was sacred because it alone preserved records of the story of the beginning of the world, uncontaminated by myths and fables.[2]

One outcome of this destabilizing episode was a precise new chronology of Genesis that pinpointed the date and even the hour of the Creation, Noah's flood, and other events. Among the many proposals was that of Irish theologian James Ussher in 1650, in a work translated from Latin into English in 1658. Ussher announced that Saturday evening of October 22, 4004 BC, was the moment of Creation. The following day—Sunday, October 23—was the "first day therefore of the world." Ussher's fastidious chronologies were printed within Bibles, and the six-thousand-year age of the planet began to circulate widely just as the Protestant Reformation delivered more Bibles into more hands than ever before. New engravings made the infant Earth seem palpably real. With a sun peeping through clouds over a watery world, images like those of Prague-born artist Wenceslaus Hollar established conventions for how Americans would portray their own origins in deep time (fig. 1.1).[3]

As data from the Americas poured into Europe, some scholars began to argue that Noah's flood had also reached the New World. The Bible being vague on the details, others counterattacked, arguing for a local flood confined to the Holy Land. But mountaintop seashells from the Andes and Appalachians offered tantalizing evidence of American involvement in the biblical flood. Scholars invented what they called a sacred theory of the Earth, sewing new American data into biblical scenarios. They added the Americas to globes and maps not just as geographical phenomena but as chapters in the story of Creation. Now embroidered with divine meaning, the New World was seen to be new because it had risen last from the receding waters of Noah's flood. The Americas' emergence was like somebody climbing out of the bathtub, still wet and shivering even as Europe had already warmed its flora and

FIGURE 1.1. This seventeenth-century illustration of the Genesis story shows the young Earth just after dry land and plants have appeared, but as yet no animals or human beings. Wenceslaus Hollar, *Creation of the Earth*. Mid-17th century.

fauna to fecund perfection. In books that pioneered time-lapse sequences of the planet changing over time, the Americas appeared still swaddled in primordial mist (fig. 1.2).[4]

Inspired by the classical idea that climate determined what life forms would thrive, some European naturalists proposed that the humid chill of the New World doomed it to inferiority. Without setting foot in the

FIGURE 1.2. The Americas rising last from the Flood on the infant Earth, as depicted in Thomas Burnet's *Theory of the Earth* (1697).

Americas, the French naturalist Georges-Louis Leclerc, Comte de Buffon, wrote that the animals and plants in the New World were degenerated versions of those found in the Old World. The cold climate shrank elephants into short-trunked tapirs and displaced gestation onto the bleak exterior of the mother opossum.[5]

The native peoples of the Americas were also inserted into the story of New World inferiority. As Secretary of the Board of Trade and Plantations and to the Lords Proprietor of Carolina, the English philosopher John Locke read extensively about the American Indians. They appear in his *Second Treatise of Government* (1689) as examples of the earliest human societies, lacking private property and contracts of government. All societies had to pass through this primitive stage on the road to civilization. "Thus in the beginning, all the world was America," Locke announced in a Genesis-like formulation. Dutch philosopher and geographer Cornelius de Pauw extended Buffon's climate thesis to the New World aborigines. It was a grim story for natives and newcomers alike. It was said that Indian men had less body hair than European men because they had been weakened by the clammy climate. Europeans settling in the New World risked relapsing into savagery.

Political independence in 1776 gave Americans the chance to reset the meaning of the New World. The stakes were high: not just scientific reputation but national survival. The Declaration of Independence had opened the United States for trade and diplomatic alliances. Powerful European nations weighed the prospects of the new nation as ally or enemy, trading partner or competitor, fleeting republican experiment or vindication of the superiority of kingless representative government.

Two paths to national self-definition opened with independence: one emphasized the newness of the New World, the other its extreme antiquity. Americans took both paths. In fact, the simultaneous declaration of extreme antiquity and extreme newness became a unique feature of American nationalism as it developed in the first century after political independence. Although Europeans also helped to forge the concept of deep time in the late eighteenth and nineteenth centuries, Americans claimed both ends of eternity. They capitalized on deep time not only as a scientific concept but as a political tool they hoped would speed their acceptance among the powers of the globe. In the uncertain first century of nation formation, political urgency led Americans to try to be old and new at the same time. This was not just a scientific revolution but a triumph of political imagination.

We can look in more detail at each of the two paths. The first celebrated the new kingless republic of the United States. As Europe's old monarchies tottered, Americans announced that their nation would be a place for fresh starts, a force of regeneration. "What then is the American, this new man?" asked the French immigrant Hector St. Jean de Crèvecoeur after the revolution. His answer: the American was Adam, a new alloy forged as Old World problems melted away. Newness and regeneration became fixtures of US nationalism. American nature would "promote National and Individual Happiness," the painter and naturalist Charles Willson Peale announced. Political freedom lay in the very ground of America, according to statesman Daniel Webster, speaking at the laying of the cornerstone of Boston's Bunker Hill Monument in 1825. "The *Principle* of Free Governments adheres to the American soil. It is bedded in it, immovable as its mountains." The invented republican goddess Columbia pointed to the western frontier, the moving engine of progress that would deliver Americans from Old World senility.[6]

Political independence also severed Americans from ancient royal genealogies. Instead, the new federal Constitution offered the eternal present of government imagined as a machine, regenerated with each electoral return. "Democracy has no forefathers, it looks to no posterity; it is swallowed up in the present, and thinks of nothing but itself," wrote John Quincy Adams about the United States when a Swiss visitor tried to sell him two small boxes of ancient coins and medals, treasures worthless to fast-charging Americans. Eternal youth sometimes felt like more of a curse than a blessing. Long past its century mark of nationhood, American democracy struck some as doomed to be forever young, never older, never wiser, its electorate swayed by fads and boorish demagogues rather than steadied by the wisdom of a stable aristocracy. This was Woodrow Wilson's diagnosis of the American condition in 1896. "The Old World trembles to see its proletariat in the saddle; we stand dismayed to find ourselves growing no older, always as young as the information of our most numerous voters."[7]

But there was also a second path to defining America, which invoked the new nation's antiquity. Europeans liked to mock the New World's lack of medieval castles and classical ruins. Patrician Americans absorbed this self-criticism. "In America we not only have no Gothic Buildings, but nothing resembling them," sighed Yale's young naturalist Benjamin Silliman during his European travels in 1805. Deep time came to the rescue, offering Americans an antiquity that preceded the most ancient civilizations. Exhuming rocks from the deepest fossil-bearing strata on the continent, American naturalists thought they glimpsed the beginnings of life on Earth. In the alien Eden of the baby planet, trilobites had trundled along a dark seafloor. Over eons, their bodies had fossilized into a lost world now exposed in cliffs and canal beds from upstate New York to the western fringe of the Great Lakes. What were the pyramids of Egypt next to this far more ancient world, pulsing with the freshness and vigor of creation itself?[8]

The nationalist assertion of US antiquity launched with political independence. By the early 1780s, Thomas Jefferson had amassed enough shells and rocks to propose that America was several hundred thousand years old. Jefferson's New World antiquity was deliberately polemical, floating phantom high numbers to annoy Buffon. Privately to his journal, the American naturalist Benjamin Smith Barton also refuted the French count. "If the late celebrated Mr. De Buffon had been acquainted with these facts, facts pretty well known to Americans, he would not have asserted so confidently as he has done, that nature, in the new-world, has formed her productions upon a small and feeble scale." Barton's private suspicions ripened into published protest. "The physical infancy of America is one of the many dreams of the slumbering philosophers of our times."[9]

The farther west Americans traveled, the more of archaic America they found. A planter in Mississippi scoffed at the "visionary philosophers, who have been pleased to amuse themselves with the pretended infantile state of our continent, compared to their trans-atlantic world." On the contrary, he added, "we must grant to it an incalculable antiquity." That planter was William Dunbar, recently arrived from Scotland. The ancient rocks near Edinburgh had provoked Dunbar's countryman, the Scottish geologist James Hutton, to proclaim that fossils were "annals of a former earth," with rocks recycled over an "immense time." Hutton's data drew largely from Europe, but American data continued to pour in and to suggest the antiquity of the new continent. "The highly primitive . . . character of the Rocky Mountains, would seem to discountenance the opinion entertained by some, that our continent has

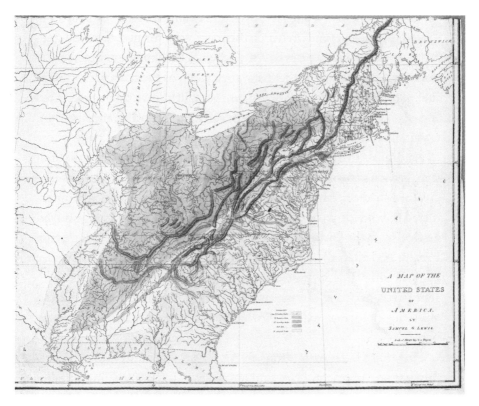

FIGURE 1.3. The first geological map of the United States, created in 1809, uses colored bands of rocks to represent the ages of the strata. From "A Map of the United States of America" by Samuel G. Lewis. In William Maclure, "Observations on the Geology of the United States, Explanatory of a Geological Map."

emerged from the depths of the ocean, at a period comparatively recent," wrote one of the members of an 1819 expedition to the Rocky Mountains.[10]

The first geological map of the United States appeared in 1809, created by another Scottish immigrant, William Maclure (fig. 1.3 and plate 2). What made Maclure's map "geological" were the colored bands representing rocks of various ages, from the oldest to youngest. The purpose of geology was to describe the "relative position" of rocks so as to see "the great and prominent outlines of nature," Maclure explained. To focus on isolated rocks would be like a "portrait painter dwelling on the accidental pimple of a fine face." Maclure declined to speculate on "the relative periods of time" in which the rocks were formed and modified. "Such speculations are beyond my range." Maclure's map opened the door for a suite of maps that used rock strata as the basic structure of the map. In contrast to earlier maps that had shown ancient features such as Indian mounds on a modern cartographic plane, time became essential rather than incidental to these newer maps.[11]

Naturalists in the United States also adopted the new idea of extinction. French naturalist Georges Cuvier had theorized in 1798 that some creatures had once lived that were no longer alive today. This contradicted the old idea of a static and timeless great chain of being arrayed by God from the lowliest lichen to man. In this older scheme, the loving Creator had filled all niches of the ladder with life. Fossils of unfamiliar animals and plants suggested that they still lived elsewhere on the present Earth, awaiting discovery. By contrast, extinction suggested the existence of former worlds in which life forms once thrived, only to perish.[12]

Had there been extinctions in America? Benjamin Smith Barton thought so. American plant and animal fossils supplied "indubitable proofs of the extinction of species," he confided to his journal. Large bones pulled from the North American mud suggested that an elephant-like animal—the mastodon—may have once stalked the land. Jefferson shipped fossil specimens of this beast to Buffon in Paris. Behold the giant American femur, Charles Willson Peale seems to say in his self-portrait, where he cradles the prized index bone that hinted at the size of the American mastodon on display at his Philadelphia museum (fig. 1.4).[13]

Peale's self-portrait captured the political triumph of deep time in the first century after American independence. Peale is both old and new: he is 83, his hair white and thinning, but he is a citizen of a republic not yet fifty years old. The bone he holds is also both old and new, pulled from primordial earth and gleaming with scientific novelty and promise. Over the next hundred years, ancient fossils kept popping out of the American ground, somehow ancient but also pristine and virginal, ongoing declarations of independence from Old World values. Other nations and regions were also turning to deep time to craft visions of

FIGURE 1.4. In Charles Willson Peale's self-portrait from 1824, the painter cradles a large American mastodon bone, evidence of the antiquity of the ostensibly New World as well as the enormous size of its native creatures.

modern life—and Americans were often in communication with leading scientific lights in those circles, especially in France and Britain, but also Mexico, Italy, and the German states. But the combination of extreme antiquity and utter novelty was pushed further in the United States. There, Americans strove to find both the ancient beginning of life and a fresh start for life.

The governing metaphor of deep time was the stratum, a layer cake of rocks. The idea was more or less invented in the seventeenth century, but it had been confined to a small circle of learned naturalists. Under the pressures of the Industrial Revolution in England and Scotland, mineralogists, miners, and fossil hunters such as William "Strata" Smith pioneered visualizations for the emerging science of geology that precisely located valuable minerals. In maps and charts, they used brilliant colors to display the many layers of rocks that lay underfoot.[14]

Strata became a new way of seeing. They suggested a place in time, a subterranean world that led from then to now. The Scientific Revolution of the seventeenth century had proposed new ideas about space and time. Instead of events containing their own meaningful time, as they had before, now events could be placed on a single axis of the timeline, a visual device that suggested that events were objective, continuous, all-embracing, and absolute. Strata sprang from a similar idea. They stacked earthly processes into a single, universally applicable sequence. But unlike one-dimensional timelines, where points on a line marked events such as wars and treaties, strata implied an x-y axis. The horizontal axis was space, and the vertical axis was time. This two-dimensional representation of strata on paper signified what in nature was a three-dimensional rock layer. The breaks between strata represented the disappearance of some kinds of fossils and the appearance of others, whether by a catastrophic event or a long, gentle process. Ambiguous about causality, both metaphors of the timeline and the stratum were explicit about direction: time flowed forward, from past to present, from then to now, never repeating.[15]

By the early nineteenth century, strata had caught on among American geologists. Surveying the chaos of nature in North America, they seized on strata as an organizing principle. Some saw strata as mirrors of their own fossil cabinets. Edward Hitchcock of Amherst said strata were "arranged in as much order, one above another, as the drawers of a well regulated cabinet." From upstate New York, his colleague Amos Eaton nodded in agreement. "Geology is chaos without systematic arrangement," he wrote, drawing strata in his notebook as though tidying the linen cupboard. That it was also divine work made it all the more

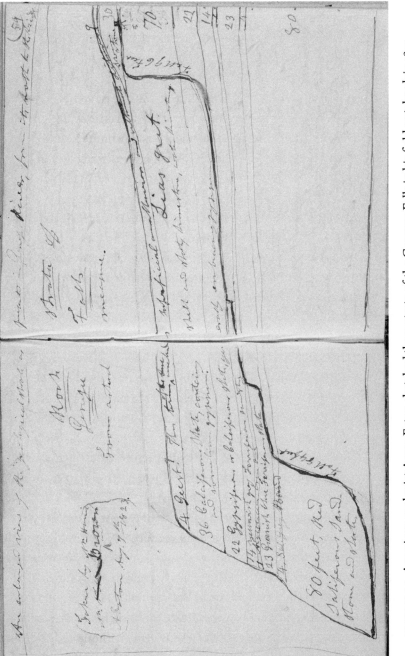

FIGURE 1.5. American geologist Amos Eaton sketched these strata of the Genesee Falls in his field notebook in 1823.

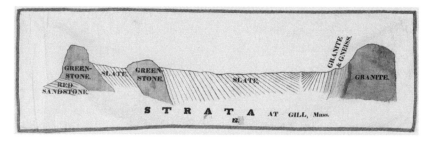

FIGURE 1.6. Orra White Hitchcock drew these Massachusetts strata onto a large linen poster for students at Amherst College, from around 1828–1840.

appealing, since the strata surely reflected God's plan for nature. Easy to draw and easy to decode, strata swept across the visual field. What began as hasty sketches in field notebooks in the 1820s became the new plan on which natural history museums were arranged in the 1830s, when the curators at the Academy of Natural Sciences in Philadelphia rearranged their fossil collection "according to Strata." Strata migrated into college classrooms by the 1840s and then into grand landscape paintings by the 1870s (figs. 1.5, 1.6, 1.7).[16]

Strata are so familiar to us today that it can be difficult to recapture the moment when they emerged as arguments about the world. Strata declared that the world is this way and not that way, deposited in an orderly fashion on a planet that was ancient and not new. Projected onto maps of the United States, strata joined other new classifications such as climatological and ecological maps to form a dense web of information about the nation.[17]

But strata were more than even that. They were statements about the power of the human mind, a projection of moral and spiritual capacities into the ever-expanding dominion claimed by the modern sciences. Scottish geologist Charles Lyell explained the higher meanings of geology and strata in the first volume of his landmark *Principles of Geology* (1830), a work that influenced generations of Europeans and Americans during the formative decades of deep time. The human mind and spirit would be the great vehicles of discovery for vast times and spaces, wrote Lyell. "Although we are mere sojourners on the surface of the planet,

FIGURE 1.7. By the 1870s, strata had migrated into American landscape paintings, such as this canvas by De Witt Clinton Boutelle titled *Trenton Falls Near Utica, New York* (1873). These sorts of detailed paintings became known as rock portraits.

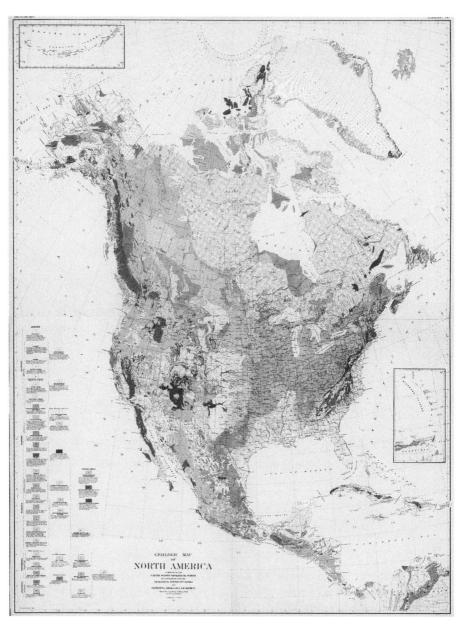

FIGURE 1.8. With undulating, colored bands representing rocks stretching to well over two billion years old, this geological map of North America from 1911 shows the transformation of the New World into a very old world in the span of a mere century.

chained to a mere point in space, enduring but for a moment of time, the human mind is not only enabled to number worlds beyond the unassisted ken of moral eye, but to trace the events of indefinite ages before the creation of our race." The human mind was "free, like the spirit."[18] Lyell's benediction infused geology with urgent moral purpose. It suggested that the strata were more than piles of rocks. Strata pointed to an unseen realm not fully legible by scientific instruments alone.

The 1911 geological map of North America unveils the outcome of the deep time revolution of the previous century (fig. 1.8 and plate 3). What had been radically new a century before was now obvious. The North American continent had an ancient history—a very ancient history indeed, so old that it predated the Old World. Ribbons of colors representing time bend and twist around North America as it rises from the cradle of pre-Cambrian antiquity. The thin grid lines of the modern nations of Mexico, the United States, and Canada snap onto the primordial background.[19]

One formation steals the show: the enormous pink blob lying over eastern Canada and the most northerly regions of the United States. This is the magnificent Canadian Shield, made of the earliest surviving surface rocks from a lost continent born billions of years before. These rocks have been scoured and blasted by time nearly to smithereens, forming a cracked pancake of barren rock. The New World is old, the map announces, and its past stretches into a dark abyss of time eons before Columbus and the *Mayflower*, before the peoples of Asia stepped eastward over the land bridge of Beringia, before dinosaurs ruled Earth. Our task is to begin at the beginning, to discover how and why the deep time revolution unfolded in the United States.

2

Beginnings

THE LONG DROUGHT finally broke with a late-afternoon June thunderstorm. The year was 1826. Rocks grayed by dust in upstate New York's arid spring now turned black and slick. Like a bucket of water thrown on a mosaic, the storm revealed the treasures cemented in the rocks: trilobites.

The trilobites looked a bit like horseshoe crabs, but they had three distinctive lobes, which is what gave them their name (fig. 2.1). They lay by the hundreds here, near a deep gash in the New York rocks called Little Falls, frozen in their death throes like the victims at Pompeii (fig. 2.2). Some had curled into balls. Others lay tattered in heaps. Urged on by their professor, the undergraduates from the new Rensselaer School in Troy pried the fossils from their tombs before heading back to camp for the night.

Everyone was tired. For the last month, on a mule-drawn packet boat named the *Marquis de Lafayette*, the students and their teacher had inched westward on the new Erie Canal. This artificial river, built with expert input from geologists, stretched over three hundred miles from Niagara Falls to Troy. On its predictable waters, wheat and corn flowed east from new western farms to Atlantic markets. The canal also occasionally ferried curious geologists and their students, and Eaton's group was finally heading back east, toward home. The name of the boat honored the French aristocrat who became an American revolutionary war hero. It was a fitting tribute in this year, the fiftieth

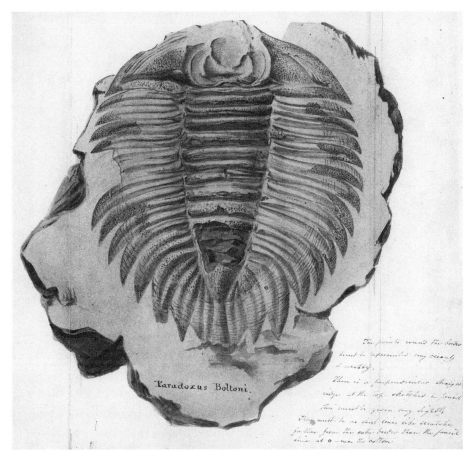

FIGURE 2.1. This is one of the earliest drawings of American trilobites, created by French naturalist Charles Alexandre Lesueur around 1825.

anniversary of American independence, the beginning of the United States.

But now the trilobites hinted at another beginning. That evening as his campfire sent sparks into the night sky, Professor Amos Eaton, one of the founders of the Rensselaer School, turned the fossils over in his hand. They were not just any fossils. They had been ripped from "the oldest rocks in which any petrifactions are found." The Erie Canal, this feat of modern engineering, had exposed a past far older than the

FIGURE 2.2. The Eaton group floated here in Little Falls on the Erie Canal in 1826, hunting for trilobites.

American Revolution, an antiquity so old that it scraped the very basement of time. No signs of life had ever been found below these rocks. This was the beginning, not of the United States, but of life on Earth. The trilobites were "the first created beings."[1]

Something else was puzzling about these trilobites. Although such fossils had been found in the Old World, they were especially abundant in North America. Could it be that the New World was as old as the Old World? In the beginning, according to the Genesis story, God created animals: whales, fish, cattle, birds. These creatures were familiar to contemporary readers, reflecting the ancient view that life on Earth today was more or less as God had made it. Ferried to safety aboard Noah's ark during the biblical flood, these familiar animals repopulated Earth.

But the trilobites looked nothing like the animals in Genesis. Wrenched from the deepest strata of Sweden, Russia, Wales, Scotland, England, Ohio, New York, and Nova Scotia, the trilobites whispered of an alien Earth that would have been unrecognizable to the animals of the Genesis story. Some were tiny, less than an inch long; others stretched to nearly two feet. Whatever their size, the trilobites were "seemingly of another world."[2]

Today, the violent metaphors of the twentieth century—the Big Bang and the Cambrian explosion, the electrical zaps of a primordial soup—shape the way we talk about beginnings, whether cosmic, planetary, or biological. They tell us that life has blasted into being through sudden, wrenching, and noisy cataclysms. But this was not the language of the nineteenth century, when the first naturalistic alternatives to the Genesis creation story gained a large public following. This was a world of silence, solitude, darkness, and mystery: of nothing becoming something by some initial mysterious force at work in the infant planet.

The great transformation of the nineteenth century was in normalizing the idea that ancient North America had teemed with Earth's first created beings, the trilobites creeping along the floor of a dark and silent ocean. The oceanic eden of the trilobites offered the first plausible alternative to the Garden of Eden in the public imagination. Soon this world became at least as palpably realistic as the unembellished, skeletal account of the creation in Genesis. With no sin and no Fall, the world of the trilobites freed Americans to imagine new stories of progress and even perfection. Investing the trilobites and their strange, primordial world with a shimmering metaphysical significance, American naturalists used words, images, and plaster models to create an ever more detailed underwater world for the trilobites that enriched the idea of a benevolent Creator rather than diminishing it. They stood back in wonder and awe. It seemed that the Creator had lavished energy even on these creatures, the lowly trilobites.[3]

The American Trilobites

A few years before the Erie Canal field trip, Amos Eaton summed up the exciting new nebular hypothesis, which gave a naturalistic explanation for the creation of Earth. Like the other planets in the young solar system, Earth had congealed from clouds of gas. "The globe has a beginning," the pious Eaton announced. It was "not eternal." This was a subtle jab at the famed Edinburgh geologist James Hutton, who had declared the continents to be endlessly recycling according to natural laws of uplift and erosion. Hutton had concluded with this elegant but inflammatory statement: "The result, therefore, of our present enquiry is, that we find no vestige of a beginning,—and no prospect of an end." But beginnings and ends were important because they revealed not just a God operating through laws of nature, but perhaps his special favor in lavishing his creative energies here and not there. The American trilobites suggested that life on Earth had a beginning—not in a garden, but in an ocean buried in the planet's deepest fossil-bearing strata. Below the trilobite-bearing rocks, all was lifeless and still. And then there was life.[4]

Eaton opened his mail one day to find a sketch from a colleague that visualized this existential drama. Strata upon strata descend in New York until finally the lowest layer, at the bottom of the page, showed only rocks "without petrifactions"—that is, without the fossils that were signs of life. But just atop this blank and lifeless rock appeared proof of life, none more abundant than the trilobites, "the first created beings." Here was a beginning, perhaps even *the* beginning (fig. 2.3).[5]

No text was more crucial for Americans' early understanding of these bizarre pioneers than Alexandre Brongniart's *Histoire Naturelle des Crustacés Fossiles* (Natural history of fossil crustaceans, 1822). Today, Brongniart has been eclipsed by the brighter scientific stars of post-revolutionary Paris, such as his friend and collaborator Georges Cuvier, with whom he wrote a field-transforming study of the geology of the Paris basin that showed that fossils in different strata were of different ages. But in his own day, Brongniart was nearly as well known as Cuvier, both as a wunderkind and as a polymath. He seemed to be good at everything from fossil hunting to mineralogy to pottery to chemistry to administration. In 1800, when he was just thirty, he became director of the great Sèvres porcelain factory outside of Paris, a post that he kept for five decades and that gave him access to Paris's highest scientific circles and to prestige publishing opportunities. His *Histoire Naturelle des Crustacés Fossiles*, co-authored with zoologist Anselme-Gaëtan Desmarest, appeared with expensive and luscious trompe l'oeil lithographs of trilobites and other fossils (fig. 2.4).[6]

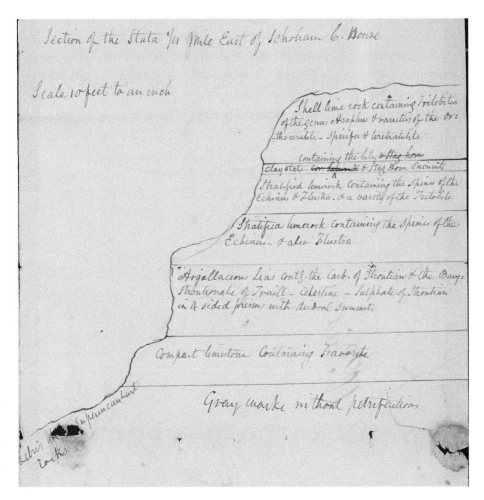

FIGURE 2.3. The lowest layer of rocks is labeled "Graywacke without petrifactions," meaning that this stratum contained no fossils, marking the boundary between a lifeless Earth and the origin of life. John Gebhard to Amos Eaton, September 22, 1834.

Just as important as the medium was the message. Brongniart's *Histoire Naturelle* advanced the provocative new thesis that these mysterious creatures were not insects (as Linnaeus had surmised) but rather crustaceans. They were like lobsters and crabs, but far older, with no apparent living analogue. Lacking feet and perhaps even eyes, they were

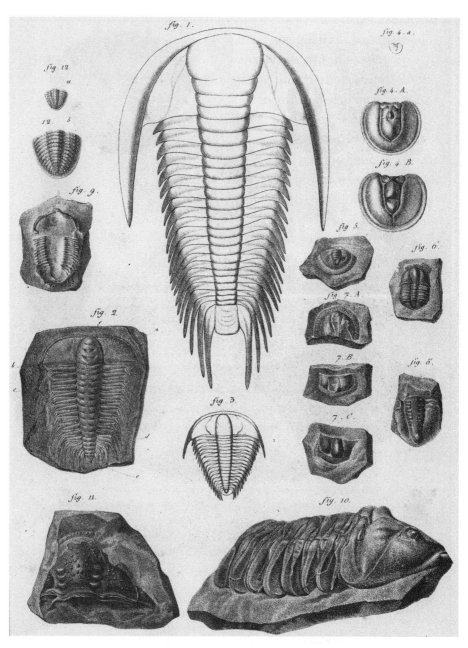

FIGURE 2.4. This magnificent image of trilobites appeared in a French book famous among American naturalists in the early nineteenth century: Alexandre Brongniart and Anselme-Gaëtan Desmarest, *Histoire Naturelle des Crustacés Fossiles* (1822).

also far stranger. They lay almost alone in the deepest rocks of the Earth, silent witnesses to the watery infant planet.[7]

Brongniart's *Histoire Naturelle* was the first major publication on trilobites to incorporate evidence from North America. This thrilled Americans who were accustomed to second-class citizenship in the republic of letters. Trilobites from what Brongniart called "le nouveau continent" bolstered his thesis that the creatures formed a distinct family. Brongniart had eagerly sought ideas and empirical evidence from his American colleagues. He thanked colleagues from Europe ("l'ancien continent") and then called out by name the American naturalists such as Yale's Benjamin Silliman, who had shipped trilobite samples to him from as far away as Ohio, and David Hosack, who presented him with a plaster model of a trilobite from Albany. Brongniart sent copies of his book to key American naturalists, for whom it became a kind of bible for the next several decades, traded among distant naturalists. Silliman thanked Brongniart for the information the book would provide for "those who would pursue this interesting but, hitherto, obscure branch of knowledge."[8]

Obscure they may have been, but now trilobites from the western fringe of the United States proved that the trans-Appalachian West was older than anyone had suspected, Brongniart maintained. William Maclure's influential geological map of 1809 had painted the region west of the Appalachians in blue to indicate that those rocks were "secondary": old but not the oldest. Brongniart now hazarded that the rocks of this region might be the most ancient "transition" rocks, so named because they were thought to have been formed when Earth was making the transition from uninhabitable to inhabitable. Transition rocks formed the lowest, earliest layers in which organic beings had ever been found. East of the Appalachians the story was the same. Brongniart ranked trilobites from Albany alongside French and Swedish specimens as among the oldest in the world. The northern half of the Appalachians was a great trilobite-filled spine, with the fossil creatures spilling out in outcroppings on either side, in New England to the east and Ohio to the west.[9]

Americans seized on the trilobites to gain entry into Europe's exclusive scientific circles. British and French naturalists often dismissed

Americans as "Cabinet men," fragmented and dilettantish compared with the "working men" of Europe, who had the money, leisure, and institutional structures to advance "Geological Knowledge." But now those American "Cabinet men" were using American specimens to press hard on European theories. They also swatted away what Amos Eaton called the "swaggering European braggadocio" who dictated "absurdities" to America's "patient enquirers after scientific truths" in the only country big enough to show strata in their proper order of superposition. A transatlantic trade in the small fossil creatures led to the growing suspicion that the trilobites of Europe and Americas were "identical," as New York naturalist James Dekay wrote. Now Europeans demanded "*very good* specimens of American *trilobites*," Silliman complained to a colleague. Famous European conchologists added American trilobites to their collections. The eminent English naturalist and shell expert James de Carle Sowerby agreed to Philadelphia paleontologist Samuel George Morton's offer of "an exchange of American for English Fossils," and promised Morton to have his trilobite paper published in an English journal.[10]

Fossil Messengers

Trilobites were messengers from primordial America. But what messages, exactly, were they sending—about the young planet and about God's plans for his creation? One American rose to the challenge of answering these questions: the Philadelphia naturalist Jacob Green. Along with his collaborator Joseph Brano, Green created colored plaster casts of trilobites that rank among the first models of extinct animals in American history. Dozens of them appeared in conjunction with his landmark book, *A Monograph of the Trilobites of North America* (1832), and the several other publications that followed it in the 1830s (fig. 2.5 and plate 4).[11]

Today Jacob Green's tiny yellow and brown trilobite models lie forgotten in the backroom storage cabinets of the nation's oldest natural history museums, such as the Academy of Natural Sciences in Philadelphia. It is the terrifying giants—the colossal T-Rexes and brontosauruses—that

FIGURE 2.5. Three of the small, painted plaster trilobite casts made by Jacob Green and Joseph Brano in the 1830s. They accompanied Green's book on North American trilobites and were intended to give distant naturalists a sense for the great variety of trilobites and the endless generative powers of the deity.

have a special place in the heart of the public, and museums featuring a small trilobite in the entrance lobby might well run into attendance problems. Modern historians have also been transfixed by the swashbuckling and publicity-hungry paleontologists of the post–Civil War "bone wars" to the exclusion of other dramas.

Yet the humble trilobites carried the mightiest message of all: that God's generative powers were endless, extending to the meekest of creatures and the earliest of days. Created in the first flush of deepening time in America, the plaster trilobite casts did not challenge the Genesis story of Creation so much as flesh out a parallel story that revealed God's benevolence through a natural world that now appeared to be far older than previously suspected. Unlike the static dinosaurs of the Gilded Age, these tiny models traveled from naturalist to naturalist, part of the mobile scientific culture of the pre–Civil War era that carried scientific information to cities and hamlets alike. Their ancient scales rubbed rough against much younger fingers, making the deep past present.[12]

Born in Philadelphia in 1790, just a few months after Benjamin Franklin's death, Jacob Green lived in a world in which science and religion formed a seamless whole. Franklin had helped to encourage a scientific infrastructure in this, America's largest city, with a population of around thirty thousand, and Green benefited from Franklin's vision. By his own reckoning, young Jacob was "a lover of natural history, in all its branches, from my youth up." He loved to drop by Charles Willson Peale's natural history museum in the Old Pennsylvania State House. By the time he was sixteen, he had fashioned an electrical machine to repeat Franklin's galvanic experiments. Smallpox when he was just two years old robbed him of most of the sight in his right eye. This may have influenced his decision to read the book of nature more intensively than the book of God. He may also have been searching for his own family niche. His father, the severe Presbyterian divine and future Princeton president Ashbel Green, openly favored his bookish older brother, Robert. Jacob collected so many plants, rocks, shells, insects and other curiosities that his parents banished the lot to the top floor of the family home, a room he promptly rebranded his "*museum*."[13]

From his father, Jacob learned that nature expressed God's benevolence—and God's wrath. A small child wandering Philadelphia's riverbanks collecting salamanders and driftwood would see first-hand the destructive power of tides and waves. As an adult, Green remained terrified by the fathomless darkness of the ocean, "with only a plank between you and death." Noah's flood was not a mythical event for him. It was the great world-shaping event that left "terrifick and awful desolation" in its wake. But the Bible had been vague on the details, and Jacob wondered whether fossils were remnants of life before the flood or part of a new creation after the flood. Given the violence of the deluge, he found it difficult to imagine that all present-day animals could be "the offspring of primeval parents." It was more likely that a new "creative fiat" had followed the flood, with either one or numerous sequential creations of trilobites and other creatures.[14]

Robert's early death plunged Jacob into a deep depression from which, according to his father, he never truly recovered. The loss launched him into a more spiritual direction, attuned to the mysterious veil separating life from death. He moved nearby to Princeton once his father had become president, and even spent a few years studying theol-

FIGURE 2.6. The American trilobite expert Jacob Green stands with his scientific apparatus in this portrait by John Sartain, after Hugh Bridport, from the mid-nineteenth century.

ogy, the family business. For a while he taught chemistry at Princeton. But he was soon back to his original love, natural history, and after a long period of dabbling finally settled down in 1825 to become a professor of chemistry at Jefferson Medical College in Philadelphia. A portrait shows him looking contented amid his scientific apparatus (fig. 2.6).

Overcoming his fear of oceans, Green embarked on a transformative seven-month trip to England, France, and Switzerland in 1828, when he was thirty-eight years old. From a modernizing Europe, he learned how to portray the most ancient America. The journey drove home two lessons that inspired the trilobite book and casts he fashioned after his return.

The first lesson was the importance of models and miniatures for the edification and moral elevation of the public. Disembarking in Liverpool in May, Green saw that England was deep into the Industrial Revolution. Hulking factories hid production processes that had previously been visible in small-time shops. In response, some factories had opened showrooms for the curious. Green saw facsimiles of the real thing, a new public-facing ethos that extended to art museums and civic celebrations, which now staged reenactments of famous events. He inspected a tiny pair of scissors the size of a bumblebee at the cutlery manufacturer in Sheffield; plaster models of future London buildings at the Royal Academy of Arts in London; an "exact copy" in bronze of an enormous Roman vase pulled from the ruins at Herculaneum; a scale model of the sphinx and pyramids of Giza in Paris's royal library. Everywhere he went in Europe, somebody was offering a facsimile of nature or human art reworked for the viewing public. Green became convinced that such demonstrations could "improve the state of society" and encourage literary and scientific institutions. More than this, he hoped that these facsimiles and models could "carry away the senses." The eminent geologist William Buckland showed Green the trilobite fossils in the British Museum, and he was awestruck by their sensuous presence. "They almost seem reanimated,—the bony fossil starts to life."[15]

A magical August afternoon found Green touring the porcelain manufactory at Sèvres with his hero, the trilobite expert Alexandre Brongniart. Green saw how art, industry, and science formed a unified whole. Eager to stay in the good graces of his American trilobite supplier, Benjamin Silliman of Yale, who often pleaded with him to give tours to visiting Americans, Brongniart now obliged by leading his Philadelphia visitor around the factory. The harried director took time to explain the complex processes of manufacturing the neoclassical melon bowls and sauceboats that had made Sèvres the leading porcelain producer of Europe. It was an encouraging omen of French national revival after the humiliation at Waterloo. The geologist in Brongniart also recognized that pottery survived in many archaeological sites, and stylistic dating permitted the creation of chronologies and comparisons across sites without absolute dates. Style and substance met in the deliberate neoclassical shapes of modern industry, announcing hopes that self-culture and national industrialization formed a unified project. "We examined minutely all the different processes for moulding, baking, painting and gilding . . . the china; and visited a number of apartments in which the finished articles were displayed to great advantage," Green recalled.[16]

Amid all the grandeur, there was much to inspire the citizen of a nation barely fifty years old. "In one room we saw specimens of china, from all the countries known to manufacture it," Green observed. But "there was a *blank* left for the United States, which was to be filled up, as I understood the guide to say, whenever we attained sufficient skill and taste to turn our attention to this subject." Green bought a set of Brongniart's teacups and saucers to show Americans how things should be done. It might have looked like the plate pictured here, manufactured under Brongniart's directorship and testament to the marriage of science, art, and industry in the service of France. The coffered interior border, evoking the neoclassical ceiling of the Pantheon in Paris, surrounds the shells that had made Brongniart famous in Europe and America (fig. 2.7 and plate 5).[17]

Green also took from Europe the lesson that the Old World, for all its beauty, was a carcass. Decaying châteaux and moldering abbeys littered the countryside, monuments to past glory. At Versailles, a few bored soldiers napped in the dusty galleries of the long-dead Sun King. Sèvres itself, churning out all that neoclassical pottery, was to Green the symptom of an exhausted society recycling ancient shapes. "Is this all," he wondered.[18]

Perhaps the United States—all future, with no vanished glory to haunt it—would light the way forward. As Green's colleague, the American geologist John Lee Comstock, pointed out, the trilobites predated any human art form. No works of art in antiquity could compare with the trilobites, Comstock announced. "Even the remains of Babylon and Egypt . . . are infants in age, when compared with these things."[19] The trilobites were a new beginning for a new nation.

FIGURE 2.7. Jacob Green may have bought a plate much like this one when he visited the French trilobite expert Alexandre Brongniart, director of the Sèvres Porcelain Manufactory outside of Paris, in 1828.

Returning to the United States at the end of 1828 flush with these new ideas, Green now brought the trilobites to life: not just in two-dimensional illustrations, but in three-dimensional, colored plaster models to be shared among American naturalists. "Multa renascentur quae jam cecidere" ("Many in disuse will now be revived"), read the Latin quotation from Horace on the title page of the book he published four years after his return, *A Monograph of the Trilobites of North America*. Green acquired trilobite specimens from dozens of private and institutional cabinets in Philadelphia, Pittsburgh, Reading, New York, Albany, and Baltimore, which he personally visited. From the cabinets he could not visit, he requested samples to be sent in the mail. Assembling information on dozens of North American species, he announced the trilobites' importance as God's first creation. "The Trilobite is supposed by many naturalists to be one of the first animated beings of our earth called into existence by the great Author of nature." God's first beings might even include not just Europe but America, since there were "identical species on both continents." Accompanied by the plaster casts, Green's *Monograph* became essential to American naturalists, who asked for Green's book and its "Indispensable" casts. From Europe arrived more approval. The Swiss naturalist Louis Agassiz praised it in his landmark book on fossil fish.[20]

The plaster casts had two intertwined purposes. One was to assist with species identification in remote areas of the United States, a chronic problem that everyone complained about since book images were often of poor quality. Colored "according to nature" by the Philadelphia modeler Joseph Brano, the trilobite casts were meant to mimic the color in which they had been found in the rock, such as the dark yellow limestone of one specimen found in Virginia. Costing ten dollars for a batch of thirty-six (a pretty penny at the time), the trilobite casts would be "*fac similes* of the objects described." Under the close supervision of the curators at the Academy of Natural Sciences in Philadelphia, Brano made casts of unique specimens in their geological cabinet. Benjamin Silliman waved them around in public lectures in Boston, telling the audience that these were among the earliest created beings: "We do not know of any that were earlier." Green shipped some to England to his new friend William Buckland, who had ushered him around the British Museum to see trilobites.[21]

The second purpose of the casts was to reveal God's plan in nature. The three-dimensional models promised to solve some of the great mysteries of the trilobites, none more momentous than the question of whether trilobites had eyes. Eyes would show that the Divine Author not only created these first life forms but was concerned that they see and perhaps enjoy the sunlit oceans of the young planet. So great were God's generative powers that they created not just life but sensation and even pleasure. The trouble was that the stony lumps on the side of trilobite heads were not unambiguously organs of vision. The question of the trilobite eyes fractured the scientific community. "Nothing that we know of the subject authorizes us to infer, that these were ever eyes," Amos Eaton told his diary. "Brongniard [*sic*] himself admits that this is not a strong evidence." Other early publications on the trilobites concurred that there was "a want of eyes."[22]

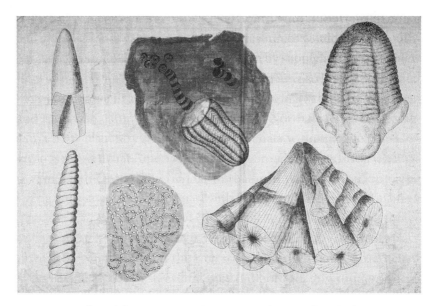

FIGURE 2.8. The trilobite is at top right in Orra White Hitchcock's linen poster made for classroom use at Amherst College, c. 1828–1840.

Jacob Green was sure they were eyes. "The conical eye-like protuberances on the head of this species, are very remarkable" he wrote, leaving "no doubt that they once contained the organs of vision." His plaster casts would spread the good news to the scientific community even as he awaited the discovery of living trilobites, a prospect much more appealing than the "blank" hypothesis that the biblical flood had wiped the little creatures from Earth. Some geologists thought trilobites still lived today, and that the modern crustaceans could settle the eye questions once and for all. Promising candidates had been unearthed on the Falkland Islands by the geologist James Eights, a student of Amos Eaton. The American naturalist Timothy Abbott Conrad was so sure that trilobites had eyes that he wrote an ode to a trilobite, "Thou large-eyed mummy of the ancient rocks." American college students learned of the divine trilobite eyes in their classrooms. Enormous paintings were being unfurled in the front of some innovative classrooms in the 1840s, such as those at Amherst College. They showed off the "perfectly preserved" eyes of the trilobites (fig. 2.8).[23]

The divine message of the trilobites was international, as England's William Buckland showed in his influential *Bridgewater Treatise* on geology and mineralogy (1836), which was reprinted in the United States. Buckland explained that trilobite eyes offered a "marvellous" example of delicate "organs of feeling" from animals that had perished "many thousands, and perhaps millions of years ago." Human beings, too, shared those organs, "through which the light of heaven was admitted to the sensorium of some of the first created inhabitants of our planet." Connecting the "extreme points" of animal creation, the ancient and modern eyes pointed to one conclusion: "the exercise of one and the same Intelligent and Creative Power."[24]

It was not only human beings who could see and feel. From the infant Earth, fossil messengers announced the intelligence and creative power of a deity who worked his wonders through thousands and perhaps millions of years. Even the lowly trilobite, inching along the shallows of the first seas, had basked in the feeble rays of a young sun.

How long ago had this happened? By the 1830s, a variety of answers had been proposed to align Genesis with geology. Some urged that the days of Genesis were twenty-four-hour days; others promoted a "gap" theory (in which a gap of time opened at certain points in the Genesis narrative to allow time for geological strata to be deposited); others proposed a "day age" theory that interpreted each day of the Bible as thousands or millions of years. Still others ignored geology and revived James Ussher's iconic 4004 BC Creation. The influential educator Emma Willard's diagram for American schoolchildren floated Ussher's date in heavenly clouds, with human history stretching forward and the birth of Jesus marked by a star (fig. 2.9). In a public lecture in Boston in 1835, Benjamin Silliman tried to smooth out any conflicts between Genesis and geology. "The audience were assured," he recorded in his journal, "that when the facts of the course are all before them that they are entirely consistent with the Mosaic account properly understood. The order of the geological arrangements corresponds with the sacred history but it must be understood as to allow sufficient time for the events to be brought about by natural laws which are only the expression of the will of God as it is recorded in the structure of the globe".[25]

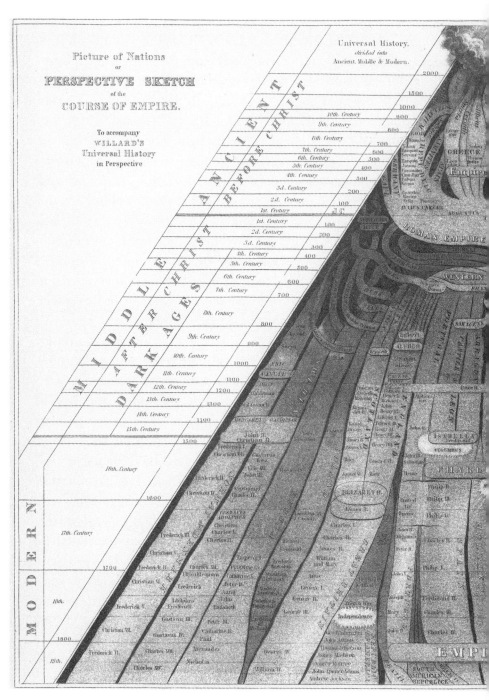

FIGURE 2.9. All history unfolds from the Creation in 4004 BC in Emma Willard's "Picture of Nations, or, Perspective Sketch of the Course of the Empire" (1836).

Ancient History.
divided by Epochs into 6 Periods.
Middle. into 5 Periods.
Modern. into 9 Periods.

Year	Event	Period
1921	The calling of Abraham.	1st.
1491	Institution of the Passover. Moses leads forth the Hebrews from Egypt.	2d.
980	Death of Solomon.	3d.
752	Foundation of Rome.	4th.
		5th.
323	Death of Alexander.	6th.
J.C.	Birth of Christ.	
		1st.
395	Division of the Roman Empire.	2d.
622	Flight of Mahomet.	3d.
800	Coronation of Charlemagne.	4th.
1100	Commencement of the Crusades.	5th.
1492	Discovery of America.	1st.
1559	Treaty of Chateau Cambresix.	2d.
1610	Death of Henry IV. of France.	3d.
1649	Execution of Charles I. of England. Peace of Westphalia.	4th.
1713	Peace of Utrecht.	5th.
1748	Peace of Aix la Chapelle.	6th.
1776	American Independence.	7th.
1803	Coronation of Napoleon.	8th.
1815	Slave trade	9th.

Explanation.

That events apparently diminish when viewed through the vista of departed years is matter of common place remark. Applying the principle to a practical purpose, we have here brought before the eye, at one glance, a sketch of the whole complicated subject of Universal History. Names of nations and a few distinguished individuals are found in the Ancient; of the most distinguished sovereigns in the Middle; and of all the sovereigns of the principle kingdoms in Modern History.

Despite the growing presence of fossils in museums, cabinets, lectures, books, and magazines, the Garden of Eden also remained a visible presence in children's spellers, history books, and paintings. The new beginning of life suggested by the trilobites did not immediately displace the traditional biblical account of creation but rather grew alongside it. French artist Claude-Marie Dubufe's enormous paintings *Adam and Eve* and *The Expulsion from Paradise* (both 1827) toured the United States in 1832 and 1833. Dubufe, who studied under the preeminent neoclassical painter Jacques-Louis David, had mastered his teacher's waxy realism. In garish colors, he brought Americans face to face with the Genesis story. In *Adam and Eve*, the serpent ejects a white spray toward a sumptuous Eve, who slips the fatal apple to Adam, fig-leafed and startled. The "moral" and "chaste" painting was suitable for both ladies and gentlemen, announced the *Albany Evening Journal* (fig. 2.10).[26]

By the 1830s, Americans had not one beginning but several beginnings to ponder. More was more: more beginnings did not challenge or eclipse Genesis (yet) but rather compiled more evidence from natural history that God had taken care to include America in the Creation. Jacob Green's trilobite eyes suggested that nature revealed the divine, just as the Bible did. The planet's beginnings were not only to be counted in numbers or parsed through biblical philology. They were to be lifted from the rocky ground in the United States. They were to seen, touched, and classified in wooden cabinets. The trilobites became a new theater for the imagination, which was propelled back in time to imagine the first moments of life. As Green's colleague Timothy Abbott Conrad explained, "the earliest physical condition of the globe" was a job not just for reason but for the imagination. These deepest fossil-bearing strata offered "peculiar attractions both for the reason and imagination," facts "colored to the eye of inexperience with all the exaggeration of romance."[27]

The Silurian System

The American trilobites were old. But was America therefore the oldest of continents? Now that geology was offering up strata as a measure of antiquity, the race was on in Europe and America to determine which

FIGURE 2.10. The tangible reality of the Garden of Eden for Americans is revealed in this French painting, Claude-Marie Dubufe's *Adam and Eve* (1827), which toured the United States in the early 1830s.

nation was the oldest of all. A new most ancient era was summoned and fully imagined: the Silurian. The Silurian became the natural Eden in which Earth's first animated beings had appeared. As nations vying for global dominance competed to be the most modern in science and technology, they lashed themselves to a new deep time. The two projects of industrialization and deep time were connected. Even though early human civilizations—Egypt, the classical world—continued to provide cultural touchstones and stories of noble national origins, science and industry required something more solid. The Silurian offered access to the earliest moment in the planet's history—a kind of Genesis in its own way—but with the added benefit of economic utility. For above Silurian strata lay the coal layers that powered modern industry. The presence of Silurian rocks indicated that the ground above might yield treasure troves; lucky was the nation that sat atop Silurian strata. Trilobites became tiny beacons, signaling the possibility of treasure nearby.

The debate over national priority of the planet's most ancient fossil-bearing deposits centered on the British geologist Roderick Impey Murchison and his three-volume work, *The Silurian System* (1839). The product of nearly a decade of fieldwork in Wales and western England, it revealed some of the world's most ancient fossiliferous strata, "a remote antiquity teeming with organic remains," as Murchison put it. These rocks were so old that they lay even "*below* the Old Red Sandstone"—that is, below the primordial rocks that had fired debates among the British scientific community for some time. The Silurian strata lay at the very limit of life, for in the layers beneath them "no traces of life have been discovered."[28]

The words *Silurian* and *system* in Murchison's book title were both significant. The conceptual power of the Silurian System was to offer a naturalistic organization for Earth's earliest days that also had real economic implications. Murchison christened these ancient rock layers *Silurian* to honor the Silures, an ancient tribe of western Britain that had fought fiercely, if in vain, against the invading Romans under Emperor Claudius. The name was nationalistic, reminding the British of an ancient era when the soil was theirs, unsullied by later foreign invaders, whether Roman, Scandinavian, or Norman. What made these Silurian strata a *system* was the perceived unity of space and time that Murchison thought they represented: the strata consisted of characteristic fossils deposited in distinct periods of geological time. Strata now joined other natural phenomena that were thought to form interconnected wholes in time and space, such as the nervous system and the solar system. The Silurian strata seemed to lie mostly below the great coal formations that were transforming Wales from a sleepy agricultural periphery into one of industrializing Britain's mining powerhouses. The Silurian was a period in Earth's history before there were large plants from which the coal seams had been formed, and so became an obvious marking point beyond which it was futile to spend precious funds to mine for coal. Murchison's *Silurian System* would distinguish where coal "*may* be advantageously sought for" from "where *it can never be found*."[29]

Roderick Impey Murchison was the ideal messenger for Britain's Silurian System. Born to the Scottish aristocracy in 1792, he embodied the ethos of British imperialism in the age of Victoria, seeing science as part of the nation's world colonization brief. He dressed the part, posing with fossils in a starched shirt whose high collar partially engulfed his muttonchop sideburns (fig. 2.11). His long career included supporting David Livingstone's journey to find the source of the Nile and the presidency of the Royal Geographical Society, the exploration arm of the British empire. His nickname, "King of Siluria," described his haughty persona (he bragged of grouse-shooting while fossil-hunting) as well as his ambition to extend the Silurian System to the world at large, including the vast domains of Russia. Murchison must have nodded in agreement when a correspondent urged "a national point of view" in British science.[30]

Citing Americans such as James Dekay and Jacob Green, Murchison conceded that among these foreign Silurian deposits were some that "unquestionably exist in North America." This was welcome news to Americans, who had been unearthing trilobite fossils for decades, and now had a system in which to place and legitimize them. Murchison's *Silurian System* immediately caught on among American geologists, appearing in their college textbooks and public lectures. The grateful Americans shipped more American Silurian fossils to Murchison, who responded with encouraging words about the progress of geology in America. But it would not, he hastened to add, be possible for him to visit the United States, in view of his busy schedule. Given the nationalistic nature of naming to which geologists had by now committed themselves, a debate soon erupted about whether there existed in the

FIGURE 2.11. Scottish geologist Roderick Impey Murchison, aka the King of Siluria, poses with fossils.

United States a local version of the Silurian—to be called "Taconic" after a local tribe of Native Americans. But Murchison's Silurian ultimately triumphed, becoming the dominant name for these early strata, even in the United States.[31]

Some Americans argued that the Silurian System was more visible in North America than in Europe. In New York alone, wrote Timothy Abbott Conrad, there were three times as many Silurian strata observable on the landscape as in Murchison's Wales. The same could be said for the United States as a whole. England could be dismissed as a good source for primordial rocks. "England is a limited theatre for the display of the order of succession, which sinks into insignificance in comparison with the colossal development of the transition in North America," wrote Conrad. The major American textbook on geology also announced the good news.[32]

Murchison soon came around to agreeing that the Silurian—while it existed elsewhere in the world—was most legible in North America. He made this clear in *Siluria* (1854), which updated his thesis about a Silurian System based on new findings from naturalists around the world. Murchison conceded that American geologist David Dale Owen's discovery of trilobites in Iowa, Wisconsin, and Minnesota showed that "the United States, does really contain the first recognizable group of fossil animals." In the United States, whose upper Silurian rocks were "the same age" as similar formations in England, many formations were of much "vaster dimensions." The King of Siluria had spoken.[33]

More Beginnings

Between 1840 and 1870, the Silurian—the trilobite-rich stratum containing the first glimmerings of life—was further embellished. Maps of North America featuring the Silurian extended the continent's history backward in time to the moment of Creation itself. François de Castelnau, a student of Georges Cuvier, toured the United States and Canada from 1837 to 1841 and published one of the earliest maps of the Silurian System in North America. Nearly the whole continent rested on a Silurian foundation, from the Atlantic northeast to the big Silurian swamp of the far west. This was not a new world but a very old one (fig. 2.12).[34]

Landscapes of the Silurian period appeared as well, first in literary form as imaginary promenades. In the 1850s, women's magazines such as the *Ladies' Repository* invited Ohio's women on a "trip to Siluria" and on a botanical walk to inspect the plants of the Silurian period. As the question of women's political, economic, and educational rights captured national attention, Americans debated what was "natural" for the female sex. The biblical Eve had long been used to understand what was primal and therefore "natural" for women. The Eve story had pinned the

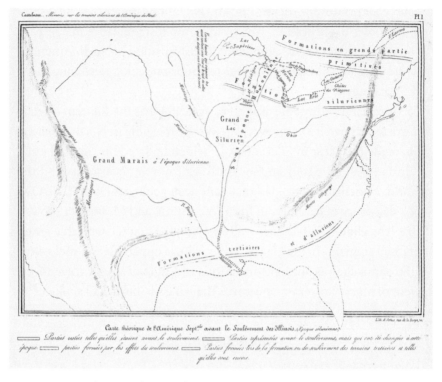

FIGURE 2.12. This French map of the Silurian regions of North America, from François de Castelnau's *Essai sur le Système Silurien de l'Amérique Septentrionale* (1843), unveils the extreme antiquity of the continent.

primal transgression onto the female sex, who now (it was said) bore the pain of childbirth as a constant reminder. But the Silurian opened happier horizons. Young women at the coeducational University of Wooster in Ohio could look at magic lantern slides of the Silurian. In one slide, the rosy dawn illuminates a primordial island world of bizarre plants, a garden so ancient that it preceded not only Eve but the serpent and the apple as well (fig. 2.13 and plate 6).[35]

Even as this first naturalistic Eden of the Silurian entered mainstream culture in the 1870s, geologists were uncovering still earlier signs of life on Earth. In the *Popular Science Monthly*, an American geologist announced the "fact" that both plant and animal life "flourished through untold ages" before the trilobite era. Canadian geolo-

FIGURE 2.13. This magic lantern slide of the infant Earth, fresh and clean in the rosy dawn light of the very early Transition Period, must have enchanted the undergraduates at the University of Wooster in the decades after the Civil War.

gist John Dawson's *The Story of the Earth and Man* (1873) popularized this Eozoic or dawn time in the planet's history. Though antitransformist, Dawson accepted the reality of deep time by interpreting the days of Genesis as very long periods of time. Dawson proposed the gelatinous "dawn animal of Canada," or *Eozoön canadense*, as perhaps the oldest known animal on Earth, the first creature in which "the wondrous forces of animal life were manifested" by God (fig. 2.14). The dawn animal had been discovered a few years earlier in eastern

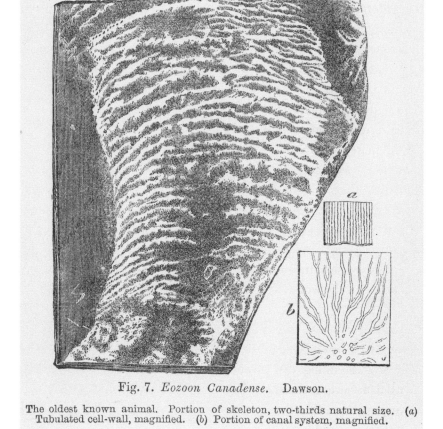

Fig. 7. *Eozoon Canadense.* Dawson.

The oldest known animal. Portion of skeleton, two-thirds natural size. (a) Tubulated cell-wall, magnified. (b) Portion of canal system, magnified.

FIGURE 2.14. Was this the first life on Earth? The tiny, gelatinous *Eozoön canadense*, or dawn animal of Canada, as it appeared in John Dawson's *The Story of the Earth and Man* (1873).

Canada, a region Dawson called the "nucleus" or oldest region of the North American continent. The planet's most ancient formations were far better exposed in North America than in the "old world," he continued. Little Eozoön offered a window into the "actual workshop" of God. And its era, the Eozoic, proved that God worked very slowly indeed, and on a vast canvas of time. "Perhaps no lesson is more instructive than this as to the length of the working days of the Almighty," Dawson concluded.[36]

Much had changed in barely half a century. What began as a few curious trilobite fossils unearthed in the 1820s had exploded into colorful scenes of primordial creatures set in strange but beautiful landscapes. Fieldwork, plaster casts, magic lantern shows, and images in scientific and general publications fleshed out this earliest Earth. Tiny trilobites washing ashore under a pale Silurian sun, it now seemed, were but late arrivals on the planet. For before these lurked an even earlier sea creature, the bizarre Eozoön, so ambiguous as to be only debatably a life form at all. All these words and objects made the young planet as believable to the mind and the senses as the Garden of Eden. For a while, the two first gardens coexisted, each offering its own set of stories and possibilities.

But along with new possibilities came new questions and dilemmas. One was the idea of a single beginning. In 1875, the Harvard philosopher Chauncey Wright argued that the meaning of the word *origin* had shifted over the course of the nineteenth century. It used to mean a first introduction into the world, as the Genesis phrase "In the beginning" suggested. But now the word *origin* often meant things merely changing in the course of events, as in the title of Darwin's *Origin of Species*. Darwin studied how species changed, not the moment of their first appearance. With "the beginning" (the moment of life's origin) now tantalizingly out of reach, it became more plausible to focus on "beginnings" (the process, such as natural selection, by which species changed through time). Nothing was solid; all was change.[37]

Some Americans began to wonder if beginnings were by definition unknowable. The geologist Joseph LeConte explained this to the students at the new University of California in 1878. Pleased that North America had revealed the very oldest rocks in the world, LeConte noted that these rocks, too, showed evidence of having been created from still older rocks, and those from even older rocks. One searched in vain for the first rocks. Everything was always made of something older. Undergraduates must have paused when they read his existentially troubling conclusion: "Thus, we search in vain for the so-called *primary* rocks of the original crust. Thus is it with all history. No *history is able to write its own beginning.*" Everything unfolded from a still earlier thing, backward into infinity.[38]

With no discernable beginning, could there be any direction to life at all? Henry Adams pondered this gloomy prospect a few years after the Civil War as he sat on a ridge in Shropshire, England, called Wenlock Edge. The nearly twenty-mile-long limestone ridge encased Silurian trilobites and other organisms from the primeval seas. Adams had been a student at Harvard in the 1850s, during the heyday of Louis Agassiz, whose lectures on geology he remembered as the only bright spot in an otherwise bleak and desiccated curriculum. Harvard had taught him to believe in "a universe of unities and uniformities." Now, among the fossils of Wenlock Edge, he thought the universe offered neither unity nor uniformity, let alone comfort. The Civil War had crushed any hope of moral evolution. And now only an "eternal void" lay prior to the Silurian ridge on which he sat. He really did not care what truth was, he realized. Adams had entered "a far vaster universe, where all the old roads ran about in every direction, overrunning, dividing, subdividing, stopping abruptly, vanishing slowly, with side-paths that led nowhere, and sequences that could not be proved."[39]

By the 1870s, North Americans had gathered compelling evidence that their continent was very ancient indeed, perhaps more ancient than the Old World. The trilobites had first come to notice as curiosities wrenched from the Erie Canal. Now they pointed backward to a planet so ancient that it was already old when the trilobites came along. "A very great lapse of time," as LeConte put it, had passed since the Silurian seas had swept the trilobites along in their currents. Along with these thrilling ideas came the unsettling possibility that there was not one Beginning but multiple beginnings receding into a near "blank" in the record of Earth's history.[40] So where did Americans come from? And where were they going? Fortunately, the next layer of rocks offered some answers, a fossil future sparkling with possibilities for the nation.

3

Fossil Futures

This is the forest primeval.
—HENRY WADSWORTH LONGFELLOW, *EVANGELINE* (1847)

FROM THE DECK of their ship, Charles and Mary Horner Lyell could see the spruce-covered hills of Nova Scotia rising before them. He was the famed Scottish geologist, author of the internationally renowned *Principles of Geology* (1830–33). She was an accomplished conchologist and geologist in her own right. The couple stood aboard the *Acadia*, one of Cunard's new coal-powered steamships. Belching black smoke, the vessel had sped the couple from Liverpool across a mild Atlantic Ocean in July in a mere eleven days—about half the usual time for sailing ships. Invited to give the Lowell Lectures in Boston, Charles Lyell had also come for a yearlong geological tour of the place he called the New World.

Within hours of docking in Halifax, the Lyells had already made their way to the local museum. They paused to look at the object that had generated the coal powering their ship: "a large fossil tree filled with sandstone, recently sent from strata containing coal in the interior." As the couple traveled thousands of miles over North America, they marveled at what Charles called "the most wonderful phenomenon perhaps that I have seen." These were the underground coal forests of North America, some with trunks still standing fully upright. The American fossil forests were vast, sometimes entombed deep underground, sometimes thrust upward as great mountainous wrinkles. "This subterranean forest exceeds in extent and *quantity of timber* all that have been discovered in Europe put together," Lyell told his sister.[1]

It was 1841, and this was the Age of Coal. The name was given by the English economist William Stanley Jevons to capture how this rock fundamentally transformed the modern world: economically, politically, scientifically. The discovery of thick coal seams stretching from England and Wales to Pennsylvania and Ohio sent both Europe and the United States on the path to the Industrial Revolution. Britain arrived first, but the United States was gaining fast. Worried British economists broke the news at the end of the Civil War: America's coal supply exceeded Britain's thirty-seven to one. By 1885, coal had become the leading source of energy in the United States. The terms *fossil fuel* and *natural resources* date from this era, pregnant with optimism about the past Earth's gifts to the present.[2]

The Age of Coal was also the first era to hitch the planet's deep antiquity not just to the present but to the future. Economists, politicians, and engineers in Europe and America summoned statistical data about population growth and coal supplies to paint scenarios about the century ahead, signaling what the old Earth offered to the new. Some forecast a sunny future of human comfort based on the exploitation of coal. In 1830, Yale naturalist Benjamin Silliman marveled at America's "Gigantic vegetables." The excavation of America's coal beds "is only begun and ages will not exhaust these immense deposits of coal, the tribute of early geological ages to modern arts and progressive civilization." But as the decades passed, worries mounted. The steep "Jevons curve" in English economist William Stanley Jevons's influential *The Coal Question* (1865) forecast empty coal beds by the year 2000 (fig. 3.1).[3]

But in the first flush of the Age of Coal, the optimists held the field. Coal had brought comforts to life that could only grow with time: warmth, light, speed, power. The bright coal future was calculated with mathematical precision from imports and exports, consumption patterns, and population projections. The total equaled national coal dominance, which equaled human happiness. This was the promise of the new fossil future.[4]

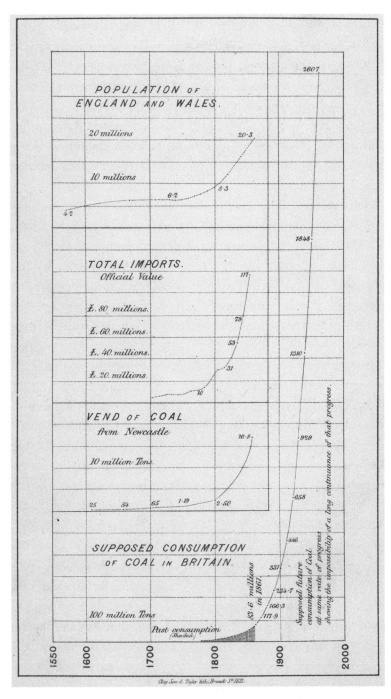

FIGURE 3.1. The alarming "Jevons Curve" projected steeply increasing coal consumption between 1800 and 2000. From William Stanley Jevons, *The Coal Question* (1866).

The fossil future did not belong only to coal economists and engineers, however. American naturalists also thought about the coal beds, but they chased different quarry than profits and efficiencies. What they cared about were the ancient plants that had produced the coal: the gigantic fossil vegetables. These stony trunks and fronds were the shredded remains of the mighty forests that once swaddled the infant Earth. Compacted into subterranean ribbons of carbon, the primeval forests powered the Industrial Revolution and gave the ancient era its new name, the Carboniferous.

Where the capitalists saw the invisible hand of the market in the triumph of coal-powered industry, the naturalists saw the benevolent hand of God. Transforming America's dark Carboniferous forests into a golden vision of tomorrow, they spun a new tale of American national glory based on a primordial gift of latent power now unlocked by modern engines. The Carboniferous became part of everyday experience. In everything from greenhouses to college curricula to lyceum lectures to paintings, Americans summoned to life an ancient era unsuspected until now, and with it a vision of their nation's divinely ordained destiny to lead the world in coal production and coal consumption. Earthly salvation lay underground, God's gift from the past to the future. As Charles Lyell himself wrote of Ohio after his first visit to the United States: "The time has not yet arrived . . . when the full value of this inexhaustible supply of cheap fuel can be appreciated; but the resources which it will one day afford . . . are truly magnificent."[5]

The Occult Science

These naturalists were, it is safe to say, an anxious group. They saw that most people were not fascinated by fossil vegetables. They knew that the ancient era whose fragments they pondered was "a world little looked into, or thought of." It was a place of steaming swamps, colossal insects, and towering ferns, a shadow world before dinosaurs or even flowering plants had appeared on Earth. They called their field an "occult science," as though they were summoning ghosts. Their scientific publications, full of grey images of fern fragments and root tendrils, look

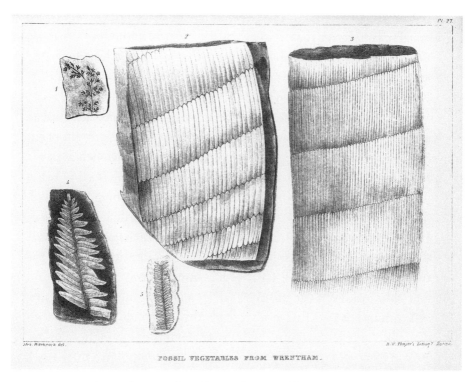

FIGURE 3.2. Orra White Hitchcock's 1841 drawing of fossil plants from the Carboniferous period.

like catalogues of wallpaper samples destined for gloomy Victorian parlors (fig. 3.2).[6]

But in their hearts, these students of coal knew their field was the most important of all. Their goal was to uncover "the early, the successive, the magnificent works of the GREAT CREATOR." They spun a new planetary history from the stony relics they collected in their fossil cabinets. As the global flood receded, scattered islands and then finally continents surfaced. Bathed in tropical warmth, plants colonized these fragile beachheads. Scraggly mosses stretched skyward to become the enormous palms, ferns, and conifers that carpeted the planet. This was God's first forest.[7]

These naturalists promised to answer the great question of the Age of Coal, asked by one exasperated American in the middle of the nineteenth century: "What is coal?" The fact was that for all its heat and light, coal remained mysterious before 1850. The discovery of the nature of coal was one of the early achievements of these naturalists, not least for being knowledge deeply desired by the coal industry. What was coal made of? How had it formed? When had it formed? Where? And why?[8]

To the last question—why was there coal?—these naturalists had an answer: God. Coal, they agreed, was nothing less than God's gift of energy from the past to the present—and the future. Words such as *energy* now crept into writings about God. In the Genesis story of Creation, the plants appear before the animals so as to provide them with food. The function of plants was to energize animals, a brilliant example of God's "creative energy," according to Edward Hitchcock of Amherst College. It was now agreed that God's creative capacities came in the form of power supplies. Coal was ancient energy stored by God and later unlocked for the civilization and happiness of humanity. "[N]othing has struck us more forcibly than the abundant supply with which Providence has furnished the inhabitants of our globe, particularly in its northern hemisphere," one geologist wrote in a thick book about coal statistics. Manpower was measured against coal power. Both were "dynamic," but the "latent strength" of coal was thousands of times greater than even the mightiest man. God had lavished coal upon his most favored nation, a divine truth rendered as a stark statistic: the superficial coal deposits in the United States were far bigger than anybody else's (fig. 3.3).[9]

Coal's dazzling promise explained what Charles Lyell was doing in North America in 1841. Perhaps on his lecture tour the great Scottish geologist could answer the coal question—and other pressing questions about the deep history of the Earth and North America. Although some British colleagues thought him a dull and long-winded lecturer, many in the general public found him eloquent, bringing Earth's ancient history to life with vivid analogies. A decade before his arrival in North America, when he was in just his mid-thirties, Lyell had published his controversial *Principles of Geology*. In readable prose studded with relatable examples, Lyell had offered a new way of reading the planet's history. It was not a tale of repeated cataclysms operating more forcefully

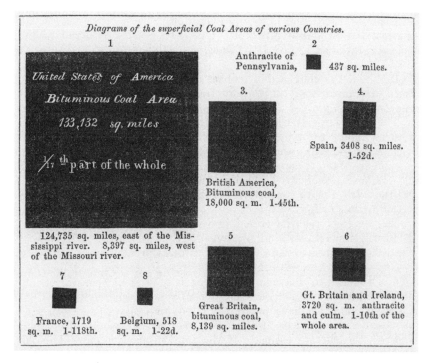

FIGURE 3.3. The size of the black boxes in this illustration from Richard Taylor's *Statistics of Coal* represent the square miles of bituminous coal in various nations.

in the past than in the quieter present, a scenario his older colleagues in Britain had proposed and that Lyell now dismissed as a "train of absurdities." Lyell argued instead for a steady-state Earth, in which the same causes that had shaped the past Earth were also shaping the present Earth, over an immense quantity of time, perhaps millions of years. The present was the key to the past.[10]

This steady-state theory became known as *uniformitarianism*, an insult first hurled against it by skeptics. But soon the term was commonly used to distinguish Lyell's theory from its opposite, now dubbed *catastrophism*. Scorning miracles and catastrophes as "mysterious agents" of geological change, Lyell insisted instead that the "fixed and invariable laws" of nature operated slowly over long periods of time to create the modern world. In his view, even catastrophes were merely the scaling up of known processes rather than a fundamental changing of the rules, as represented by miracles.

Lyell's view linked geology to human cognition. Earth's history could be told only through the relatively few rocks and fossils that had survived to be discovered by human beings. Most of the planet's past could not be known, since the evidence of rocks and fossils had been lost. All this still implied for Lyell the existence of a divine author of nature, who created "the reasoning powers of man," as he explained in one of his public lectures. Reasonable human beings would decode the meaning of fossils and thereby write Earth's history. Subterranean deposits of coal were made of plants belonging "to species which for ages have passed away from the surface of our planet." Any nation harboring coal deposits must be very old indeed. Uniformitarianism also suggested that the future would be somewhat knowable, since "all times, past and future" had been and would be driven by the same unchanging laws of nature.[11]

Lyell did more than just affirm the antiquity of North America by highlighting its coal deposits: he gave the continent a divine purpose. North America's history stretched from a swampy past into a happy present and an even more glorious future ordained by God. Coal was the thread of a story that also justified imperialism. The God-given tropical forests of today in the world's equatorial belt were merely "future beds of lignite and coal." As he toured the nation in 1841 and 1842, Lyell became the itinerant preacher of the new coal creed.[12]

Ohio

Lyell was especially eager to visit Ohio, where what he called "prodigious" coalfields forecast an "inexhaustible supply of cheap fuel." Admitted into the Union as a free state in 1803, Ohio was a flat slab whose eastern third crumpled into the coal-filled Appalachian mountains. With slavery legally cabined below the state's southern border on the Ohio River, and with a series of treaties and wars pushing Native peoples such as the Shawnee westward beyond the Mississippi, Ohio was emerging as a coal-powered, free-labor industrial leader. By the middle of the nineteenth century, Cleveland led the nation in iron and steel production. New railroads put the state at the heart of the nation's westward-moving transportation infrastructure.[13]

Ohio was also home to a group of naturalists who clustered in the state's coal-rich eastern towns: John Locke of Cincinnati, Caleb Atwater of Circleville, Ebenezer Granger of Zanesville, and Samuel Hildreth of Marietta. Treated by East Coast professors as frontier bumpkins to be acknowledged only when useful, the Ohio group was eager to "to impress haughty Easterners ignorant that Ohioans are not quite Barbarians," as one Ohio naturalist confided to another. In the Ohio coal formations—"almost as extensive, as all Europe," according to Benjamin Silliman—they seized their chance to shine on the world stage.[14]

For answering the coal question, the major figure in the Ohio group was Samuel Hildreth. Born in 1783, Hildreth was a small-town doctor with no academic position but a passion for natural history pursued as a gentlemanly hobby. He lived in Marietta, a hamlet nestled at the confluence of the Ohio and Muskingum Rivers cutting through the eastern Ohio coal country. Moving to Ohio from Massachusetts in 1806, Hildreth parlayed his strategic location into the nexus of an international network of fossil plant exchange. Ohio miners, farmers, and naturalists sent him reports of "Carbonized wood" and "Impressions of Ferns" from local mines. When he died in 1863, his natural history collection numbered four thousand specimens. Ohio's riches attracted the attention of coal experts on the East Coast. Edward Hitchcock of Amherst College promised to ship boxes of specimens and minerals from all over the world if Hildreth would only send "a suite of specimens from your bituminous formation."[15]

A more macabre traffic mixed in with the fossil plants: the skulls of Native Americans. White settlers culled Native skulls from the ancient mounds dotting Ohio, as well as from the ongoing modern wars they waged to push the Indians off the land to make way for farming and industrialization. Digging for coal in deep time was a way to undercut—quite literally—the claims of more recent humans to habitation in a particular region. Native skulls and fossil plants could be used to show the inferior cultural capacities of the Indians, exemplified by their failure to exploit the natural resources that lay just beneath them in the coal fields of Ohio. It was "the ordinance of Divine Providence that he [the Indians] must be exterminated," wrote one of Samuel Hildreth's correspondents. Hildreth sent skulls to many a "learned Craniologist" on the East Coast, such as the Philadelphia physician Samuel George Morton. The skulls piled up in Morton's cabinet and also appeared in widely known illustrated works as Morton's *Crania Americana* (1838) and Ephraim George Squier and Edwin Davis's *Ancient Monuments of the Mississippi Valley* (1848). The human skull and fossil plant networks reached from the swampy past into the violent present, as scientific knowledge began to be used to assert the primacy of white settlement, economic exploitation, and cultural values. Geological conceptions of deep time, in Ohio as elsewhere, began to be brought to bear on pressing social questions.[16]

Hildreth's major fossil plant agent on the East Coast was Benjamin Silliman at Yale. Ever since his youthful visit to a coal mine in Newcastle, England, in 1805, Silliman had been pushing his New England colleagues to investigate the American coal formations.[17] Silliman's letters to and from the father-and-son fossil plant experts Alexandre and Adolphe-Théodore Brongniart of Paris also alerted him to the fact that Americans still lagged far behind Europe in the study of fossil vegetables and coal formations. Silliman saw that Hildreth, whom he ranked as one the "brighter stars" of the West, could help catapult the United States to the forefront of research in this area. Beginning in the 1820s, he peppered Hildreth with requests for "impressions of plants & of any organized bodies on your coal strata" that Silliman could then ship to European savants. In exchange, Silliman offered Hildreth access to his rarefied scientific world: to Charles Lyell (supplied by Silliman with a letter of introduction to visit Hildreth in Marietta); to East Coast audiences (by mentioning Hildreth's coal plant findings in his own public lectures on geology); and to new scholarship from Europe (by sending Hildreth the important new books on coal plants).[18]

The most important of these European books was Adolphe-Théodore Brongniart's *Histoire des Végétaux Fossiles* (History of fossil vegetables, 1828–37), the first volume of which Silliman duly shipped to Hildreth.

Just twenty-seven when he launched the multivolume project, Adolphe-Théodore Brongniart was the newest generation of the prominent Brongniart family of Paris. Their numerous scientific competencies over three generations extended from architecture to ceramics to fossil shells. The history of fossil vegetables—launched when he was only seventeen years old—immediately established this youngest Brongniart as the undisputed world expert on coal plants.

To a scientific world infatuated with animals, Adolphe-Théodore Brongniart explained the importance of studying plants—and especially fossil plants. Unlike animals, which (as their name suggested) wandered about, plants were rooted in place, doomed to endure local conditions or perish. This meant that plants were better guides to past atmospheres and other clues about the early Earth. Brongniart larded his work with new classifications for fossil plants to rival the modern botanical classifications of Linnaeus. He dedicated his work to Georges Cuvier, the world's authority on fossil animals, because he saw his plant studies as the necessary companion to Cuvier's investigations. Just as Cuvier's success had hinged on comparing bones sent to him from around the world, Brongniart put out the call for plant fossils from outside of France. The main supplier of American coal samples would be none other than Benjamin Silliman, funneling fossil plants from Hildreth and others in the coal-rich interior of the continent. Brongniart thanked Silliman publicly in the *Histoire des Végétaux Fossiles*.[19]

These intensive studies of ancient plants allowed naturalists in Europe and America to answer the coal question at last. Coal was made not of minerals but of plants—in fact very old plants—a truth that until then had been "rather inferred than demonstrated," according to the Edinburgh naturalist Henry Witham in his book *The Internal Structure of Fossil Vegetables* (1833) and in an article that had appeared a few years earlier in Silliman's *American Journal of Science and Arts*. Witham's microscope unveiled the "intimate organization" of fossil plants. His beautiful illustrations showed petrified "traces of organic tissue ... in which the woody layers are distinctly seen." Witham confirmed that coal was not only of plant origin but also of "extreme antiquity," a sultry forest swaddling the young planet (fig. 3.4).[20]

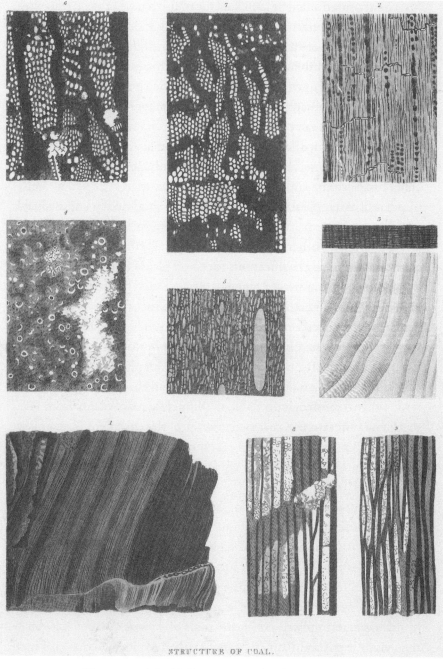

FIGURE 3.4. "Structure of Coal," illustrating Witham's argument that coal was made of ancient plants, from Henry Thornton Maire Witham, *Internal Structure of Fossil Vegetables* (1833).

This international group also confirmed how the coalfields had arrived at their present location. The fossil plant shreds had led some naturalists to argue that calamitous violence—perhaps the "mighty burstings" of the biblical flood—had carried them from far away. But their own microscopes suggested otherwise. Coal was merely old peat, deposited in the past as it was now: by slow, gentle, repeated accretions over eons rather than catastrophic flooding. As the planter Rush Nutt of Mississippi argued, "To account for these strata of coal we must suppose that . . . successive movements took place, in which the bogs were sunk, and raised again at distant intervals of time."[21]

If the coal had never moved, then the modern United States could claim God's primordial fuel as its own. That was more or less the argument in the most important American publication on coal to appear before the Civil War: a long 1836 article by Samuel Hildreth in Silliman's *American Journal of Science and Arts*. Rounding up the European and American work of the last decade, Hildreth proposed first that coal was of vegetable and not mineral origin; second, that it dated from "some remote period" in a process "still going on" in modern peat bogs; and finally that it had been deposited not by a violent flood but through "gradual" and "tranquil" conditions that pressurized and heated it into the various varieties of coal (less pressure producing bituminous coal, more pressure producing anthracite coal). He pointed to Ohio's Grotto of Plants as proof of this slow, gentle deposition. In the accompanying illustration, two smartly dressed gentlemen converse in a cave whose ceiling shows fossil vegetable impressions deposited by a "moderate" current of water operating "age after age." The "perfect state" of the fossils proved that the ancient plants had grown "on the spot where they were buried." Long ago, Europe and America had shared a climate favorable for the formation of peat bogs "on the surface now occupied by coal." The coal had not moved; it now stood at the place of its creation, awaiting extraction by the nation that had grown around it (fig. 3.5).[22]

The modern implications for the coal industry were clear. American coal was American in a primordial way. British coal was British in a primordial way. National greatness was locked into the very ground on which these industrial nations stood. Hildreth explained: "The coal

FIGURE 3.5. In this image from Ohio geologist Samuel Hildreth's internationally influential article from 1836, two gentlemen in top hats examine Ohio's Grotto of Plants, where impressions of fossil plants suggested their gentle deposition by water long ago. This theory was a strike against the catastrophic-flood deposition hypothesis.

deposits of Britain, by nourishing her manufactures, which have raised her to her present proud attitude among the nations, are the principal source of her present greatness." God smiled on these lucky nations, Hildreth went on. "How beautiful and how valuable are the means by which the all wise Creator has provided for the comfort and the happiness of man. Vast magazines of iron, salt and coal (the latter indispensable to the population of many parts of the globe) were laid up in store for his use, before he was 'yet formed from the dust of the earth.'" All this coal had been "here deposited by the Creator, from the earliest ages."[23]

The fossil plant network sprang into action to publicize Hildreth's article. Silliman praised it as "by far the most valuable contribution which up to that time had appeared on the subject discussed." He sent the article to Europe, where it was read by Brongniart and Lyell. Brongniart then asked for fossil vegetables from Hildreth in 1837. Hildreth duly obliged with a box for Brongniart sent through Samuel George Morton in Philadelphia as intermediary.[24]

It paid off: Hildreth's article was the magnet pulling Lyell to obscure Marietta in 1842. The publication had shown Lyell's uniformitarian system at work, not just for thinking about the formation of coal over long expanses of time, but for how this could be used to justify the modern national exploitation of the coal fields. Hildreth had confirmed what Lyell had merely ventured in his *Principles of Geology*: that present national greatness hinged on ancient coal deposits. The "subterranean deposits of coal," Lyell had proposed in 1830, showed that "commercial prosperity, and numerical strength of a nation, may now be mainly dependent on the local distribution of fuel determined by that ancient state of things." And now, thanks to Hildreth, Lyell had proof from America that this was so. "So great are the facilities for procuring this excellent fuel," Lyell wrote of Ohio's coal, that "almost every proprietor can open a coal-pit on his own land."[25]

Amherst

While on his American tour, Charles Lyell paid a visit to another coal expert: Edward Hitchcock of Amherst College, a tiny new college sitting near the banks of the Connecticut River in western Massachusetts. At Amherst, the Carboniferous would emerge as a star in a new science-centered curriculum.

Born in 1793, Edward Hitchcock suffered a childhood case of the mumps that robbed him of much of his sight and ended his dream of becoming an astronomer. Ordained as a Congregationalist minister in 1821, he served as a pastor until 1825, when he took the job of professor of chemistry and natural history at Amherst College. Nearly ten years before, he had opened a geology-centered correspondence with Benjamin Silliman of Yale, eventually going to Yale to study under him. The two became some of the earliest American coalfield explorers. Hitchcock's work on the Massachusetts geological survey in the 1830s had led to his optimistic report about the abundance of what he called "fossil fuel" in the United States. From the front of his Amherst lecture hall Hitchcock preached the new dispensation to his students: "Probably no part of the world contains such immense beds of coal as the central parts of the United States," he told them.[26]

Now on April 15, 1842, the great Lyell arrived in Amherst. Hitchcock had read his *Principles of Geology* from cover to cover. It convinced him of something his own fieldwork had also suggested: that Earth was very old, much older than a traditional reading of Genesis might suggest. In his younger days, flush with his ministerial training, Hitchcock had seen rocks as a "cosmogonical chronometer" and judged by their rapid erosion that the time of their creation could not have been "vastly remote." And initially he had been skeptical of Lyell. Did not uniformitarianism amount to "infidelity"? Was it not better to say that God occasionally reached into time, speeding up some processes and slowing others down, just as he suddenly annihilated some of his creations and breathed life into others? Did not uniformitarianism make God superfluous to Creation? His friend Silliman had agreed that Lyell "pushes the operation of causes now acting much too far."[27]

But fifteen years of fieldwork in New England had driven Hitchcock to accept a far longer chronology. By the mid-1830s, he had become convinced not only that Earth had been formed over vast periods of time but that eastern North America was probably older than Europe. Fossils buried in hundreds of feet of strata, formed "at least in part" by mechanical agencies, could only have occurred in an "extremely slow" process. Lyell's long chronology was certainly intriguing. "How immense the period requisite for the production of such vast masses!" Hitchcock wrote in the theological journal *Bibliotheca Sacra*. His students received the same message in the lecture hall: "The time was immense for all these rocks ... to have formed," one student jotted in his notebook. In fact, God seemed only grander if his wisdom extended to millions of years rather than a mere six thousand, Hitchcock explained. "The common interpretation limits the operations of the Deity ... to the last six thousand years. But the geological view carries the mind back along the flow of countless ages.... Can the mind enter such an almost boundless field of contemplation as this, and not feel itself refreshed, and expanded, and filled with more exalted conceptions of the divine plans and benevolence than could possibly be obtained within the narrow limits of six thousand years?"[28]

Hitchcock drove Lyell through the leafy spring countryside to the banks of the Connecticut River to see the curiosities that had startled Lyell's colleagues in Britain and made Hitchcock famous in its geological circles: thousands of fossil footprints scurrying over the blood-red sandstone. Hitchcock thought they had been made "many thousands of years ago" by birds—some monstrous, twice the height of an ostrich, bounding with four-foot strides after their prey, which defecated in terror. This explained the fossil dung, dropped "hundreds of thousands of years ago" when Massachusetts had been a tropical delta. Hitchcock hauled blocks of rock footprints back to the college, where students could lay their hands on the curious relics (fig. 3.6).[29]

Hitchcock's fossil footprints set English high society gawking when a sample was displayed in London. Prince Albert and the prime minister leaned in to inspect the curious American antiquities. Charles Darwin flattered Hitchcock with a letter calling the extraordinary scene on the Connecticut River "highly important." And now, standing on the very spot, Lyell agreed that the rocks encasing the prints were of an "acknowledged antiquity."[30]

Still older American wonders awaited Lyell. Returning with Hitchcock to campus as dusk fell, Lyell immediately saw that his companion had another claim to fame besides the fossil footprints: Hitchcock had transformed Amherst College into a temple of deep time for his students. Fossil specimens scattered in several rooms and even outdoors displayed the terrifying bird footprints and the leafy coal plants. Collecting them on field trips notorious for the boisterous antics and even "immoralities" of the students, Professor Hitchcock lugged the fossils back to campus. In front of Hitchcock's lecture hall appeared another marvel, the gigantic and colorful canvas paintings of fossils and strata made by his wife, Orra White Hitchcock. She was a skilled botanical illustrator who had once taught painting and drawing to young women at Deerfield Academy, where Edward had been a teacher. Beginning during the years of their courtship and extending to their death, the couple collaborated. Edward wrote a popular geology textbook and Orra created the hundreds of illustrations to accompany it, though the

FIGURE 3.6. A photograph of the tracks left by what Edward Hitchcock believed were gigantic ancient birds scurrying through mud long ago in the region near what was now Amherst College.

FIGURE 3.7. Orra White Hitchcock and family greet her husband Edward Hitchcock as he returns from a journey, perhaps a geological field trip. Attributed to Robert Peckham, *Professor Edward Hitchcock Returning from a Journey* (c. 1838).

works did not advertise her contributions. All the while, Orra Hitchcock ran their home life. A painting shows the grumpy geologist returning home to be greeted by a large family spilling from a tidy home and Orra holding the youngest child in her arms. From the stagecoach a man has unloaded a trunk, perhaps full of rocks and fossil bones. The octagonal fossil cabinet Hitchcock later built on campus to emulate the "harmony and beauty" of European architecture is heralded here by the octagonal building looming in the background (fig. 3.7).[31]

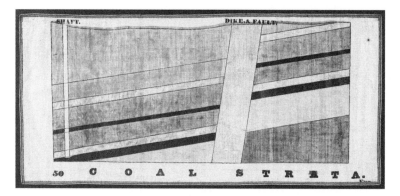

FIGURE 3.8. Orra White Hitchcock's poster of coal strata, for use in Professor Edward Hitchcock's classes on geology and natural history at Amherst College.

Many of Orra Hitchcock's drawings showed the Carboniferous, reinforcing the message that coal was relevant to modern life. In one canvas for the lecture hall, a mineshaft pierces through layers of earth to reach a thick black seam of coal lying deep underground (fig. 3.8 and plate 7). Her drawings of Carboniferous trees appeared in his textbook (fig. 3.9).

In a culture saturated by Protestantism, it was difficult to be a strict uniformitarian, and Edward Hitchcock was soft around the edges. Mostly the primordial world chugged along according to the dictates of natural law, he said, just as Lyell had proposed. But there was always the "possibility" of a miracle, even if the "probability is denied." God had performed miracles more often in Earth's early history, Hitchcock decided, and it was likely that God "did formerly act with greater energy than at present." He explained in an article in the theological journal *Bibliotheca Sacra*: "The law of miracles is a force occasionally manifesting itself to counteract, intensify, or diminish the power of natural law."[32]

To show how deep time could accommodate both uniformitarianism and occasional miracles, Edward and Orra Hitchcock devised an innovative diagram: their "Paleontological Chart," one of the earliest biological tree diagrams in the United States. Orra created the colored chart for Edward's popular textbook, *Elementary Geology* (1840), and it appeared in thirty-one editions of the work until 1860, when Edward

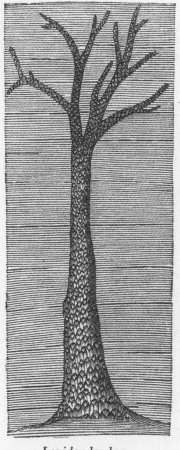

FIGURE 3.9. Orra White Hitchcock's drawing of the extinct vascular plant Lepidodendron, which could reach heights of 150 feet in the coal forests swaddling ancient North America. From Edward Hitchcock, *Elementary Geology* (1840).

withdrew the chart so as not to appear to support Darwin's theory of evolution by natural selection rather than divine direction. Hitchcock claimed his chart was the first to represent the "leading facts of Paleontology." This was not strictly true—he was somewhat discombobulated to realize that the German naturalist H. G. Bronn had created a similar one slightly earlier—but it was still close (fig. 3.10).[33]

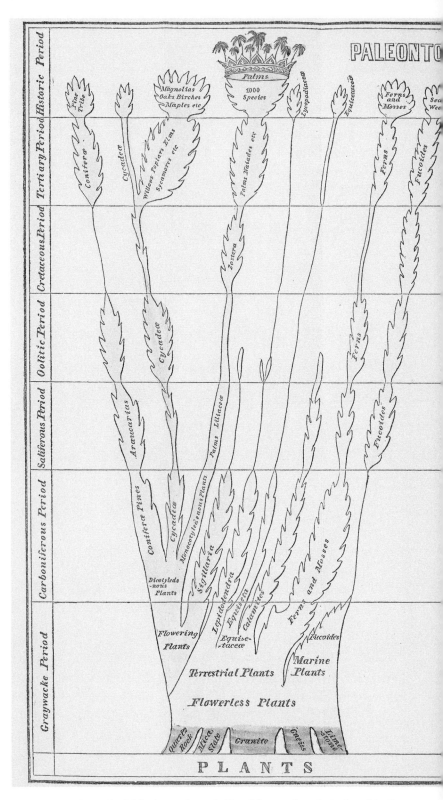

FIGURE 3.10. In this colorful "Paleontological Chart" from Edward Hitchcock's *Elementary Geology* (1840), the separate plant and animal kingdoms progress upward through seven immensely long geological eras that mirror the seven days of the Genesis creation story.

L CHART.

Periods (right column, top to bottom): Historic Period · Tertiary Period · Cretaceous Period · Oolitic Period · Saliferous Period · Carboniferous Period · Graywacke Period

Top labels: Sharks, Fishes, Crinoidians and Cycloidians, Lizards, Birds, Man, Mammalia, Microscopic Shells, Insects, Corals

Tertiary / Cretaceous labels: Extinct Mammalia, Nautilus, Ammonites, Nummulites and Melilites, Fresh Water Shells, Marine Shells, Serpulæ, Spiders, Polyparia, Articulata

Cretaceous / Oolitic labels: Bones of Birds, Ammonites, Scaphites etc., Reptiles, Hetoloites, Terebratula, Shells, Estuary Shells, Insects, Pentacrinites

Oolitic / Saliferous: Saurians, Phascolotherium, Ammonites, Belemnites, Macreurds

Carboniferous: Sauroid Fishes, Sharks, Ctenoidians, Placoidians, Terebratula, Spirifer, Producti, Sigillaria, Encrinites etc., Insects, Echini etc.

Graywacke (base): Fishes, Ammonites, Orthoceras, Terebratula, Producti, Spirifer, Chambered Shells, Trilobites, Hydrade Phytolithes, Nautilites Shells, Crustaceans, Trilobites, Annelidans, Serpulæ, Crinoidians, Polyparia, Corals

Base categories: Vertebral Animals · Marine Shells · Molluscous Animals · Articulated Animals · Radiated Animals

Bottom rock layers: Quartz Rock · Mica Slate · Granite · Gneiss · Mica Slate

ANIMALS

The Paleontological Chart pinned the story of life to the geological strata representing an "immensely long" span of time. Two trees represent the biblical idea that God created plants and animals as separate projects. They grow through seven long time periods that equal the number of days in Genesis. Students were instructed to imagine Earth gradually cooling over time, with each of the seven regimes of life "exactly adapted to the varying physical conditions of the globe," thanks to the deity's benevolent design. While the "precise age" of each stratum was unknowable, wrote Edward, each long geological period was probably of roughly equal duration. "Extremely slow change of climate" and occasional "sudden catastrophes" brought each geological era to a close, with the "creative energy of the Supreme Being" generating life anew in the next period. The idea of a divine re-creation of life did not necessarily contradict Lyell's idea that long-extinct organisms could recur in similar forms if the same environment reappeared, and Orra's spindly stalks puffing out in later eras seems to leave that possibility open.[34]

The Carboniferous period played a starring role in the Paleontological Chart since its role was "to prepare fuel for man." The plant kingdom culminated in the most "perfect" group of plants, the palms, which encircle a crown. The formerly hotter Earth had nourished the abundant plant life of the Carboniferous, created as "a very slow process," during a period "indefinitely long." The Carboniferous had produced the magnificent palm family, which had grown in number so that now in the modern era a thousand species flourished as "the CROWN of the vegetable world," just as Man crowned the animal world. The palms were endlessly useful to modern humans, yet another gift from antiquity to modernity.[35]

The "landscape" of the Carboniferous, however, must have presented a "uniform and somber picture," Edward Hitchcock wrote, not a place for Man's "happy dwelling" but rather a preparation for his future happiness. His use of the term *landscape* suggest that he was groping toward the idea of a landscape representation of this ancient time as no human had ever seen it—what Martin Rudwick has called a "scene from deep time," a new idea in the nineteenth century. For now, however, Orra Hitchcock's image of an underground forest showed this dark under-

Subterranean Forest : Isle of Portland.

FIGURE 3.11. As though standing in front of a cliffside, American students in the 1840s could gaze at a subterranean fossil forest exposed on the southern coast of England, in Orra White Hitchcock's illustration for her husband's geology textbook.

world as it appeared to modern eyes (fig. 3.11) and not as a fully formed scene from deep time.[36]

With its two separate trees of life and seven uniformly long periods of geological time, the Hitchcocks' paleontological chart would not survive the nineteenth century. But in its day, the chart showed how the concept of deep time expanded the creative power of the Deity rather than challenging it. With the palms crowning the plant kingdom for the use of Man, the chart also explained why coal was one of the major signs of God's special favor to America.

The Coal Sublime

By midcentury, Americans understood that progress rested on the primordial coal deposits given by God for humanity's use and happiness. Geologist Henry Rogers saw coalfields "approaching the sublime." Coal, in short, was not just science, but art. We might term this aesthetic vision "the coal sublime," since depictions of coal in this first flush of the Industrial Revolution suggested the lofty spiritual dimensions that this carbon block could reach. Artists now fashioned landscape paintings of coalfields and coal-powered trains, three-dimensional models of coal seams, and coal strata diagrams. As coal mining became the very definition of labor-intensive industrial capitalism, the coal sublime veiled this raw power in a benign origin story in which Americans reaped the God-given fruits of his first forests. And not just any Americans. University of Michigan geologist Alexander Winchell stressed that coal was laid down in America to prepare "for the occupancy of the Caucasian race."[37]

Coal is the invisible progressive force in George Inness's *The Lackawanna Valley* (c. 1856). Commissioned by John Jay Phelps, the president of the Delaware, Lackawanna and Western Railroad, the painting unveils coal's positive end products. Harmless puffs of steam billow from a train gliding through a meadow cleared of trees to make way for civilization. A resting picnicker takes in the harmonious scene—nature and civilization happily married by the coal industry (fig. 3.12).

Geology books also illustrated the coal sublime. No miners labored, no smoke blackened the sky, no foreman barked commands. At the Brownsville coal seam along the Monongahela River in eastern Pennsylvania, the coal appears to have extracted itself from the cliffs under a balmy sky. From the smokestacks, sinuous black wisps quickly vanish. No miners are present, yet a railway car has somehow deposited coal in a departing barge (Figure 3.13). In an image from Henry Rogers's *The Geology of Pennsylvania* (1858), four tiny workers dwarfed by an immense and cloudless sky dot the summit of the Lehigh coal mine, a scene of "impressive" but "simple grandeur."[38]

Seen as the modern version of the sublime ancient coal deposits, present-day tropical forests were charged with meaning. Their abundant products invaded many aspects of American life, from oils and resins to umbrellas and cane furniture. Tropical forests offered romance and reverie. Joseph Meeker's painting of the Louisiana bayou is nearly indistinguishable from the Carboniferous landscapes appearing in geology textbooks at the same time (figs. 3.14 and 3.15).[39]

These steaming ancient forests even offered adventure. American readers loved Scottish geologist Hugh Miller's *The Old Red Sandstone* (1841), which took readers on an adventure into the Carboniferous. "We

FIGURE 3.12. In American painter George Inness's *The Lackawanna Valley* (c. 1856), coal-powered industrialization has begun to transform the American landscape. Stumps are all that remain of a forest clear-cut to make way for trains and factories that belch smoke into the skies. Yet the wooded hills receding gently toward a golden horizon are so picturesque that a farmer has paused to admire the scene.

FIGURE 3.13. In this idealized image of a Pennsylvania coal seam, from Charles Lyell's *Travels in North America* (1845), the coal appears to extract itself without human assistance, as white clouds billow across pollution-free skies.

FIGURE 3.14. In Joseph Meeker's painting *Bayou Teche* (1874), the sultry swamps of a Louisiana bayou look very much like images of the Carboniferous age produced at this same time.

FIGURE 3.15. Many American college students would have seen this image of the muggy Carboniferous age, which graced the front of James Dwight Dana's popular college textbook, *Manual of Geology* (1863). A lone but cheerful amphibian creeps out of the swamp, following the footsteps of an unseen companion. A similar scene would again be interpreted for the benefit of American college students in the 1870s: see figure 7.10 and plate 21.

have entered the Coal Measures," he announced as he spotted a tangled forest in the twilight gloom. "Land, from the mast-head! land! land!—a low shore, thickly covered with vegetation. Huge trees, of wonderful form, stand out far into the water. . . . A river of vast volume comes rolling from the interior, darkening the water for leagues with its slime and mud, and bearing with it, to the open sea, reeds, and fern, and cones of the pine, and immense floats of leaves, and now and then some bulky tree, undermined and uprooted by the current."[40]

Greenhouses and conservatories brought modern tropical foliage to private gardens and public parks. Americans could see living, modern versions of the ancient Carboniferous forests as they went about their day. Built of iron and glass, heated by coal, and filled with ferns and palms, some greenhouses were owned by the same industrial magnates who had made their fortunes mining coal and iron. On his European tour in 1850, Edward Hitchcock visited the botanical garden in Regent's Park, London. Better even than the higher classes of England, who were swanning about on the sunny day, were the "living tropical species that throw light on those dug out of the rocks," such as the cycads and date palms. By midcentury, American landscape architects recommended greenhouses even for the humble. Gardening treatises explained how to arrange the greenhouse plants to capture feeble sunlight and to rest the eye. Philadelphia's Centennial Exhibition of 1876 displayed the largest conservatory of its time in Horticultural Hall. It was made of glass, heated by coal-powered steam pipes, and filled with lush tropical ferns and palms. Thousands entered the tranquil gloom of a modern-day Carboniferous (fig. 3.16).[41]

Americans of the nineteenth century began to paint a vision of their national future by gazing into the remote past. Turning backward to the Carboniferous period, they looked forward to the Industrial Age. Toggling between the two eras, Americans imagined primordial trees and ferns powering a present of railroads and factories. The fossil future promised ease and light, a liberation from the gloom and fatigue that had always been humanity's lot. The Carboniferous reassured Americans that God had smiled upon the United States, lavishing the nation with his gifts of natural resources.

FIGURE 3.16. Modern ferns and palms recall the primordial Carboniferous age in the Horticultural Hall at the Philadelphia Exposition of 1915. Earlier images show that this installation was in place as early as 1876.

But the fossil future was also a Trojan horse, bringing an outcome no one wanted. Unlike the Christian eternity, the fossil future was bounded by the coal supply. This coal future would end some day, plunging Earth back into dark, cold, and hunger. And so with the fossil future came anxiety: anxiety about the limited fuel supply and anxiety about the skies blackening with the airborne particulates released when the coal was burned for fuel. This was not what the first coal boosters had hoped for when they looked into the forest primeval and saw a bright future. But it is their fossil future that we are reaping today, as we cope with the unintended consequences of turning Carboniferous fossil forests into fuel. To us, the ancient forests often seem less like a gift than a curse, leading us down a dark path of our own devising.

4

The Oldest South

IN THE FIRST decades of the nineteenth century, American slavery expanded from the Atlantic seaboard into Alabama, Mississippi, Louisiana, and Arkansas, a region roughly the size of France. As slavery began to disappear in the North and the importation of captive Africans was prohibited by the Constitution, the Deep South transformed into one of the most dynamic slave systems in the world. Fed by a massive importation of slaves, the Deep South's annual production of cotton grew rapidly after 1800, displacing earlier sources of cotton such as the West Indies, the Ottoman Empire, and Brazil. It soon dominated the global cotton market. For most of the era before the Civil War, 80 percent of the cotton imported to Britain's manufactories came from the United States. In the Deep South, cotton was king.[1]

The speed of the region's transformation is easy to overlook. After the Civil War, new origin myths soon papered over its novelty. Defeated Confederates, traumatized by their collective loss, fabricated the Lost Cause myth of the Old South. In this telling, Southern whites were not new arrivals but heirs to a noble Anglo-Saxon past, upholding a vanishing tradition of chivalry and honor in the face of the machine age and modernity. Through the Lost Cause myth, Southern whites found a history that could mobilize large numbers of people to act politically and even violently in the name of a higher truth.[2]

But before the Lost Cause, before the Civil War itself, some newcomers to the Deep South were already promoting a vision of the region's history, one so ancient that it stretched into the planet's infancy. They used this history to lash the region to the rest of the nation and to make its emergence as a slave society and cotton-based economy seem natural rather than intentional. This was the oldest South, a place in time crowned with the authority of nature itself.

The new story of the oldest South was written by a group of planters, geologists, and paleontologists in the early years of the region's political and economic consolidation. Fanning out among the cliffs lining the wide southern rivers that drained the continent's belly into the Gulf of Mexico, they plucked marine fossils from a thick, chalky ("cretaceous") layer of rock. Some of these creatures were tiny, like the belemnite, basically an ice cream cone with tentacles. Others were monstrous sea serpents with teeth as big as a human hand. An ancient ocean, it seemed, had once blanketed what were now cotton fields. Soon these naturalists and planters applied the term "Cretaceous" not just to the rock layer itself but to the whole primordial era during which this American ocean had teemed with life. Lying silent in their rocky Southern tombs, these fossil creatures were now unearthed and celebrated as the bedrock of Cotton Kingdom's prosperity.

Proposed in the age of slavery, industrialization, and empire, the Cretaceous was not a brute fact existing in a natural world outside of society. It was a human invention, a set of signs and symbols shared among distant naturalists who wove regional data into a tapestry of modern meanings. As they had with the Silurian and the Carboniferous, the sciences of geology and paleontology offered a way of surveying a region's landscape so as to locate modern projects in a particular moment of deep time. The antiquity of the Cretaceous was so great that it undercut all human habitation of the Deep South, allowing arriving white planters and naturalists to lay claim to the region and its natural resources. Naturalists recorded this Cretaceous antiquity in field notebooks and letters crowded with sketches of strata. The region's Cretaceous era was formulated with remarkable speed. Naturalists had proposed and essentially completed it within the decade that ended in 1835, when sectional lines were hardening and Northern opposition to the South's peculiar institution was growing.

The Cretaceous charged the Deep South with important national and global errands. The region's natural resources would enrich Southern planters while also benefiting the nation by supplying cotton to New England and British textile mills. The Cretaceous also linked the United States to Europe in a new planetary history being written by European and American naturalists. No longer would North America be the laggard on the planetary stage, its clammy climate withering life. The Cretaceous showed that America, too, had vigorous life in its history, beings that in death created the rich earth that now made cotton king. The Cretaceous was also a shared past with Europe, erasing the radical break between Old World and New. The Cretaceous made all these things seem natural. It was a small step to asserting the natural growth of a slave plantation economy from the rich soils deposited eons ago. Ancient rocks seemed to suggest, even if they did not ordain, a modern social order. The increasingly popular metaphor of the stratum authorized a link between the geological antiquity of the region and the racial status of its Native peoples and the arriving labor force of African slaves.

So came the term *black belt*, which sprang from that blend of rocks and people unfolding westward, from Georgia into Alabama and Mississippi. Coined in the middle of the nineteenth century, the term originally referred to the dark Cretaceous earth that sustained an especially abundant cotton crop. A map published by a New England naturalist during the Civil War showed the dark apron wrapping around Cretaceous North America (fig. 4.1).

As the population center of US slavery shifted to the Deep South, the term *black belt* began to be applied to the people working the region's plantations. Half of Alabama's slaves labored within ten black belt counties, making the region one of the wealthiest in the antebellum United States. The former slave Booker T. Washington defined the original meaning of the black belt in his autobiography, *Up from Slavery* (1901). "So far as I can learn, the term was first used to designate a part of the country which was distinguished by the colour of the soil. The part of the country possessing this thick, dark, and naturally rich soil was, of course, the part of the South where the slaves were most profitable, and consequently they were taken there in the largest numbers." Only

North America in the Cretaceous period; MO, Upper Missouri region.

FIGURE 4.1. This map published during the Civil War shows the Cretaceous "black belt" wrapping around ancient North America. From James Dwight Dana, *Manual of Geology* (1863).

after the Civil War, he went on, did the term acquire a political meaning, designating counties where black people outnumbered white. Then the black belt became a new sepulcher, no longer the graveyard of Cretaceous beings but of the former slave plantations, a "strange land of shadows," as W.E.B. Du Bois called the black belt in *The Souls of Black Folk* (1903). A photograph taken around 1860 showed black workers, possibly slaves, picking cotton in Alabama's Cretaceous lands (fig. 4.2).[3]

The two black belts—earthy and human—proved irresistible to naturalists seeking order in nature. At the very time and place of peak slavery in the United States, the Deep South's rich Cretaceous soils formed an early chapter in a planetary history dependent less on Genesis

FIGURE 4.2. Black workers picking cotton in the "black belt" near Montgomery, Alabama, in the 1860s.

than on fossils. At the end of that long history came human beings, who built societies atop these natural resources. The Cretaceous naturally led to cotton, and cotton naturally led to slavery, according to the early geologists and paleontologists working there. "The fertile lowlands of that territory can only be worked by blacks, and are almost of illimitable extent," one of them wrote. The American naturalists digging in the Cotton Kingdom invented a new story of America's oldest South that influenced naturalists from Mississippi all the way to the scientific capitals of Paris and London. The Deep South's Cretaceous opened a vast theater of possibility: of economic success, imperial expansion, and racial and social hierarchy apparently based in nature.[4]

The Paleontologists of Philadelphia

The Americans who did the most to establish new knowledge of the ancient fertility of the Deep South were a group of young naturalists in Philadelphia in the 1820s and 1830s. Coining a new term, they called themselves *palæontologists*. They included Timothy Abbott Conrad, Isaac Lea, Thomas Say, Richard Harlan, and Samuel George Morton. All had been born near Philadelphia at the turn of the century, in the long shadow of a new federal Constitution that had swept away monarchy and its patronage apparatus while remaining silent on where exactly funding for science would now come from. So the group stayed in Philadelphia, what one Dutch naturalist in 1835 called "that scientific city of the United States." Philadelphians benefited from the most extensive network of privately financed scientific institutions in America: the American Philosophical Society, the Academy of Natural Sciences, Charles Willson Peale's museum, several medical schools, and a robust culture of anatomical dissection.[5]

Pocket maps in tow, the paleontologists took field trips into New Jersey, just across the Delaware River from the elegant gridded streets of Philadelphia. This was prime fossil-hunting ground. The calcium carbonate–rich soil, mined from so-called marl pits, was such a rich source of fertilizer that it supplied farms in other states. New roads and canals cut deep into the earth, bringing the marl to market and exposing a strange underground world of fossil creatures. Fifty miles southwest of Philadelphia, the new Delaware and Chesapeake Canal revealed "great numbers" of fossil shells. Closer to the city, in the marl pits near Woodbury, the paleontologists found fossil teeth unlike any "existing genera." Into the paleontologists' satchels went ancient shells called ammonites, so named because their whorled shape looked like the ram's horns of the god Ammon. There were also the skinny belemnites and other marine fossils that soon filled specimen cabinets around the nation and bedecked a stream of new publications.[6]

If they noticed the complexion of New Jersey's marl pit workers and canal diggers, the paleontologists never mentioned it. By 1830, most of the state's black inhabitants were free (though stripped of the right to vote), but over two thousand black women and men remained enslaved. Some were now also being funneled into the Deep South to feed the growing demand for labor in the cotton and sugar fields.[7] For however profitable New Jersey's marl pits might be, they were no match for the earthy riches that awaited planters in the southwestern annex of the Cretaceous geological formation that the Philadelphia group would soon uncover.

New Jersey's fossils soon put the Philadelphia group in the spotlight. Visiting naturalists dropped in for field trips, bringing news about similar European formations at a time when Americans chafed at their inferior scientific status. One visitor was Lardner Vanuxem, who had just returned from three years studying alongside French naturalists Alexandre Brongniart and René Just Haüy at the École des Mines in Paris. While in Paris between 1816 and 1819, Vanuxem had examined the fossils of the Paris basin, made famous by Georges Cuvier and Brongniart because they contained the fossil remains of now-extinct animals. Paris seemed to be composed, like New Jersey, of ancient marine deposits. It was exciting to imagine that Paris was the New Jersey of the east.[8]

The Philadelphia fossil hunters embodied the spirit of another French paleontologist, whose motto was *C'est ce que j'ai vu*: this is what I have observed. The motto's simplicity concealed complex assumptions. It suggested that careful observation and measurement would reveal facts. Facts in turn would unveil God's plan for the world—and for America. The paleontologists were engaged in a project that was both natural and religious, in an era when no sharp lines distinguished the two realms.[9]

The axis of the Philadelphia fossil group was the mild and methodical physician Samuel George Morton. Educated in medicine in Philadelphia and Edinburgh, he had by the 1820s established himself not only as a busy physician but as a renowned fossil shell specialist—so respected by his colleagues in Europe that one scientist there called him "the real founder of invertebrate palæontology in America."[10] In Switzerland, the eminent fish expert Louis Agassiz knew of Morton's fossil marine finds by the mid-1830s, thanks to English colleagues who connected the two men through letters.[11] Most important is what Morton called himself: "a geologist." His first published illustration was on a geological topic, a cross-section of the Perkiomen lead mine in Pennsylvania, from 1826.[12]

But with a growing family, a heart condition, and a demanding day job tending to the sick, Morton struggled to find time to join the local field trips. Instead, he established himself as ground zero for fossil shipments from around the world. Working in his few "leisure hours," he received the boxes that arrived damp and moldy from Ohio, Louisiana, Florida, England, Scotland, West Africa, France, Germany, Canada, India, Brazil, Mexico, and China. In the international fossil shell economy, Morton made himself into a central bank. He collected specimens from everywhere and then published new findings, linking far-flung people who would otherwise labor in solitary obscurity. "I am indeed highly indebted to you": some version of this sentence appears dozens of times in Morton's voluminous correspondence. By 1828, Morton had accumulated thousands of fossil shells, an "extensive" collection even by the exalted standards of Lardner Vanuxem, whose own cabinet was one of the largest in the United States. As each specimen arrived, Morton unboxed, categorized, labeled, and shelved it. When his home cabinet become too crowded, he carried the shells and bones a few blocks away to the Academy of Natural Sciences, where he held several leadership positions from 1825 until his death in 1851. There they joined other fossil and modern specimens in the growing collections of the academy: armadillos, shark fins, elk antlers, snakes, a tiger skull from Java. On Sunday evenings, Morton's Mulberry Street house became a scientific salon. Morton opposed sectarian religious and political views in science. So the salon was attended by locals and visiting stars, such as John James Audubon and the Scottish geologist Charles Lyell, representing a rainbow of intellectual and political opinions.[13]

In the early 1830s, a new interest took shape: skulls. Morton began to collect "the Crania of all Animals, from Man to Beasts, Birds & Reptiles." To a fellow physician in Venezuela, he called this "comparative Craniology." He remained especially interested in the diverse creatures of the Americas, collecting skulls of "our indigenous animals."[14]

Morton was especially fascinated by human crania, and he began to collect them in 1830. He eventually amassed the largest collection of human crania in the United States, with over a thousand specimens. By 1835 he was planning to publish a *"Liber craniorum,"* or cranium book. Two such books soon appeared, *Crania Americana* (1839) and *Crania Ægyptiaca* (1844), establishing Morton as the leading American expert on craniology as a science of human racial hierarchy at the moment when sectional tensions over slavery were hardening between North and South. In both books, Morton argued that whites had larger cranial capacity and therefore higher intelligence than Native Americans and Africans. Later in life, Morton promoted the doctrine of polygenesis, of separate creations of unchanging human beings rather than single descent

FIGURE 4.3. The physician and paleontologist Samuel George Morton poses around 1850 with a human skull on the bookcase behind him. Portrait by Paul Weber.

from a common ancestor, as the Genesis story suggested. He sat for a portrait in front of one of the human skulls. As his own large cranium glistens in the light, Morton poses in front of a bookcase that holds a skull in its dark recesses, a memento, perhaps, to the importance of knowledge and the uncertainty of life (fig. 4.3).[15]

Today, Morton is notorious for these craniological studies. He has been vilified by modern historians as a major architect of modern racial hierarchies.[16] He has also been condemned by modern scientists as a textbook example of professional bias.[17] Interest in Morton continues to dwell almost entirely on his human craniological studies. And yet placing them in the context of his earlier and ongoing geological and paleontological research shows Morton's significance to be even greater than we have realized. Working with other American paleontologists and geologists, Morton used the Cretaceous fossils of the United States to write a new planetary prologue to the more recent human story. The history of the Cotton Kingdom did not begin in the early nineteenth century, according to the view propagated by Morton and his contemporaries. It began long ago in the Cretaceous, when an ancient ocean deposited a fertile layer of earth that now upheld the plantation economy of the United States. The new cotton plantations were best worked by certain kinds of people—people with allegedly smaller crania. An ancient geological fact could be used to support a modern racial one. The paleontological contributions of Morton and his American contemporaries could be read to suggest that the slave-heavy cotton-growing South was not merely a human invention but a natural fact embedded in the region's remote antiquity. Linking geology and ethnology, they formulated a link between Earth's antiquity and the racial status of its inhabitants. By the 1850s, this seemed to point to an irreparably vast and primordial distinction between the slave South and the free North.

These unsettling conclusions lay decades in the future. In the 1820s, Morton and his young group of Philadelphia fossil hunters were interested in fossil shells because these tiny creatures promised to answer one of the fundamental scientific questions of the day: the origin of species, what Charles Darwin later called "that mystery of mysteries." Many naturalists believed species to be the fundamental unit of biological classification, though they sparred bitterly over how precisely to define it. Now geology was adding more questions by revealing fossil species with no living analogue. The ancient Earth, it appeared, had been profoundly different from the modern world, inhabited by now-vanished creatures. How could naturalists explain that process? Did strata show species changing continuously over time, or did the strata show discrete, disconnected layers of species? To invoke the metaphor of a novel, did characters develop and change over the course of the book, or did each chapter introduce a whole new cast of characters? The question of species change, also called transmutation, was among the most hotly debated of the nineteenth century, since it promised to unlock the mysterious history of life on Earth.[18]

One answer to the puzzle had come from the French naturalist Jean-Baptiste Lamarck, one of the many French naturalists whose work Americans read with interest. Today Lamarck is remembered for one thing only: giraffe necks. He is taught in biology classes as the misguided person who imagined giraffe necks lengthening over the generations as the hungry giraffes strained to reach treetop leaves and then passed on those acquired changes to their descendants. Dismissed today, Lamarck's ideas were widely known and debated in his own era because he offered one of the first major explanations of how life on Earth had changed over time. Rejecting the idea that species died out during extinctions, Lamarck argued instead that species merely changed into other species little by little over vast periods of time, and that life formed a great continuity from the deepest antiquity to the present. As Morton explained it, "Lamarck and Geoffroy St. Hilaire insist upon the uninterrupted succession of the animal kingdom—the gradual mergence of one species into another, from the earliest ages of time; and they suppose that the fossil animals whose remains are preserved in the various geological strata, however different from those of our own time, may nevertheless have been the ancestors of those now in being."[19]

Morton rejected this view, as did many other British and American naturalists. Instead, he followed the eminent French anatomist Georges Cuvier in arguing that species were unchanging. Cuvier's attempts to reconstruct animals from fragmentary bones had convinced him of their staggering complexity, and the unlikelihood that any species could somehow transform into another one over time. God must have created the natural world as an exquisite, complex, and unchanging perfection. As Morton phrased it, species were "*a primordial organic form,*" unchanged from their origin in "the night of time." In occasional catastrophes, for reasons beyond mere human capacity to discern, God annihilated some species, only to then repopulate Earth in a subsequent new Creation, a demonstration of his endless generative powers. Morton's friend Timothy Abbott Conrad, also a fan of Cuvier, imagined "strange revolutions" caused by "the mysterious operations of Nature" during a "past eternity." These revolutions included "the sudden elevation of mountains, irruption of seas, and destruction of various races of animals and plants." Far from being made irrelevant by the new geology and its deepening time, God was now enlisted to provide "a long succession of relatively petty and definite miracles," as Arthur Lovejoy has put it.[20]

Morton's view of unchanging organisms also implied a vision of the social order. Change should not come from within the organism, as Lamarck's theory implied, but from outside, steered by a great and superintending power. Organisms did not have agency, in Morton's view. On its home turf of France, the debate between Lamarckian transmutation and Cuvierian fixity was also a declaration of political allegiances. Lamarck's theory implied that changes in an organism could come from within the creature itself, impelled by desires to do this or that, and therefore dynamically interacting with its environment to cause potentially revolutionary change. It was understandable that such views about individual agency might be greeted with alarm in a nation where memories of the guillotine were still fresh. Unsurprisingly, Napoleon endorsed Cuvier's view of fixed species. Here was a model of nature in which even the smallest change awaited the action of a powerful outside force.[21]

The stakes therefore were high: fossil shells promised to unlock the very secrets of Creation. By 1830, Morton's decade of fossil shell research—the field trips, the late nights in the cabinet, the letters exchanged with other scholars, the fossils shipped to him from around the world—was pouring forth in a series of influential articles in scientific journals that used the New Jersey fossils near his home to overturn the prevailing consensus that this region was a single, recent deposit "of one age."[22] This had been the theory of the respected naturalist William Maclure, whose influential 1809 map (fig. 1.3 and plate 2) had shown the United States wrapped in a U-shaped deposit of alluvial rocks younger than all the other rocks on the continent.

Instead, Morton proposed that New Jersey had a very long history indeed. It was a layer cake of deposits, each containing characteristic fossils of different ages. In some of the lower layers—the "secondary" layers that lay under the even younger "tertiary" strata—the Philadelphia gang had found marine fossils so similar to those in France and the Netherlands that these fossils could only be "contemporaneous" deposits of "former ages." In these secondary layers lay distinctive fossil shells such as ammonites and belemnites, and even the remains of the giant marine reptile christened Mosasaurus because its fossil bones had recently been unearthed in the Meuse River near Maastricht in the Netherlands. Morton promptly shipped plaster casts of the New Jersey Mosasaurus tooth to Georges Cuvier and Alexandre Brongniart in Paris and to Gideon Mantell and Charles Lyell in England. "How extraordinary that this wonderful reptile should have been an inhabitant of both the old & new world!" gasped Benjamin Silliman. He agreed that the giant animal offered further proof that formations in New Jersey were not all the same age but in fact "really different." It seemed that the same strange ancient animals had once lived on continents now separated by thousands of miles of ocean, a scenario the Genesis story of Creation had said nothing about.[23]

In 1830, Morton unveiled his revolutionary proposal: at some point in the remote past, Europe and America had been the same age. This was a heretical proposition given the European insistence that the Americas were younger and therefore inferior continents, risen last from the biblical floodwaters and hobbled by a cold and humid climate that shriveled all life forms whether animal, plant, or human. Morton announced that New Jersey's marine fossils were not from the young alluvial formation, nor even from the older tertiary formation, as some American colleagues were insisting. No, these marine fossils lay underneath both of those formations, in an even older stratum also shared by both Europe and America.

The fossils had spoken from deep in the past to suggest that some parts of the New World were as ancient as some very old parts of the Old World. In his careful, understated, even boring style, Morton ventured his radical theory in print:

> We know that there is at present a remarkable generic accordance between the living mollusca of the eastern and western shores of the

Atlantic ocean.... Is it not reasonable to suppose that this accordance was formerly as great as at present? And with existing analogies before us, may we not with confidence resort to similar data to ascertain the contemporaneous deposits of former ages?[24]

Eminent scientists in Europe took note, a special honor for Americans who yearned for recognition in those exalted circles. Alexandre Brongniart, the expert on the chalky fossils of the Paris Basin, praised Morton's discovery in the French journal, *Annales des sciences naturelles*. Morton was thrilled, bragging to Benjamin Silliman that the great Brongniart had endorsed his theory. "That distinguished Geologist unhesitatingly considers the Fossils of the American Marl region as Secondary, and equivalent to those of the Great Chalk Formation. It is of course highly gratifying to me to receive this corroboration of my sentiments, especially from a man so universally respected for Geological acumen."[25]

The international recognition put wind in Morton's sails. He had put together the puzzle of ancient New Jersey, showing that its pieces locked into the White Cliffs of Dover and the chalky riverbeds of the Netherlands. All had once formed part of the same primordial ocean. But now he wondered: what about the rest of America? How far and in what directions did the secondary formations extend? Was just a small sliver of North America as old as ancient Europe, or did the antiquity of the continent reach much further? Put in economic terms, how far west did the fertility of the United States reach?

Tantalizing evidence was already arriving from elsewhere in the New World. Fifteen hundred miles to the south of Morton's Philadelphia cabinet, in the British colony of Jamaica, the young British geologist Henry De la Beche was claiming that this fertile island, a sparkling emerald in Britain's plantation empire, was also made up of ancient secondary strata.

Born in England in 1796, De la Beche had inherited a large Jamaican sugar plantation of 207 slaves when he was just six years old. Named Halse Hall, it stood on the southern part of the island. De la Beche was an absentee planter, preferring to live handsomely in southern England on the profits of his Jamaican sugar sales and to putter among the giant, sea-serpent-like marine fossils just pulled from seaside cliffs by local fossil hunters. One of these locals was Mary Anning, just three years his junior but born on the opposite end of the social ladder, to a struggling cabinetmaker. Like other poor people in her coastal hometown of Lyme Regis, Mary Anning's family had begun fossil hunting among the Dorset cliffs, selling the curios to wealthy visitors (inspiring the tongue-twister, "she sells sea-shells by the sea-shore"). Anning eventually became internationally famous for digging up whole skeletons, such as an Ichthyosaur when she was only twelve years old. As a woman, Mary Anning was always treated as an outsider by the scientific community, but she and De la Beche became friends in adolescence because of their common fascination with fossils. They liked to troll the cliffs together, digging up the lost world of fossils. But finally duty called, and De la Beche sailed for his Jamaican plantation in 1823. His mind brimmed with idealistic plans to improve the treatment of his many slaves in "a land of physical and moral pestilence," notorious for its brutal slave regime and large slave revolts. To reward his slaves' work, he commissioned a good conduct medallion that featured a cluster of tropical trees and sugar cane sprouting luxuriantly without the help of human hands.[26]

Leaving much of the day-to-day operation of Halse Hall to his overseers, De la Beche spent the next year making geological expeditions around the island accompanied by other planters and "man-servants" (probably slaves, in the gentlemanly code of the era). De la Beche's survey convinced him that Jamaica had similar and "contemporaneous" limestone formations to those in Europe. The Cretaceous might be here too—possibly even on his own plantation, whose sugarcane and Guinea cornfields yielded the same fossil shells Morton was finding in New Jersey. De la Beche published a geological map of the eastern part of Jamaica that showed the bands of strata responsible for the island's fertility.[27]

After his return to England, he drew an imaginary scene of Mary Anning's fossil finds in Dorset, titled *Duria Antiquior* (a more ancient Dorset). It is perhaps the first landscape scene of prehistoric life, showing Dorset's smiling ancient creatures cavorting in shallow seas. The scene has

FIGURE 4.4. Robert B. Farren's *Life in the Jurassic Sea* (1850) is one of the many reproductions of Henry De la Beche's *Duria Antiquior* (A more ancient Dorset, 1830) that were circulated in both Great Britain and the United States. It is one of the earliest landscape representations of prehistoric life, showing recently discovered marine reptiles devouring each other with blood-thirsty abandon. The tropical landscape of palms and warm seas may recall De la Beche's time in Jamaica tending to his slave plantation and visiting sites of geological interest. © 2023 Sedgwick Museum of Earth Sciences, University of Cambridge. Reproduced with permission.

been celebrated for its half-subaqueous viewpoint, showing above and below water at the same time, perhaps a result of the new aquarium craze. Lurking in the background may be De la Beche's memory of his time in Jamaica, also memorialized in his good conduct medallion: the tropical foliage sprouting from sun-drenched islands (fig. 4.4 and plate 8).[28]

By 1830, in the Caribbean and on the North American mainland, plantation slavery drew geologists and paleontologists into new regions. In the southwestern United States, the Cotton Kingdom now also beckoned the paleontologists. Morton was thrilled that the greats of

Europe—De La Beche, Brongniart, Mantell, and others—accepted his theories about the age of New Jersey's fossil-rich secondary formations. But these formations wrapped the whole length of the "Atlantic frontier," extending to the Mississippi River. A deadly cholera epidemic trapped Dr. Morton in Philadelphia. But he saw what was needed: a paleontologist on the ground in the "remote southern sections of this continent" to confirm his radical theory that some strata in the New World might be as old as the Old World, and that the fertility of New Jersey might unroll westward along the black belt of the Cretaceous.[29]

Alabama

That paleontologist arrived in 1833 in Claiborne, Alabama. He was none other than Timothy Abbott Conrad, one of the young fossil hunters in the Philadelphia group. Situated ninety miles north of the Gulf of Mexico, Claiborne was a new cotton plantation in the new state of Alabama. It was home to US senator and federal judge Charles Tait, his family, and the many slaves who worked his cotton fields. The plantation mansion overlooked the Alabama River, the winding artery that ferried slaves and cotton between the Gulf of Mexico and the American interior. In its 200-foot bluffs, like dried fruits in a cake, lay millions of fossil shells embedded in thick layers of ancient rock.[30]

Alabama boosters were advertising the new state as the garden of North America. This garden was in fact a region of rolling, pine-covered hills that was home to Creek, Choctaw, and Chickasaw Indians, who claimed millions of acres between Georgia and the Mississippi River during the early nineteenth century, with a population outnumbering the white and black population. Turning Alabama into a productive agricultural region—a garden—required a federal policy of extinguishing Indian title and exterminating or relocating the Indians. This "civilizing" process converted mobile Indian hunting grounds into settled cotton plantations, with some Choctaw and Chickasaw themselves owning plantations and black slaves. Indian depopulation occurred quickly. Rush Nutt, a cotton planter and naturalist living several hundred miles west of Claiborne in slave-heavy Jefferson County on the Mississippi River, commented on depopulation in the early 1830s. "At a former period the inhabitants were exceedingly numerous," he wrote. Now it was likely that the southern Indians would "become extinct."[31]

In place of the Indians arrived white planters and thousands of enslaved people from the exhausted fields of the Atlantic seaboard. Foreign geologists did not miss the contradictions involved in this demographic churn. The English geologist George Featherstonhaugh expressed doubt in his private journal that "Americans as a people" could ever pass into history as "a good People," since they claimed to love freedom while keeping "near two millions of human beings in Slavery, merely because they are of a different colour from themselves, and drive from the unsettled parts the original occupiers of the Soil." He concluded that the words liberty and freedom in the United States meant "being free to do every thing they choose, without regard to the rights of others."[32]

The decade of the 1830s was also when the human skulls entered the picture. A new scientific market opened for Native skulls from the Deep South. They appeared from a variety of processes in this unsettled region. As the Choctaw and others were forcibly resettled, they brought along the bones of their dead to rebury them; some of these were seized and shipped East. Other Native skulls came from the violence that engulfed the region, and the looting that was its constant companion. Some of the skulls in Morton's cabinet came from the Deep South region now, shipped by American soldiers, naturalists, and physicians in boxes that often included fossil shells. A Natchez skull from Alabama, for example, appeared in Morton's *Crania Americana* (1839), an example of how his early fossil shell investigations laid the groundwork for his later craniological research (fig. 4.5).[33]

Judge Tait's plantation lay squarely in some of that dispossessed land, now US territory. But without good soil there would be no cotton, an economic imperative that drove the judge to Philadelphia to learn more about soil chemistry and fertility from the naturalists there. Embarking with his wife in 1826 on the long overland journey, Tait met Samuel George Morton and the other paleontologists at the Academy of Natural Sciences. He took courses with the mineralogist Solomon Conrad at

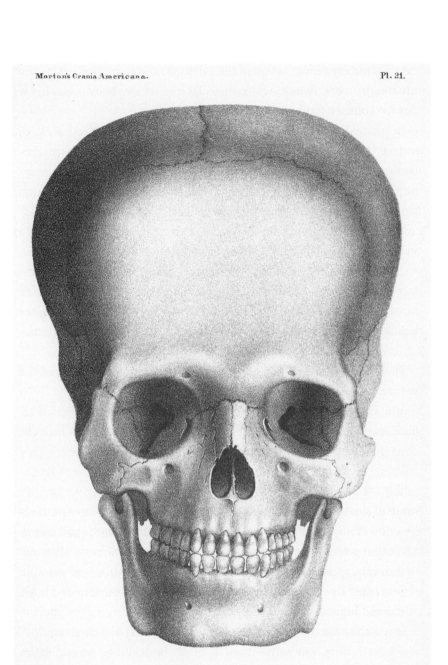

FIGURE 4.5. A Natchez Indian skull shipped to Dr. Samuel George Morton in Philadelphia from the Deep South later appeared as this life-sized engraving in Morton's influential *Crania Americana* (1839).

the University of Pennsylvania. Returning to Claiborne in 1827, and retired from the bench and politics now that he was fifty-eight, Tait devoted himself to natural history, opening a fossil-centered correspondence with Morton, Isaac Lea, and other Philadelphia paleontologists, sending them box after box of the most beautiful Alabama shells. He had just been elected to the American Philosophical Society. Proud to be "enrolled in the same catalogue" as Thomas Jefferson and Benjamin Franklin, he wanted to pursue more natural history research from afar. Would it not be possible for the Philadelphia group to send one of their number to Claiborne, Tait wondered?[34]

The call was answered in December 1832 by Solomon Conrad's son, Timothy Abbott Conrad, one of the young Philadelphia fossil gang. Just thirty, Conrad was by his own admission shy, awkward, often late, and largely lacking in ambition. Even his closest friends called him "a man of singularly absent mind & immethodical habits." He wrote poetry to combat the bouts of depression that dogged him. Perhaps an isolated plantation such as Claiborne would be just the thing, offering both the opportunity for scientific discovery and relief from tiresome social obligations. "Mr. Conrad and myself are resolved that the Geology of this country shall be thoroughly explored," Morton explained of Conrad's plans for Alabama.[35]

His letter of introduction from Dr. Morton to Judge Tait packed into his luggage, Conrad set off for Alabama in December 1832, eager to escape the cholera epidemic ravaging Philadelphia. Morton told him to keep an eye out for various scientific specimens on his journey, including any skulls from *Homo sapiens* or other animals. But Conrad's eye was preoccupied by other things: a slave society on a scale that he had never witnessed before. Everywhere Conrad saw poverty, sickness, and suffering. Returning late from a beach excursion in search of shells in black-majority South Carolina, he lodged in a tattered "negro cabin, with a dozen little black devils nearly naked."[36]

After the grueling overland journey, Conrad arrived at Claiborne in February 1833 and settled in for a year of fossil hunting. He was thrilled: here was "a perfect El dorado of fossils." Soon came a poem entitled "Claiborne." But sickness and slavery ruled here as well. Claiborne had by now become one of the most populous plantations in the state, with 115 slaves by 1835, stretching over five generations in a constant cycle of births and deaths, many dying from the fevers that swept over the region. Conrad was also stricken, mornings often finding him too feverish to rise from his bed.[37]

Still, over many months, Conrad collected hundreds of fossils. The routine varied. Some days Conrad clambered down the steep wooden stairs that led from the clifftop plantation to the Alabama River. From the shade of the magnolia trees, he pecked at the fossils that jutted from the looming bluffs. Other days, he hopped aboard one of the slow, low-slung cotton steamers to gather shells upstream or down. Churning along the Alabama River a decade later, Charles Lyell called the cotton steamers "the paradise of geologists," since with their shallow drafts they could press right up to the fossil-studded riverbanks. Judge Tait's many contacts along the river unlocked social connections, allowing Conrad to take side trips to collect more fossils, journeys in which he was often accompanied by a few of Tait's "servants" (slaves), a common practice in southern geology. One spot down the river was a "Golgotha," a field full of gigantic "saurian" or reptile bones (the word "dinosaur" would not be coined until 1841 to describe these extinct reptiles). Conrad dragged home the hulking vertebrae and even the "nose" of a marine saurian. The work was dangerous. One slave, his bag laden with fossils, almost fell off a cliff. By the end of 1833, they had harvested nearly two hundred new fossil shell species from Claiborne.[38]

At night, Conrad filled letters to Morton about his finds at Claiborne, unrolling a story about the great antiquity of the region. He sketched a geological stratum suggesting that perhaps the Cretaceous was here too, in the deepest strata of this new state, whose fossil shells resembled those of New Jersey and France. The modern water line is at the bottom, with the cliffs rising above, and the Claiborne plantation imagined at the top (fig. 4.6).[39]

Conrad's sketch of the cliffs looming over the high waterline of the Alabama River shows the Southern past culminating in the black belt of the Cotton Kingdom. The source of Claiborne's fertility lay at the bottom, at the beginning of the story. Labeled "Secondary?" (Cretaceous)

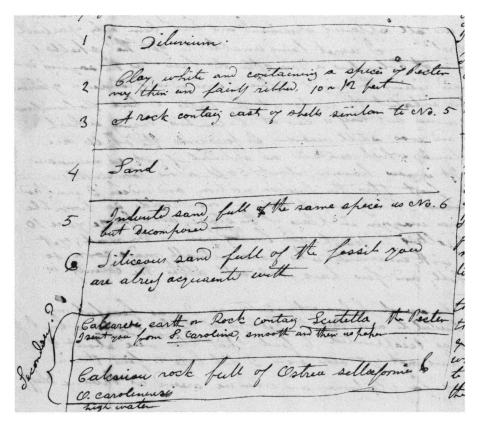

FIGURE 4.6. Timothy Abbott Conrad sketched these strata near the Claiborne plantation overlooking the Alabama River in 1833. The modern water line is at the bottom, with the cliffs lining the river rising above. Conrad wrote "Secondary?" at left to suggest that these lowest layers might date from the Cretaceous period.

in the left margin, that shell-rich layer was literally the bedrock of the Southern economy. The snowy cotton growing on the Claiborne plantation had roots near the very bottom of time.[40]

The strata imagery made its way back to Philadelphia. As they received boxes of Alabama fossils from Conrad, Morton and the curators at the Academy of Natural Sciences totally reorganized their own fossil collection "according to strata." Morton was thrilled. "When we shall be able to arrange this series, in its relative position, with the fossiliferous

& other overlying beds, what a magnificent and instructive series it will present!" Knowledge would be revealed to "the eye & the ear," and through them "the understanding," he continued. "With us instruction consists in demonstrating to the senses, by means of objects arranged according to their natural affiliation." The museum would make real to the human senses the abstract new concept of deep time, how the enormous antiquity of the land culminated in the apparently natural realities of the present day.[41]

The big payoff of Conrad's year of digging was to establish that Alabama, New Jersey, and parts of Europe were all the same age, a great band of contemporaneous strata linking the Old World and the New. Conrad fired off his Eureka moments in breathless letters to Morton. The Alabama Cretaceous fossil shells were "the same age as Jersey!" Conrad permitted himself a romantic reverie about the dark abyss of time in America's Cotton Kingdom. "Claiborne is a most interesting spot to the geologist," he wrote in 1834. "No where will he find in greater profusion those mute historians of the ancient revolutions of our globe, who seem as if by the wand of a magician to come forth from the eternity of the past." Similar findings about the Cretaceous origins of the Deep South's fertility also emerged from farther west, along the Mississippi River. The geologist George Featherstonhaugh reported finding fossil-rich "Secondary" strata as far north as Saint Louis. At Walnut Hills (now Vicksburg) along the Mississippi, the French artist Charles Lesueur drew one of the first stratigraphic images of that river. It shows the bank cut away to reveal fossil shells, which are magnified in the surrounding frame: here were many more mute historians to the ancient revolutions of the globe. To the white planters of the Deep South, the word "revolution" would have carried ominous implications. Sometime in the remote past, revolutions had rocked the region, exchanging one regime of life for another. But the planters could take comfort in the fact that this geological revolution had come from the outside, guided by the superintending power of the deity and not emanating from the mute creatures themselves.[42]

About seventy miles south of Vicksburg, the area around Natchez on the Mississippi River proved to be another treasure trove of fossil bones.

Natchez was by now one of the chief cotton emporia of the Deep South, its growers sending forty thousand bales two hundred miles downriver to New Orleans, the doorway to the Atlantic and the markets beyond. The geologists announced that the rich land sustaining Natchez cotton were the remains of an ancient sea. At his Laurel Hill plantation, which overlooked the Mississippi River near Natchez, Rush Nutt thought he saw the strata of what looked like a former ocean. "Such are the appearances of all this scope of country, that if we can suppose the sea to have once inundated one or two miles of the region nearest the coast, we see no reasons why the whole extent should not have been submitted to a similar influence." Scoffing at the Mosaic story of Creation because it rested neither on "an abundance of facts" nor "our experience," Nutt theorized a deep time for the region that extended to an era when an ocean blanketed Mississippi. The resulting fossils could not have occupied "a shorter period than several hundred thousand years." Nutt's insights about Natchez built on earlier reports from the area of bones and oyster shells from "some very distant period." They must grant to the Mississippi River "an incalculable antiquity," wrote planter William Dunbar in 1809.[43]

The fossil bones left by the ancient sea once covering the Mississippi were captured in the remarkable *Panorama of the Monumental Grandeur of the Mississippi Valley*, painted by John Egan around 1850. Moving panoramas were popular during the middle of the nineteenth century, and this one offered over twenty separate scenes of sites along the Ohio and Mississippi Rivers, from the perspective of someone churning along on the deck of a steamboat. Many of Egan's scenes show the different antiquities of the vast region, whether human or animal. One shows the Natchez plantation of William Feriday. Just downhill from the plantation, workers directed by Philadelphia archaeologist Montroville Wilson Dickeson are excavating the mounds left by the Native peoples of the region (fig. 4.7 and plate 9). A close-up of this scene shows modern alligators sunning themselves next to the fossil bones of extinct marine reptiles (fig. 4.8).[44]

Back in Philadelphia, Morton took stock of all the Deep South marine fossils: the shells, the saurian noses, the monstrous vertebrae. It was

FIGURE 4.7. In this scene of a Natchez plantation on the Mississippi River, workers directed by archaeologist Montroville Dickeson excavate mounds left by vanished Native peoples. John J. Egan, scene 18 from *Panorama of the Monumental Grandeur of the Mississippi Valley* (1850). Saint Louis Art Museum, Eliza McMillan Trust 34:1953.

FIGURE 4.8. In this detail of Figure 4.7, fossil marine reptiles are visible on the shore, to the right of the modern alligators sunning themselves. John J. Egan, scene 18 from *Panorama of the Monumental Grandeur of the Mississippi Valley* (1850). Saint Louis Art Museum, Eliza McMillan Trust 34:1953.

an extraordinary trove, collected in the space of just a few years. And now, in 1834, he made the final leap, connecting Europe and America in the crowning scientific achievement of his young career: a short book about extinct creatures of ancient America, titled *Synopsis of the Organic Remains of the Cretaceous Group of the United States* (1834). "[A]ll my character as a geologist rests upon this one effort," he confided anxiously to a friend.[45]

The book's unglamorous title concealed its revolutionary thesis: the Cretaceous was an ancient rock belt linking now-extinct animals across the United States and Europe. "[T]here is no perceptible difference between the teeth of the Mosasaurus of Europe and that of New Jersey," Morton wrote of the giant sea monster that had prowled the Cretaceous oceans. "From these and other data scattered through this work, I arrive at the conclusion, that when the chalk fossils were living inhabitants of the seas of Europe, the organic relics of this synopsis were alive in the ocean of America; in other words, that they were contemporaneous beings." That was it: Europe and America were the same age. The process was not a sudden, catastrophic "rush of currents." It was a gentle process, either the sea subsiding or the land rising, which had left the fossils in their pristine, unbroken condition.[46]

Europeans took note. Now the European and American Cretaceous could be joined, an ancient time connecting two spaces sundered by a modern ocean, a past synchronicity fueling the economic projects of the present. Mindful of his debt to European fossils of the Cretaceous, Morton dedicated his book to his English friend, the geologist Gideon Mantell, who had long advised him to find American "equivalents of the chalk formation of Europe." And now these equivalents had been found, stretching like a belt of time from the mid-Atlantic coast all the way through the Deep South. "Your works will form a grand epoch in American geology," Mantell wrote to his American friend. As his candle guttered in the evening darkness of his Brighton study, Mantell sketched a stratigraphic diagram of the Cretaceous in a letter to Morton. He included European Cretaceous strata in Maastricht and elsewhere, and then what must have thrilled Morton: "& probably those of the United States."[47]

American paleontologists rejoiced. Europe had noticed them. "I am now reading Mantell's geology of the South of England & am much gratified to see the high value he places on the contributions of the Philadelphia savans, & especially those of yourself," wrote a colleague from Ohio. Morton was onto something. In a letter to an English colleague, Scottish geologist Charles Lyell deemed Morton's theory persuasive: "The analogy of the genera, and even of the species to the European chalk [in New Jersey], is most striking." And then Lyell blessed the theory publicly in a book published in 1845: "Morton was right." The New World and the Old World were the same age. The Deep South, so new to the United States, now had a deep history that revealed its utility for modern projects.[48]

Aftermath: From Shells to Skulls

Over the next two decades, from about 1840 to the outbreak of the Civil War, Americans built on the findings the paleontologists had first proposed in the pages of their scientific journals. Earth itself, long ago, had produced a great band of fertility that underwrote the booming cotton economy of the American Deep South. Some naturalists added data from the new science of climate, which also linked humans to Earth's processes. They stated that only people from Africa were fit to work in that torrid region and on that black soil. Some physicians, too, claimed against the teaching of Genesis that an unbridgeable gulf of difference separated the physiology of white and black people. Their skins, their crania, everything was different and had been since the Creation itself, a doctrine that became known as polygenesis.

Political maps of the United States reflected what also seemed to be a primordial reality, locked into the depths of the continent itself: that the South was a distinctive region, made of Cretaceous soil and for African slavery, a black belt of soil and people, the engine of American prosperity. John Mallet's map of the "Cotton Regions" of North America, published in the midst of the Civil War, showed the Cretaceous fertile crescent of the Deep South in green against a pale pink and yellow background, making it pop out of the map. A chemist from Ireland

RACES.	No. of skulls.	Mean internal capacity in cubic inches.	Largest in the series.	Smallest in the series.
Caucasian.	52	87.	109.	75.
Mongolian.	10	83.	93.	69.
Malay.	18	81.	89.	64.
American.	147	80.	100.	60.
Ethiopian.	29	78.	94.	65.

FIGURE 4.9. Samuel George Morton's ranking of human races by cranial volume from his *Crania Americana* (1839).

who had immigrated to the United States, Mallet used the American cotton boom as a blueprint for British cotton cultivation in India. He cautioned the British that cotton could not be grown just anywhere with a tropical or semi-tropical climate. The crop also required "conditions natural and social," of which the American South had "mastery."[49] In other words, economic prosperity required more than geological luck: it needed a servile labor force.

Examining Samuel George Morton's early geological and paleontological research on fossil animals brings the full scope and significance of his infamous craniological program into view. The young Dr. Morton had launched his side career as a paleontologist, an expert on the fossil animals of the Cretaceous that proved that North America and Europe shared a common past deep in the abyss of time. Fired by rising sectional tensions in the later 1830s and the 1840s, Morton then built on his expertise in fossil animal anatomy to craft a new story about human anatomy. He would measure human crania shipped to him from those very regions of North America whose geology seemed to prepare it for slavery: the Deep South. From the geological he moved into the ethnological, the political climate encouraging him to imagine human hierarchy where once his studies had been confined to the geological hierarchy of strata. Morton now ventured the theory of innate black and

Native American inferiority that he thought his human crania supported. He published this theory in articles and books that established him as the leading American voice in the new science of separate and unequal human creations or polygenesis, what his admiring colleagues termed "the American School of Ethnology." In books such as *Crania Americana* and *Crania Ægyptiaca*, Morton published his human cranial measurements to propose that racial inequality was written into nature. Caucasian crania were larger than Ethiopian and American Indian crania, by Morton's reckoning (fig. 4.9).[50]

Until his death in 1851, Morton never stopped publishing about the menagerie of exotic creatures that had once glided through the sparkling primordial ocean blanketing North America and Europe in the Cretaceous. But his new line of controversial ethnological research finally obscured his revolutionary geological theories. Today, archives housing the papers of Samuel George Morton direct us toward his craniological research and notorious racial theories, concealing the earlier moment in which Morton helped to write a new ancient history of the planet. But that earlier moment is essential to our understanding of the rise of race as a science of human difference in the nineteenth century. The new sciences of geology and paleontology, pursued by Morton and his colleagues, became the crucial preamble to the story the South began to tell itself about racial difference in the decades before the Civil War. Geology and ethnology were both written into nature, according to Morton and his colleagues: ancient shells and human crania proved that hierarchy was natural and God-given, and not a human invention.

In short, Samuel George Morton was even more significant than we have suspected to the rise of racial science in nineteenth-century America. His theory that the oldest South underwrote the fertility of the modern South vindicated the harshly unequal economic, social, and political projects over which the nation eventually shattered in civil war.

5

Mammals, the First Americans

FROM THE PRAIRIE in South Dakota and Nebraska, the ground suddenly drops away into a gash about a hundred miles long and fifty miles wide. These are the Badlands, a region of earthen spires and steep gulches so dry and forbidding that farming is impossible and even grazing is difficult. Early French visitors called these *mauvaises terres*, lands that were bad because they refused to sustain human life. The shaded gully walls offer no relief from the heat. They yield instead a different reward: the fossil bones of extinct mammals. Some are eerily familiar, like the horses that look modern but for the fact that they are tiny, the size of dogs. Others are as big as rhinos, with fearsome horns and crushing hoofs.

These are the extinct creatures of the Age of Mammals, a new term coined in the second half of the nineteenth century that was central to the mythification of the American West as a land of fresh starts. Engineers laying railroad tracks on the prairie stretching from the Missouri River to the Rocky Mountains brought along geologists and paleontologists. These naturalists found mammal fossils lying in strata beneath the railroads but above the rocks holding dinosaur fossils. The scientific name they gave to the Age of Mammals—Cenozoic, or age of new life—underscored the dramatic changing of the guard in the fossil record, when the mighty dinosaurs of the Mesozoic yielded to horses, camels, and the rhinoceros-like Titanotherium. The Cenozoic stretched through thick strata, suggesting a long era—perhaps millions of years—that had culminated in the Age of Man, the greatest mammal of all, according to these naturalists.

These ancient mammals seem to have sprung from the American soil itself. Horses, camels, and other helpful creatures had thrived for millions of years in their land of origin before migrating to Asia and beyond. Scientists in the United States concluded that the true native Americans were not the Indians but these far more "ancient Americans," as one paleontologist called the fossilized mammals of the West. In this longer history of the American West, the Indians became immigrant latecomers, arriving eons after the truly indigenous creatures of the Age of Mammals. Catapulted to a starring role in the national drama, the extinct mammals of the Cenozoic appeared in dioramas, murals, and skeletal reconstructions in natural history museums from Washington, DC, to Los Angeles. College classrooms featured courses on mammal evolution. They created a thrilling backstory for the modern West, dramatizing large-scale processes of emergence, migration, and extinction that whispered lessons to the present.[1]

Creating the Ancient West

To citizens of post–Civil War America, the prairie West shimmered with promise. Dreams of prosperity sent thousands of white Americans across the Mississippi River. Soon the transcontinental railroads crossed the region, promising to mend a nation shattered by civil war by linking distant markets and suppliers at unheard-of speeds. From the perspective of the many Native groups inhabiting the Great Plains, matters looked quite different. Many Indians had been pushed westward onto the forbidding prairies from the Great Lakes region over the previous century, joining the tribes already there. They found ways to survive and even thrive in grasslands that became ever more arid the farther west they stretched. Capitalizing on the strength and speed of the horse reintroduced to the Americas by the Spaniards, the Lakota and other Native peoples in the northern Plains centered their societies on the bison. A series of treaties governing Indian-white relations from the early 1850s onward established forts and trading posts throughout the territory. The Union Pacific Rail Road, founded in 1862 and originating in Council Bluffs on the eastern side of the Missouri River, eventually cut straight through bison-rich and fossil-rich Lakota lands along the North Platte River in what is today

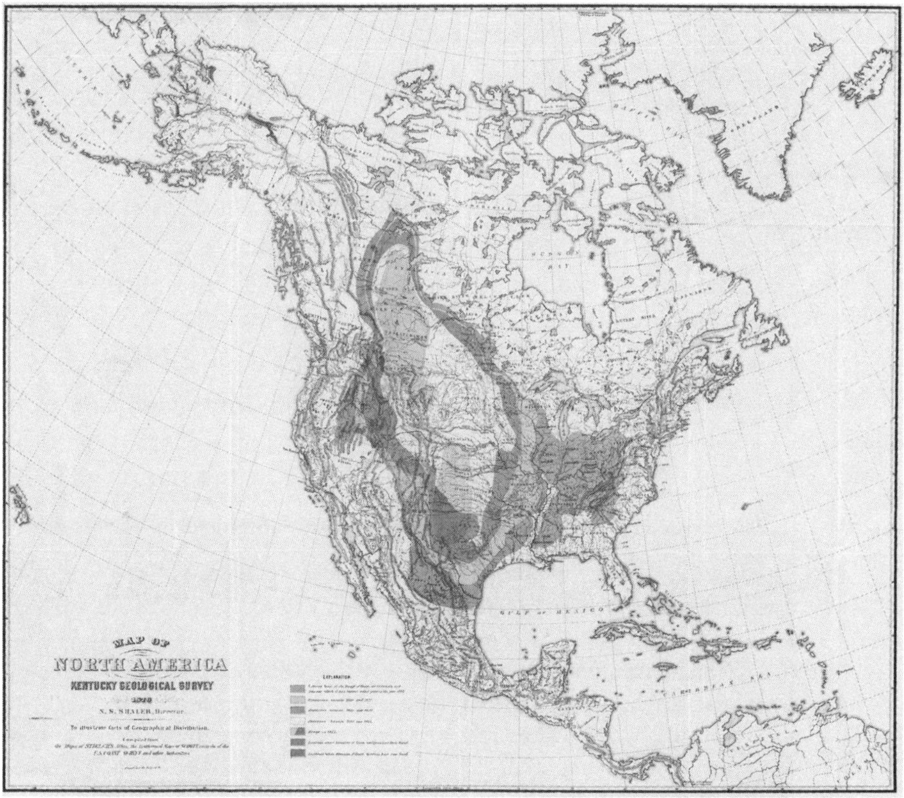

FIGURE 5.1. In this 1876 map of the shrinking habitat of the American bison, the smallest colored area at center is where the bison "still exists," while the largest area shows its range before 1800. J. A. Allen, "The American Bisons, Living and Extinct."

South Dakota, Nebraska, and Wyoming. This cut off the roving bison herds, whose numbers declined rapidly, a process exposed in maps unveiling a time-lapse tragedy of shrinking habitats (fig. 5.1 and plate 10).[2]

As the dramas of the modern West unfolded on the surface, a new antiquity stretched out below. By midcentury, American geologists had ventured across the Missouri River in search of mineral wealth. David Dale Owen, standing at the rim of the Badlands in 1849 with his band of geologists, marveled at "the former existence of most remarkable races" (fig. 5.2). He was not speaking of the many Native peoples who still lived there, some of whom accompanied his expedition, supplying it with information and executing some of the sketches that later decorated the published geological

FIGURE 5.2. The Badlands or Mauvaises Terres in the Nebraska Territory, where many fossil mammal bones jutted from strata laid down during the Age of Mammals. Note the Native people at right; Owen and his companions cited them frequently as informants about geological matters in the West. From David Dale Owen et al., *Report of a Geological Survey of Wisconsin, Iowa, and Minnesota* (1852).

report. Instead, Owen was speaking of the fossil remains of extinct mammals embedded in the Badlands, what he called "a city of the dead."[3]

The end of the Civil War brought more scientists to the American West and added more millennia to the region's presumed antiquity. "The vast extent of our country west of the Mississippi seems to have been the arena on which were enacted, during the Mesozoic and Cenozoic times, some of the most important events in the geological history of the American continent," wrote geologist Ferdinand Vandeveer Hayden in 1869. The Reconstruction policies that reunited the modern nation were mirrored by the geological processes that had first united the ancient continent. The great interior sea dividing North America in the Age of Reptiles "was abolished, and the continent became one" in the Age of Mammals, wrote one geologist at the close of Reconstruction.[4]

Many of these excavators had been sponsored by East Coast universities and museums, who filled their galleries with Western fossil mammals. Personifying the Gilded Age of industrial capitalism and unvarnished greed, these paleontologists—Edward Drinker Cope, Othniel Marsh, Joseph Leidy, John Bell Hatcher, Fielding Bradford Meek, William Berryman Scott, and Henry Fairfield Osborn, among others—competed to unearth, name, and publish as many new fossil mammals as they could. They erected towering skeletal reconstructions, drew wall charts for new college courses on the Age of Mammals, fashioned life-sized models, and painted full-color landscapes featuring a cast of extinct mammals. Although they hired women such as Cecilia Beaux as illustrators for some of their landmark publications (since drawing was taught in schools for girls as an appropriately genteel skill), they diminished women's contributions. They bragged to Europeans that the American West yielded more mammal fossils than anywhere on Earth. "The Tertiary of Western America comprises the most extensive series of deposits of this age known," wrote Othniel Marsh. The eminent English scientist Richard Owen confessed his astonishment to Cope. "I could not foresee that your virgin soil would be so productive of new, strange" forms. One American paleontologist hoped that the Age of Mammals would rid the world of Eurocentric geological "time standards" once and for all and make time truly international.[5]

These hopes and dreams unleashed the fossil rush of the 1870s. Reconstruction policies attempting to reunite a nation sundered by the Civil War were now joined by the ideological program of enclosing a very brief US national history into a much more ancient history of the West. One paleontologist dubbed this "the long history of the great West." The middle of the North American continent, paleontologists decreed, was not just the site of the current Indian Wars by which the United States was attempting to wrest lands from the Native peoples who had occupied it for thousands of years. It was the stage on which far more stirring episodes had unfolded, dramas of epic significance involving giant beasts prowling land and sea. These awesome creatures now lay quietly in their rocky graves, offering themselves as contributions to knowledge for the expanding nation. According to paleontologists, the modern-day Indians did not understand the significance of these fossils, instead wrapping them in legends of mythic thunder beasts. American scientists had other plans for the fossils: to write a new history of the American West, one that began long before any human being had walked the land. In this telling, the West was not a frontier but a center of elemental processes that led to the modern United States. This long history of the great West was like an enormous book consisting of three enormously long chapters that stretched from the dinosaurs to today.[6]

The first chapter opened with the fearsome monsters of the Mesozoic era, or Age of Reptiles, as it was popularly known in a coinage of the early nineteenth century. Before the prairies, before there was even a single landmass of North America, an ancient sea corridor had stretched from the Gulf of Mexico to Canada. It had blanketed the infant nubs of what became the Rocky Mountains. American paleontologists called it "our great interior sea" and the "Cretaceous mediterranean." This ancient sea eventually became a graveyard for its primordial inhabitants, whose fossilized skeletons underwrote the fertility of the modern Great Plains. "This is the great source of its wealth in nature's creations of vegetable and animal life, and from it will be drawn the wealth of its future inhabitants," explained Edward Drinker Cope of the ancient sea. Speeding across the prairies, modern trains would realize the potential of the primordial sea by linking Atlantic and Pacific markets. "[S]o long as peace and steam bind the natural sections of our country together," Cope continued, "so long will the plains be one important element in a varied economy of continental extent."[7]

Images of this western Age of Reptiles began to appear in both scientific and popular publications. These illustrations showed sea reptiles cavorting in the West's ancient sea as dinosaurs patrolled the beaches. In "The Sea Which Once Covered the Plains," from William E. Webb's popular book *Buffalo Land* (1872), marine monsters paddle swan-like, their necks held high above water, in what was now the new state of Kansas (fig. 5.3). Webb, an eccentric, mustachioed agent for the National Land Company promoting western settlement and the building of the Kansas Pacific Railway, was also fascinated by the Mesozoic marine reptiles then being unearthed around him. *Buffalo Land* was his fictional reconstruction of actual events of the late 1860s that mixed in episodes from the ancient West. The illustration by the fun-loving Kansas artist Henry Worrall depicts an epic battle in the future farmlands. The chunky Liodon at center squirts water toward the sinuous Elasmosaurus, which squawks in protest. Two pterosaurs hover safely out of reach. The pond-like flatness of the water suggests the gentle, gradual transformation of this ancient sea into the modern farmland that fed the nation's growing population. The fossils of these primordial Kansas inhabitants were shipped on the new Kansas Pacific Railway to East Coast luminaries such as Louis Agassiz and Edward Drinker Cope, who published plates of their vertebrae and skulls in their scientific periodicals.[8]

The Elasmosaurus proved that American fossils from the Age of Reptiles were bigger and more frightening than their European peers—according to American scientists, that is. Its fossil bones, first found in Kansas in 1867, showed that America possessed a marine animal with a neck length "far exceeding" that of the famed Plesiosaurus unearthed half a century before in the English coastal town of Lyme Regis by the fossil hunter Mary Anning. Compared with the English Plesiosaurus, the Kansas Elasmosaurus "possessed a much longer neck,—one indeed that exceeded that of all known animals," according to Joseph Leidy. It was also a fierce predator, "at one moment darting its head a distance of upwards of twenty feet into the depths of the sea after its fish prey, at another into the air after some feathered or other winged reptile, or perhaps, when near shore, even reaching so far as to seize by the throat some biped dinosaur."[9]

Then the first chapter of Western history closed. The strata told the story: rock layers brimming with giant reptile bones stopped. Above those layers lay the fossil remains of mammals of all kinds, from tiny shrew-like creatures to horned beasts. Reptiles there were, but they were small and outnumbered by the mammals. As the American naturalists put it, the Age of Reptiles had ended and the Age of Mammals had begun. "Modern history commences here," one geologist observed.[10]

The big question was this: what had happened to end the Age of Reptiles? Today, many people imagine the end of the Age of Reptiles (or Mesozoic) as a sudden cataclysm, when a giant asteroid plunged into the Yucatan Peninsula and instantly ended the reign of the dinosaurs. The asteroid theory was proposed in 1980 in the journal *Science* by several Berkeley scientists, including former Manhattan Project physicist Luis Alvarez. According to this theory, the asteroid impact ejected so much debris into the atmosphere that sunlight was blocked for many years to

FIGURE 5.3. In "The Sea Which Once Covered the Plains," an epic battle unfolds in the future Midwest farmland of America. This is one of the earliest landscape illustrations of North America during the Age of Reptiles, when a large ocean split the continent in two. From William E. Webb, *Buffalo Land* (1872).

ERED THE PLAINS.

come. What plants and animals were not instantly vaporized were felled by the cold gloom of the long post-impact winter. In a nuclear age, when it seemed plausible that Armageddon might come from the skies, the asteroid theory captured the public imagination. With its image of fiery calamity followed by what resembled the much-feared nuclear winter, it offered an alternative to the gradualist explanations based on other evidence—such as the Deccan Traps vulcanism thesis proposed by Princeton geoscientist Gerta Keller and others. Only in the wake of the 1980 asteroid theory did the term *extinction event* come into popular usage. This further spread the idea that extinctions were sudden bombshells rather than long, drawn-out, multifactorial processes. The cultural preoccupation with so-called *mass extinctions* began. Finally, the asteroid theory also—not incidentally—focused attention on the American West as the site of one of the most important events in the history of life on Earth.[11]

In short, the asteroid theory's neo-catastrophism helped to bury the century of gradualism that preceded it in a collective amnesia. But for a full century before 1980, gradualism was the dominant explanation for ending the Age of Reptiles. Rudolph Zallinger's Yale mural from 1947 (fig. I.1 and plate 1) displayed that gradualism with its trio of volcanos gurgling ominously behind the Cretaceous dinosaurs. Many nineteenth-century naturalists agreed that there had indeed been "a time of disturbance" in the West involving the "sudden disappearance of great numbers of giant reptilian forms at the close of the Cretaceous epoch, and their replacement in the early Tertiary by numerous larger mammals," as Yale's influential naturalist James Dana put it in 1863. Yet fossil strata did not necessarily reveal what counted as "sudden" when the time frame was possibly millions of years. "But there is no reason to believe that the revolution was the result of an instantaneous movement," Dana continued, speaking of the Cretaceous-Tertiary boundary. "It was probably slow in progress . . . and may have occupied a long age." Osborn agreed that the causes and duration of the extinctions were more likely uniformitarian than cataclysmic. Well-defined "lines of separation" between geological ages were impossible because they were at some level "arbitrary," geologist David Dale Owen concurred. There was "blending" that occurred because of a "gradual dying off of orders, genera, and species." Should a line between the two eras even be drawn? And if so, where? "[W]here the end of the cretaceous? where the beginning of the tertiary?" asked Edward Drinker Cope exasperatedly of the boundaries between the Age of Reptiles and the Age of Mammals in 1873. Charles Darwin expressed skepticism in the *Origin of Species* about the evidence for rates of extinction. Extinction was not so much an event, he argued, as a process rolling out over "wide intervals of time."[12]

The gradualism maintained by these late-nineteenth-century American geologists helped them to embed the West into a national story. The long twilight of the Age of Reptiles, they argued, prophesied the region's agrarian future. They imagined the ocean floor rising into a plateau, and the new Rocky Mountains forming a barrier that blocked moist winds from the Pacific Ocean. In a "gradual change," said one paleontologist, the inland sea corridor of the Mesozoic now shriveled into a mosaic of freshwater lakes and rivers watering grasslands and forests, the ideal environment for grazing animals. Naturalists saw this slow, long-term desiccation continuing into their own day. The Great Salt Lake was the puddle remaining after the evaporation of its massive ancestor, Lake Bonneville. One need only go to the African savannahs to see keyholes to North America's past, they concurred.[13]

And so opened chapter 2 of the West: the Age of Mammals, which linked America's geological antiquity to the modern era of human beings. Its first epoch was termed the Eocene, literally the dawn of the new era. But the Eocene was by no means recent, Americans were assured. Rather, it formed part of the vastly ancient history of the American West. "Recent?" asked Darwinian popularizer William D. Gunning in his *Life-History of Our Planet* (1876). "We do not think of the pyramids as recent, but during the Eocene period the very rock of which the pyramids are made was not itself made." Gunning likened the Eocene to the Franklin stove, the ancestor to all modern American stoves. The mammals of the Eocene were "of commanding interest," showing how simple and "generalized" mammals went on to specialize in later eras as the forces of progress sped them up the evolutionary tree. Gunning featured these simple mammals in a tropical forest clearing in his illustration, "Wyoming in the Later Eocene" (fig. 5.4). Drawn by his wife, Mary Gunning, it may rank as the first landscape scene of prehistoric American mammals. Hiding in the forest at left

FIGURE 5.4. Strange creatures assemble in a forest clearing during the Age of Mammals. This is another early landscape illustration of prehistoric life in North America. Titled "Wyoming in the Later Eocene," it is the frontispiece to William D. Gunning, *Life-History of Our Planet* (1876).

is a tiny pre-lemur or "half-ape," meant to suggest that in North America might even lie the deep origins of the human lineage. Primeval mammals stand about, prophetic of their later versions: pig, camel, horse, and a ferocious feline that would evolve into the bloodthirsty saber-toothed cat. Dominating the center is the bizarre horned creature that was small of brain but massive of body: Uintatherium, the beast of the Uinta Mountains. Like the plutocrats of the Gilded Age, this ancient mammal ruled its world. American paleontologists mixed the modern class metaphors of their era with the monarchical metaphors of Linnaeus, who had arrayed all life into kingdoms. Mammals were now "the dominant class," Americans decreed, whereas shrunken reptiles could only cower before the "new race of lords." Edward Drinker Cope announced the discovery of this enormous, elephant-like mammal in 1872. Cope called it the "monster of Mammoth Buttes," which succeeded the dinosaur and "reigned in his stead." Crowned with curious horns, this "ancient king" ruled the new

FIGURE 5.5. The king of the Age of Mammals in North America was the curious, horned Loxolophodon Cornutus Cope—according to paleontologist Edward Drinker Cope, that is. "The Monster of Mammoth Buttes," *Penn Monthly* (August 1873), frontispiece.

mammal kingdom. Cope's version shows the beasts standing majestically, with two fighting in the center back (fig. 5.5).[14]

The ancient American mammals actively directed their own evolution, a progressive version of Darwinism that American paleontologists termed *orthogenesis*. Through the "active exercise" of a body part, animals ascended to a more advanced stage of evolution. Animals attained "superior organization" by moving from generalized body parts to ones adapted specifically to the animal's needs. These "laws" of evolution toward growing specialization led the march toward "higher forms of life." As in a modern factory, specialization was the goal—but not too much specialization, which exposed the animal to changes of climate.[15]

While hoofs, skulls, and horns revealed progression over time, the brain now became the crowning example of "the real progress of mammalian life in America" and its magnificent culmination, Man. American scientists became preoccupied with brain and skull size, whose growth

FIGURE 5.6. The plodding and very stupid American mammal Uintatherium, as depicted by William Berryman Scott in his *A History of Land Mammals in the Western Hemisphere* (1913). Visual representations of prehistoric North American mammals changed rapidly, from the simple, cartoon-like versions of the 1870s (figs. 5.4 and 5.5) to highly naturalistic renderings such as this one by the early twentieth century.

amounted to a law, according to Othniel Marsh. While the lumbering reptiles had relied on "instincts," the mammals forged a new path by focusing their evolution on the brain. "The prevailing stupidity began to yield to a mental condition a step nearer intelligence," announced one naturalist in 1895. Early mammals had brains of a low, almost reptilian type epitomized by the ponderous Uintatherium, which developed in a half century of American illustrations until it appeared in all its lumpy-skulled splendor in 1913 (fig. 5.6). Naturalists marveled at its "stupidity," its elaborate skull encasing the smallest brain of any warm-blooded animal. "Brain power" improved as mammals progressed toward the modern era.[16]

America's primordial horses embodied another prized Gilded Age trait: originality. The West's fossil record suggested that America was "the true home of the Horse." The horse had sprung from the American land, a thesis that undermined ongoing European skepticism about the New World's capacity to generate large, vigorous life forms. The horse had "originated on this, instead of on the Eastern continent," they concurred. Beginning with spry little Eohippus, the tiny "dawn horse," these native American animals grew in size, intelligence, and speed over millions of years. Eventually, the horses headed west across the Bering land bridge to the Eurasian continent. Americans had colonized the Old World, US naturalists observed with obvious relish. The Bering land bridge was "the path by which many of our ancient mammals helped to people the so-called Old World," wrote Marsh.[17]

The upward progress of the American horse was on display in Marsh's classroom at Yale. Reconstructed fossil skeletons and posters showed students a complete, linear sequence of horse evolution that culminated in the modern horse (fig. 5.7). No links were missing, an attempt by Marsh to refute skeptics of Darwin, who seized on gaps in the fossil record to cast doubt on the basic idea of evolution. (Some of those opponents taught at Princeton, where a rival schema emerged at the same time, as chapter 7 shows.) The American horse evolution sequence at Yale became internationally famous. It was visited by luminaries such as Thomas Huxley, known as Darwin's Bulldog for his fierce defense of evolution by natural selection. One classroom painting of the Badlands revealed the arid site in which the fossils had been exhumed, while a poster of the geological time column helped students to place the Age of Mammals in a new, deep American history stretching to the origins of life on Earth.[18]

The artist Charles Knight enshrined the ancient horses among America's vigorous export products in his colored paintings and models, which appeared in magazines and museums from the 1890s onward. To a public enchanted by dinosaurs but still hazy about the significance of ancient mammals, Knight offered the first full-color landscape images of the Age of Mammals in the American West. He showed the horses evolving in America from frail, dog-sized creatures nibbling on leaves to the noble grassland grazers of a later era. Knight worked from models of his own making to better show the "moving and feeding creature." His watercolor of Eohippus showed the delicate animals ambling through a

FIGURE 5.7. The poster in the left-hand corner of the Lecture Room in the Peabody Museum at Yale in 1879 shows the linear evolution of the horse in North America.

shaded glen, alert but not afraid (fig. 5.8 and plate 11). "He was the animal for the times," Osborn declared approvingly of little Eohippus.[19]

America's ancient horses also explained the cultural inferiority of the American Indians, according to paleontologists. Originating and going extinct in America long ages "prior to the advent of man," early horses antedated not just the Spaniards but the American Indians. The Indians had migrated to the New World eons after the extinction of the ancient horses. Arriving in a horse-less land, the Indians fell behind the peoples of the Old World, who benefited from the primal equine gift of ancient America. As Marsh explained of the Indians, "his slow progress towards civilization was in no small degree due to . . . the absence of the Horse." The Spaniards had been able to reintroduce the genus *Equus* in America so successfully in the sixteenth century because its ancestors had evolved there, in the "vast pastures" of the American West.[20]

FIGURE 5.8. Eohippus, the little dawn horse of North America, as painted by Charles R. Knight in 1905. American paleontologists liked to advertise the fact that the horse was native to North America, and therefore antedated Indian claims to being the true natives of the Americas.

America's fertility was further confirmed by the "reigning plutocrat" of the Age of Mammals: Titanotherium (or Titanothere). The name meant "large animal," and indeed it was, rivaling the modern elephant in size. Charles Knight's illustration for the popular *Century* magazine in 1896 unveiled a nativity scene for Americans. The armored colossus of the Titatnotherium father guards his nuzzling child as the mother rests (fig. 5.9). This two-ton brute was no "parvenu or upstart," Osborn assured American readers. Boasting a family tree stretching back to some half a million years before, it was a true native American. The Titanotherium went by several names (such as Brontotherium, or thunder beast), a result of the competition among scientists for naming rights. Whatever the name, America's gigantic mammals "far exceeded in size" any other living or extinct mammals in its order, which included horses, zebras, and rhinoceroses.[21]

Paleontologists saw Titanotherium as antiquity's bison, a thesis that linked the archaic and modern prairies of North America into a single

TITANOTHERE FAMILY—BULL, COW, AND CALF—OF THE SOUTH DAKOTA LAKE BASIN.
From a mounted skeleton and skulls in the museum.

FIGURE 5.9. A prehistoric nativity scene near the Rocky Mountains features the giant North American Titanotheres—father, mother, and baby. Drawn by Charles R. Knight, this image appeared in the widely circulating *Century* magazine in 1896. Its similarity to the bison family in the next image was probably intentional, showing that both species were doomed to extinction.

linear sequence. Like the modern bison, these large grazing animals "lived in great numbers on the eastern flanks of the Rocky Mountains." And like the bison, Titanotherium was a creature of "intense stupidity," housing a tiny brain in a vaulting skull that swelled over eons, according to supposed laws of cranial development. But this overdevelopment may itself have doomed the Titanotherium, which went extinct just as it reached its largest size. American paleontologists offered overspecialization and oversensitivity to climate change as causes of its gradual extinction. Titanotherium entered the popular imagination along with the modern bison, imperiled by industrial-scale overhunting and habitat degradation caused by the railroads. William Hornaday's influential manifesto, "The Extermination of the American Bison" (1887), included images of affectionate bison mothers and fathers who knew their "duty"

FIGURE 5.10. A modern bison family, in a scene similar to the previous one with Titanotheres. From William Hornaday, "The Extermination of the American Bison, with a Sketch of Its Discovery and Life History" (1887).

to ensure the safety of the calves (fig. 5.10). These widely dispersed images may have influenced Knight's posing of Titanotherium.[22]

The new story of the primordial links between ancient and modern plains mammals reinforced the idea that extinction was not a catastrophic event but rather a long process. It put part of the blame for extinction on external factors such as climate change, but—significantly—also on internal failures within the organism, such as overspecialization and small crania. The fault for extinction lay in part within the organism, it seemed. Displaced onto the Native peoples of the plains, this extinction narrative had important consequences.

The Age of Man

The third and final chapter of the new deep history of the American West was the present, or "Age of Man." The dinosaurs of the Age of Reptiles had yielded slowly to the furry beasts of the Age of Mammals. Now all had culminated in the glorious present—also called, with the distinctive self-congratulatory flavor of the Victorians, the "Era of Mind." James Dana's *Manual of Geology* (1863) explained to the future leaders now sitting in his Yale classroom that Man was "archon" of the mammals, distinguished from all others by his spiritual nature (fig. 5.11).[23]

Not all human minds were equal in the Age of Man, according to naturalists. The geological deep time of the North American continent brought the Indians' allegedly primitive condition into sharper relief. Categorizing all human societies on a linear sequence ascending from savagery to civilization, naturalists saw the Indians' apparent inability to conceptualize deep time as one measure of their primitive condition. Although the Indians lived in the Age of Man, they were living fossils, representing the past from which Euro-Americans had progressed. Paleontologists contrasted the Indians' cyclical or circular sense of time, in which there was no dichotomy between present and past, with the progressive, nonrepeating time of so-called civilized peoples embodied in the timeline and strata. Some saw Natives' cyclical time as a sign of effeminate weakness, a constant relapse to an unruly past. By contrast, linear time was seen as masculine because it was forward-looking and ever innovating. In his influential *Principles of Psychology* (1855), the British theorist Herbert Spencer explained that the brain's grasp of time became more sophisticated with evolution. Primitive people imagined space in terms of time, describing distances in terms of the time it took to get from here to there. By contrast, civilized people imagined time in terms of space, using devices such as sundials, clocks, and watches.[24]

The Plains Indians' circular sense of time was captured in their calendars, according to the US ethnologists who traveled to the West in the post–Civil War era. Known as winter counts today, Indian calendars were rows or spirals painted on animal hides in which each image represented the noteworthy events of a single year. Usually counted from first snow to first snow, some Indian calendars covered up to a hundred years.[25]

Among the first scholars to study the Indian calendars was Garrick Mallery, a US army colonel and ethnologist working for the Bureau of American Ethnology, founded in 1879 as part of the Smithsonian Institution. Like other ethnologists of the time, Mallery categorized human culture along an evolutionary staircase that ascended from primitive

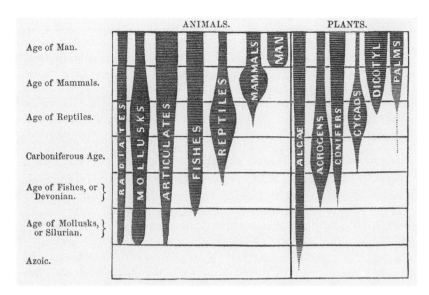

FIGURE 5.11. This chart from Dana's popular *Manual of Geology* (1863) taught American college students that Man was a mammal unconnected to other mammals.

savagery through barbarity and then to civilization. To Mallery, the Indian calendars represented "the interesting psychologic fact that primitive or at least very ancient man" used picture writing. By contrast, civilized nations possessed alphabetic writing, "the great step marking the change from barbarism to civilization." Seen as documents of savagery, many Indian calendars were classed among the ethnographic collections of natural history museums rather than the art museums dedicated to the artifacts of civilization.[26]

In 1877, Mallery published the Calendar of the Dakota Nation. Animals, people, and objects form a spiral of yearly happenings stretching from 1799 to 1870 (fig. 5.12). Mallery saw the spiral as a serial record of "trivial" events. "If it had been a complete national history for the seventy-one years, its discovery would have been far more valuable; but it is the more curious and unique, because it is only an attempt, before unsuspected among the nomadic tribes of American Indians, to form a system of chronology." Mallery also belittled other aspects of the Indian calendars. They were "not probably a very old invention," nor were they

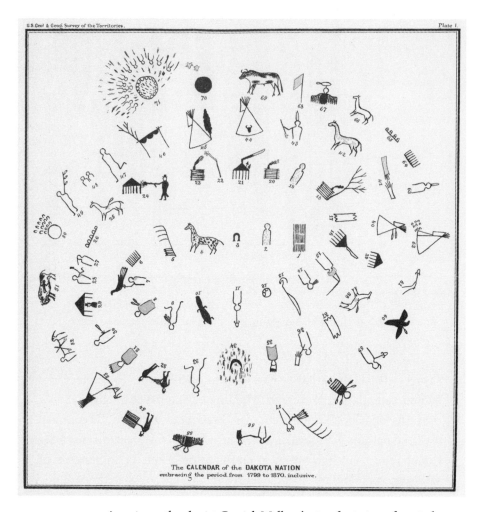

FIGURE 5.12. American ethnologist Garrick Mallery's 1877 depiction of a spiraling Dakota Indian calendar. Known as winter counts today, these calendars were used by white Americans to argue that Native Americans lacked a concept of deep time.

even a Native invention, since it was likely they were first produced in the late eighteenth century when "civilized intercourse" with Europeans began and writing rather than oral traditions became normalized. Mallery saw the calendars as further evidence that the Indians were not native to the Americas. Their spiraling calendar system was entirely dif-

FIGURE 5.13. Two men overlook the Grand Canyon in William Henry Holmes's "Panorama from Point Sublime" (1882). With its strata piling upon strata, this image accentuates the new idea of the canyon's profound geological antiquity, asserted just a few decades before it was designated as a national park. The idea of deep time helped white Americans to justify Native exclusion from the new national parks. They argued that these cathedrals of nature were so ancient that they preceded the comparatively recent immigration of the Indians into North America.

ferent from any used by the white people in the region, and therefore must be evidence of the Indians' membership in the "Mongolian race."[27]

Some images essentially evicted American Indians from deep time. Photographs and landscape paintings of the West by Thomas Moran, William Henry Holmes, and others in the last decades of the nineteenth century emptied the nation's ancient canyons and mountains of human habitation. Instead, primordial rocks were the main characters in these images. In Holmes's "Panorama from Point Sublime" (1882), two men in suits and hats observe the rocky stripes of the Grand Canyon descending into the deepest time (fig. 5.13 and plate 12). The seated man is busy drawing the landscape, as though the scene had become art rather than productive land in which people might eke out a living. Scrubbed of any other human presence besides the observing artist and his companion, the canyon becomes grand by virtue of a temporal imperialism that justified US dispossession of Native lands.[28]

By the early twentieth century, the American West had acquired a very ancient history. It was written by geologists, paleontologists, and ethnologists—but also by railroad builders and land speculators. Like all histories, the new ancient history of the American West served present purposes. Sewing the primordial creatures of the Age of Reptiles and the Age of Mammals into the national history of the United States, white Americans relegated the Indians to a temporal limbo trapped somewhere between the continent's deep antiquity and its modernity. Latecomers to the Americas when compared with the dinosaurs, the Indians became immigrants rather than native-born inhabitants such as the horse. Their spiraling calendars failed to record deep time, further evidence of their failure to progress up the ladder of civilization. In the new ancient history of the United States, in which ancient dinosaurs and mammals were the true first Americans, the exclusion of the Indians seemed to be a truth written into the Earth itself. It would not be until the 1960s that the term "Native American" emerged in popular usage to attempt to repair some of the damage that deep time had done. Today, the question of when the Americas were first peopled remains not just open but highly controversial, given the paucity of evidence and the high stakes of the answer.

6

Glacial Progress

ON A BEACH in Massachusetts lies Plymouth Rock, a shrine to the landing place of the Pilgrims. In the nineteenth century, Americans also began to celebrate that rock as something else: a glacial erratic, ferried there by ice so long ago that it made the Pilgrim landing of 1620 seem modern by comparison, less a birthplace than a waystation in a much longer American history. "Plymouth Rock is a bowlder from the vicinity of Boston, having accomplished its pilgrimage long before the departure of the Mayflower from Holland." So wrote an American geologist in 1883 as he tore down one history and installed a new one in its place: the story of the Ice Age. In this version of events, Plymouth Rock was just one of the many surface features of the North American landscape that now fused into a pattern for those who knew how to see. Boulders, gravel, lakes, the low hills called moraines—these were the ghostly footprints of the Ice Age, a shocking new idea with important meanings for nineteenth-century Americans.[1]

Today, the Ice Age is the uncontroversial setting for children's movies. But at its birth in the 1830s, the idea that the whole Earth had once been gripped by a long winter was almost too strange to be believed. Proponents of the new Ice Age theory claimed that the planet had once been plunged into a long winter. In North America, the heavy sheets of ice smothering the continent had compressed the bedrock itself, like a body on a mattress. Fingers of solid ice reached as far south as Cincinnati and New York City, dropping boulders that Central Park's builders had to work around. The glacial face, taller than the tallest building, was scoured by relentless winds that dropped the fine powder that now anchored America's vast prairies. In their final retreat, the dying glaciers melted into a ribbon of lakes stretching from Walden Pond to Lake Superior to Lake Tahoe.

This had all happened long ago, but not *very* long ago. For Americans, the Ice Age became the doorway to the modern world, their own era. Its new scientific name, the Pleistocene ("most new"), reflected that proximity to the present day. Unlike the more ancient strata that lay deep underground, largely hidden from view, the glacier's footprints marched along surface rocks, plain as day. They were the final traces of the long journey of planetary change directed by a God who chiseled messages into the hardest granite from the Atlantic to the Pacific. Americans had only to learn to read "the glacial manuscripts of God," as John Muir called Yosemite. Muir knew a divine message when he saw one. Raised in a Presbyterian family in glacier-carved Scotland and Wisconsin, he needed no translation to read God's words among the orphan ice sheets still clinging to California's highest peaks. He was just one of a group of American explorers, scientists, essayists, and painters who created a vision of glacial progress to lead the nation forward after the cataclysm of the Civil War. That this political vision centered on the North was not incidental but central. It was regionalism disguised as healing nationalism, legible in the very rocks of the United States.[2]

The Ice Age was provocative for many reasons, among them its insistence on climate oscillation. This was an abrupt shift from the one-directional climate change in the eighteenth century's nebular hypothesis, in which an infant Earth had steadily cooled from its origin in the hot gases of the parent nebula. By contrast, the Ice Age theory (or glacial theory, as it was first called) suggested that the world's climate had fluctuated, going from warm to cold to warm again, felling and resuscitating life as it went. This conceptual shift from one-way cooling to climate oscillation opened the door to our modern idea of climate change, or as it was first called in the nineteenth century, "change of climate."[3] The idea of climate as an actor was not new. Since the time of the ancient Greeks to the Enlightenment, climate had been thought to operate as a shaping force in human societies. Montesquieu's *Spirit of Laws* (1748) had influentially argued that climate determined political structures by weakening or activating the body. Hot climates wilted human energies, making people too fatigued to resist despotism. Cooler climates were energizing, resulting in the freer polities exemplified by Great Britain, where vigilant citizens checked the king's despotic sallies.

But now, with the idea of the Ice Age, climate's agency swelled across the vast abyss of geological time, driving species extinctions and repopulations from the origin of life long ago to the present. As the agency of climate swelled across deep time, naturalists wondered what agent in turn drove climate. Was it a natural agent or a still greater Agent? In addition to expanding God's agency, deep time's longer angle of vision stretched the political horizon into near infinity. Montesquieu had reached back to the ancient Greeks and Romans to find his universal laws about how societies, economies, and governments operated. But the Ice Age burrowed into the roots of the Earth itself, entraining peoples, animals, plants, and rocks in a colossal narrative that ended at the very door of the present day. What had been political became planetary.

This was why the Ice Age became the first geological era to be explicitly tied to a political program in the United States. After the Civil War, victorious Northerners decreed that with the great refrigeration of the Ice Age, God had blessed their region with the rich soils and brisk weather that energized the white population for national leadership. He doomed the South to backwardness by denying it the Ice Age, consigning it to a sultry climate and ancient Cretaceous soils that were good only for cotton and sugar production. In short, the Ice Age in the United States became a geological era proposed as much to describe the geological record of the past as it was to activate human beings in the present. It was the first time geologists wrapped politics in science, but it would not be the last. The Ice Age of the nineteenth century opened the door to the Anthropocene of the twenty-first century, when geologists have once again introduced themselves into politics—but with a crucial difference that we will discover later in the chapter.[4]

That Fatal Cap of Ice

The first American naturalists to think about an Ice Age lived in New England. Beginning in the 1830s, they became convinced that their region had once been plunged into a long winter that made the bitterest modern winter seem like a cold snap by comparison. In the rocks and hills around them, they began to glimpse the faint traces of vanished glaciers, to see not the present but the past. This was not their usual past, the snug fireside of flinty Yankee virtue that New Englanders had woven into a regional origin story that they promoted as national history. The Ice Age was far older than that, challenging New Englanders to find new ways to direct it into their preferred channels of regional self-celebration.[5]

Among these first glacialists (as they called themselves) was Benjamin Silliman of Yale. His student Samuel Morse (who went on to invent Morse Code) made a portrait of him in 1825 that shows what New Haven looked like to people without an idea of an Ice Age (fig. 6.1 and plate 13). Outside the window looms West Rock, a local igneous formation, capped by green grass and a swollen cloud that suggests water rather than ice as an Earth-shaping force. The real focus of the painting is the giant professor holding forth at the lectern with his tiny mineral samples scattered before him. Silliman "has a very lofty conception attached to the stupendous title of Professor of Yale College," one student complained. Two short decades would utterly change that perspective, shrinking the human and magnifying the rocks. During those decades, European ideas about glaciers came to the attention of American geologists, who looked around their nation and wondered how these exciting new ideas could be applied to North America.[6]

Silliman was among the first to usher in the American glacial vision by publicizing European theories about ice as a surface-shaping agent. Beginning in the eighteenth century, European naturalists had begun to consider ice as an explanation for erratic boulders and parallel striations on rocks in Sweden, Scotland, and Switzerland. Some proposed icebergs, an idea that gained a broader following when it appeared in Scottish geologist Charles Lyell's *Principles of Geology*. The iceberg theory was a classic example of uniformitarianism, in which observed pro-

FIGURE 6.1. Professor Benjamin Silliman of Yale overlooks his mineral collection as a red curtain unveils West Rock outside the window. Painted by Samuel Finley Breese Morse in 1825, the image shows how landscapes were often depicted before the idea of the Ice Age had been invented. Yale University Art Gallery. Gift of Bartlett Arkell, B.A. 1886, M.A. 1898, to Silliman College.

cesses in the present accounted for past events. To those who had sailed the North Atlantic and encountered icebergs, it was easy to imagine that these icy towers might transport boulders or scrape rocks as they crashed onto shore. In 1835, Silliman conveyed the iceberg theory to Boston audiences in a public lecture. He explained how boulders "believed to have been carried by ice bergs" had been found all over the world starting at 35 degrees of latitude north and south.[7]

Suddenly in July 1837, a provocative new idea was aired in Switzerland: the glacial theory of the Swiss naturalist Louis Agassiz. Just thirty years old and brimming with charisma and self-confidence, Agassiz used his presidential address to the Société Helvétique des Sciences Naturelles to announce that great sheets of ice had once shrouded large parts of the planet. Only a glacial epoch ("une epoque glacée") of massive ice sheets—rather than lone icebergs—could account for the variety and extent of phenomena visible in Europe, from the moraines to the erratics to the polished rocks. An English-language version of his glacial theory appeared in Edinburgh's *New Philosophical Journal* in 1838, and soon Americans took notice.[8]

Agassiz's glacial theory hinged on a total outlook from which he did not deviate over the course of his life. It derived in large part from his mentor, the French comparative anatomist Georges Cuvier, with whom he had worked in Paris, and who taught him to focus on the complex details of anatomical structure. Like Cuvier, Agassiz insisted that this anatomical complexity suggested a basis for natural classification: species were fixed and unchanging, linked in their majestic stillness to the eternal mind of the Creator. In Agassiz's view, the fossil record revealed not changing or transmuting species but a series of extinctions that were then followed by wholly new creations authored by the deity. God knew all of Creation at the outset, and merely unfolded his Idea over time. An animal might look very much like another animal in an earlier stratum, but that resemblance signaled the animal's obedience to the unchanging divine typology, not a species transmuting over time.[9]

In the ice sheets of the glacial epoch, Agassiz found a new agent for these catastrophic extinctions. In addition to floods, volcanoes, and other cataclysms, now came a winter so sudden that it froze animals in their tracks: the Ice Age. The undecayed mammoths unearthed from Siberian and Alaskan tundra hinted that this might be plausible. So did the ice world that formed each winter in Agassiz's native Switzerland, where an immense ocean of ice (*mer de glace*) blanketed the Alps. Agassiz explained that these ice oceans were the cradles of the glaciers that crept down the mountain flanks and then melted to form the largest rivers of Europe, which in turn nurtured the farms and forests with life-giving water. Most educated people merely looked at the Alps and marveled, he noted with disapproval, while also dismissing the observations about glaciers from local shepherds, farmers, and others. He would show that the glaciers were, in their own way, agents of change and indeed a new Creation in their own right, sandwiched between the original flood and the present age. Scratches and scrapes in Neuchâtel, a city less than a hundred miles from the Alpine glaciers (and where Agassiz had been appointed professor of natural history), suggested that it, too, had once been blanketed by a glacier. The glaciers must have once reached much farther than they did today, as though all the world had once been the Swiss Alps, an ocean of ice encasing the Earth.[10]

The Ice Age envisioned by Agassiz ended just as suddenly as it had arrived. Kissed by the warmth of a global spring as fragrant and fertile as an Alpine meadow in May, the ancient glaciers melted. This was a new Creation—the Modern Period, the Age of Man—known by God at the outset of time and fashioned by the glaciers of the Ice Age. Agassiz's singular and global Ice Age was akin to the biblical flood in its radical contingency: it was a unique event in Earth's history. And the Ice Age, like the flood, was closely linked to a divine plan, though enacted through the agency of climate rather than directly by God.[11]

Agassiz's glacial theory found a receptive audience among American naturalists. One early convert was Timothy Abbott Conrad, fresh from his field trip to Alabama. Now working as a geologist for the state of New York, he turned his sights to the landscape of the American North. Looking around at the polished rocks of western New York, Conrad thought that Agassiz's theory of "periodical refrigeration" better explained the distinctive flat, striated landscape than the rushing waters of a flood. He saluted the "genius of Agassiz."[12]

American naturalists had also been primed to believe Louis Agassiz's rash Ice Age theory by his well-received *Recherches sur les Poissons Fossiles* (Investigations of fossil fish).[13] Published in five illustrated volumes from 1833 until 1843, the work was the most extensive study of fossil fish to date and had established young Agassiz as a naturalist of the first order, akin in stature to his mentor, Georges Cuvier. American naturalists took notice. Get it "without delay," the English geologist Gideon Mantell urged the Philadelphia paleontologist Samuel George Morton soon after the first volume of the fossil fish investigations appeared.[14]

Hinging on the same catastrophist, nontransformist viewpoint that would inform the Ice Age theory, the fossil fish research groomed Americans to accept Agassiz's far flimsier glacial theory. Agassiz's fossil fish—gathered from around the world and deposited in Europe's natural history collections—dated from Earth's very oldest fossil-bearing strata, where Agassiz located a "Reign of Fish." His four-part categorization of fish included one group, the fearsome armored Ganoids, whose remains could be found in every rock stratum "from the most ancient to the most recent." To Agassiz, these fish proved that species did not transform over time. The Ganoids prowling the primordial seas were of the same "type" as modern Ganoids: they bore a typological rather than familial or genealogical resemblance, having been annihilated in one of the many catastrophes that had befallen Earth, and their type later re-created by God according to his unchanging Idea. The ancient Reign of Fish and the far more recent Ice Age together established Agassiz as a big-picture guy, the master of huge swaths of planetary history.[15]

Agassiz's glacial theory gained widespread attention in the United States once it appeared in book form in his landmark *Études sur les Glaciers* (Studies of glaciers, 1840). This work did more than any other to establish Agassiz in the American public mind as the inventor of the Ice Age. The stunning companion atlas unveiled modern glaciers in the Alps as keyholes to an icy past. Most Americans had never seen a glacier—there were none left east of the Mississippi River—and now they could imagine the sheer size and shaping power of these mighty ice rivers. In one illustration of a glacier near the town of Zermatt in Switzerland, two women in sunhats watch ant-sized hikers clamber

FIGURE 6.2. In this image from Louis Agassiz's revolutionary *Études sur les Glaciers* (1840), tiny hikers clamber around the towering glacier near Zermatt, Switzerland.

high upon the glacier, working their way around the characteristic glacial features: the polished stones, the *roches moutonnées* (rocks polished into an embossed pattern that looked like flocks of sheep or sacks of wool), the moraines made of rocky debris shoved to the sides and front of the glacier. Tumbling boulders and snapped trees suggested the constant groaning and cracking of the ice, as though the glacier were alive (fig. 6.2).[16]

Agassiz immediately shipped his new book to Benjamin Silliman, after years of hinting in letters that he wished to compare European and American formations. Silliman, as always, was cautious. In his inaugural Lowell Lectures in Boston in the summer of 1841, he talked up Agassiz's "new views with his drawings" of lands north of the thirty-eighth parallel (roughly the latitude of Virginia), which "excited a strong interest." But he told the audience that he was "not disposed to go quite as far as he." Privately he also worried. His English friend Gideon Mantell complained to Silliman of the "rash enthusiasm" of the British scientific community, too eager to adopt every new theory "that has the charm of novelty and strangeness." Still, Mantell conceded, in a modified form the Ice Age could explain "many hitherto obscure phenomena."[17]

There were indeed many questions about Agassiz's theory. How could such a chill have come on so suddenly? And ended just as rapidly? And a block of ice thousands of feet high? This seemed quite out of ordinary experience. Glaciers might slide easily down the steep Alps, but how could these giant ice sheets move on vast continental flats like those found in North America? Did this Ice Age kill off the existing fauna—mastodons, giant sloths—or did these warm, furry animals in fact thrive in its cold climate, only to perish afterward when warmth returned? Even Agassiz's friend, the world-renowned naturalist Alexander von Humboldt, worried about "That fatal cap of ice which frightens me."[18]

But it was precisely the catastrophism that made the theory appealing to pious American scientists, none more than Edward Hitchcock of Amherst College. Hitchcock became one of the first American advocates of Agassiz's glacial theory, though he modified it to allow for the action of water as well, a theory he termed "glacio-aqueous action." He had once studied with Silliman at Yale, and now lived a hundred miles to the north, in a similar landscape of low, rolling hills that Agassiz now convinced him were the traces of a vanished ice sheet. Silliman had whispered his doubts about Agassiz's glacial theory to Hitchcock, writing that he was "unable as yet to give an opinion." Hitchcock was game, however. With its "sudden" glaciations, Agassiz's theory gave the divine Creator a new mechanism for annihilating and then repopulating Earth in sequential new Creations, thereby disallowing the hated Lamarckian doctrine of transmutation of species. Where once there had been a flood, now there was also a Great Refrigeration. God was still with us, close by, but now working over a much longer canvas of time through various modes of extinction and new birth.[19]

Hitchcock endorsed Agassiz's glacial theory publicly at the 1841 annual meeting of the Association of American Geologists in Philadelphia and in his report on the geology of Massachusetts. Most Americans had never seen a glacier, Hitchcock explained, so Agassiz's Ice Age might seem "fanciful, and even puerile." But he thought it solved interpretive problems of the American landscape stretching from New England to Ohio. The scratches, moraines, and powdery dirt known as drift could have been created by floods or even rock-laden icebergs scraping the floor of a former ocean, as Lyell had proposed. But all these phenomena together were far better explained by glaciers such as now existed in the Alps, as revealed in the "splendid plates" in Agassiz's atlas.[20]

Hitchcock also used his classroom teaching over the next two decades to promote Agassiz's glacial theory, thereby spreading it—and the longer time frame over which it had occurred—to hundreds of young people. In 1841, he revised his textbook, *Elementary Geology*, to recommend Agassiz's new glacial theory as the only one among five rival theories most likely to be true. The suddenness of the Ice Age hinted at its divine origins. The remains of animals now living in tropical climates—hippopotamuses, rhinoceroses—had been found in the glacial drift of the northern hemisphere. Undecayed elephant-like corpses pulled from the Siberian tundra suggested that some animals had not even had time to rot before the invading cold froze them. Warming must have occurred just as suddenly. How else to explain the many lakes and erratic boulders dotting the northern hemisphere than as testaments to torrents of melting water that accompanied the sudden "return of heat"?[21]

To Hitchcock, Agassiz's glacial theory suggested a new way of seeing the landscape. College textbooks seem like unlikely candidates to be the artistic avant-garde, but Hitchcock's *Elementary Geology* unveiled new portraits of the American landscape that emphasized its final sculpting by icy action. They were created by his wife, the artist and naturalist Orra White Hitchcock, and may have served as one of the preliminary inspirations for the flowering of glacial landscape painting that arrived in the 1860s. "Glacial action" must now be considered a major force in shaping the American landscape into a "work of art," Edward Hitchcock argued. The contours of the United States were particularly shaped by the rocks beneath it, producing "landscapes abounding in beauty and

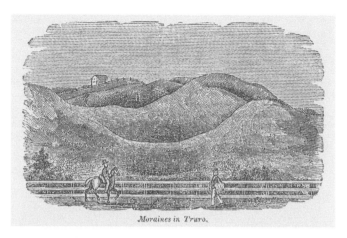

FIGURE 6.3. Edward Hitchcock revised his popular college textbook in 1841 to reflect the revolutionary new idea of the Ice Age. Moraines in Truro, Massachusetts, offer mute testimony to the towering hills of debris left behind by the retreating glaciers thousands of years before.

FIGURE 6.4. In this image from Hitchcock's 1841 textbook, a horse and carriage pick their way through a maze of glacial erratics in Gloucester, Massachusetts. Erratics are large stones left behind by the melting glaciers. These images helped New England students to imagine that their landscape was many thousands of years old.

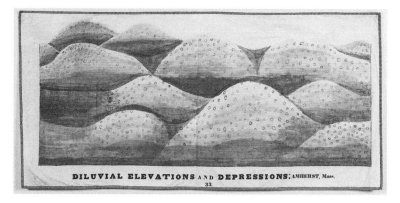

FIGURE 6.5. Orra White Hitchcock's large canvas drawing of glacial moraines hung at the front of Edward Hitchcock's lecture hall at Amherst College. Labeled as "Diluvial Elevations and Depressions," the drawing reflected Edward Hitchcock's belief that both ice and water shaped the American landscape.

sublimity," dressed by the Creator. He called this "scenographical geology." Orra Hitchcock's illustrations revealed familiar New England landscapes as the end products of ice, perhaps acting also with water: large moraines in Truro, at the tip of Cape Cod; a maze of boulders in Gloucester impeding a horse and carriage (figs. 6.3 and 6.4). Tiny New Englanders went about their business, dwarfed by the giants of a former age. The glacial moraines two miles from Amherst also appeared in her enormous canvas paintings, which she created to hang at the front of the lecture hall (fig. 6.5). As he lectured on the rivers of ice pushing boulders forward "pell mell," Professor Hitchcock gestured to these drawings and to the classroom windows, where outside the boulders and moraines still lay "almost exactly as they were left by the ice." Hundreds of undergraduates learned to see the American landscape not as present but as past.[22]

As a result of Louis Agassiz's glacial theory, New Englanders could see their landscape as a document of the powerful shaping action of the Ice Age. Painters responded by reversing the proportions of humans and rocks. George Henry Durrie's painting of New Haven, created

FIGURE 6.6. The glaciated landscape near New Haven is captured in George H. Durrie's painting of low, smoothed hills and glacial erratics. *Summer Landscape Near New Haven* (c. 1849). Brooklyn Museum, Dick S. Ramsay Fund, 46.162.

around 1849, unrolled a glacier-flattened plain leading to West Rock and its twin, East Rock, both worn smooth by heavy blankets of ice that melted and disappeared long ago (fig. 6.6 and plate 14). West Rock had previously had a theological meaning, as the hiding place for Puritan regicides fleeing King Charles II. But now, nearly two centuries later, the rocks acquired a new kind of divinity embedded in the majesty of nature. Humans were mere crumbs in this cosmic drama, miniscule in comparison with the giant professor that Samuel Morse had rendered in his painting of Benjamin Silliman just twenty years earlier (fig. 6.1). In Durrie's painting, a tiny farmer and her child amble past small cows lying next to the glacial erratics in the foreground, stony witnesses to an age when God had wrapped Earth in snow to prepare it for the fertile

Age of Man. In the Ice Age, the hidden God revealed himself, the invisible behind the visible.

This Great Agent

New England had been glaciated. But how far did this glacial landscape extend into North America? None other than Louis Agassiz himself would help to answer this question. After sailing across the Atlantic in 1846 to give the Lowell Lectures in Boston, the Swiss naturalist made the United States home until his death in 1873. During that quarter century, he helped not only to establish the plausibility of a continent-wide glacial epoch in North America but to invent a glorious national meaning for that glaciation. He sketched a vast temporal and spatial canvas for the national origins and destiny of his adopted nation. The Ice Age was America's door to modernity, the Age of Man. "This great agent," as he liked to call the Ice Age, was God himself, acting through the ice sheets to shape the modern United States. His tombstone in Mt. Auburn Cemetery in Boston captured this achievement. It was a glacial erratic, shipped from his beloved Swiss Alps (fig. 6.7).[23]

From the moment he disembarked in North America, Agassiz began to reshape his Ice Age theory. The sheer scale of the continent challenged him to think more expansively about glacial ice to account for the "universality of its action." Immediately upon his arrival in the fall of 1846, he had observed the effects of glaciation. His steamship, the *Hibernia*, docked in Halifax, Nova Scotia, before heading to Boston. "[E]ager to set foot on the new continent so full of promise for me," he recalled later, "I sprang on shore and started at a brisk pace for the heights above the landing.... I was met by the familiar signs, the polished surfaces, the furrows and scratches, the *line-engraving* of the glacier ... and I became convinced ... that here also this great agent had been at work."[24]

New England, too, showed signs of the great agent. "But what a country is this!" he gushed to his mother in Switzerland the next fall. "[A]ll along the road between Boston and Springfield are ancient moraines and polished rocks. No one who had seen them upon the track of our

FIGURE 6.7. The tombstone of Louis Agassiz and Elizabeth Cary Agassiz in Boston was made from a glacial erratic shipped from Switzerland.

present glaciers could hesitate as to the real agency by which all these erratic masses, literally covering the country, have been transported." He believed he had already converted several of the most distinguished American geologists to his theory. He saw more glacial evidence as he traveled around the region that fall: erratics, drift, and polish in a trip up the Hudson River from New York to Albany. "Glacial phenomena," he explained to a colleague in Europe in 1847, were "complicated by peculiarities" in North America that were "never brought to my notice in Europe." From Boston to Niagara Falls he found that "the whole rocky surface of the ground is polished." Nowhere in the world were there "polished and rounded rocks in better preservation or on a larger scale." Unlike his native Switzerland, where it was easy to imagine glaciers flowing down steep valleys, here in North America he had to account for glacial progress across flatlands.[25]

The success of the glacial theory in the United States had a lot to do with Agassiz's sheer personal magnetism. Many thought him handsome, his French accent adding continental flair to a pushy and relentless persona. Within a few years of his arrival, he had become the most famous naturalist in the United States. He had bottomless energy for the lucrative national lecture circuit, where the competition was scientifically respected but oratorically weak. Charles Lyell could be dull, Benjamin Silliman priggish. Photographs document Agassiz's sheer physicality as he served up the latest science, pausing at the end of key sentences to allow time for applause and ambidextrously covering the chalkboard with anemones and starfish. People from all over sent him specimens. One admirer in Maine shipped him a live bear. Specialists, too, were smitten, at least early on. "[W]e have all fallen in love with you," wrote James Dwight Dana of Yale in 1847. "[Y]our thoughts are all wisdom to the rest of us."[26]

With the help of well-placed friends in Boston, Agassiz accepted a chair in zoology and geology at the new Lawrence Scientific School associated with Harvard. There he created an enormous collection of animal and plant specimens in an abandoned bathhouse on the banks of the Charles River. His laboratory remained his primary classroom for the rest of his life. He liked to hover next to his students, clenching a cigar between his teeth as he explained some finer point of zoology. Many former students found him overbearing, even dictatorial, especially as they attempted to establish careers of their own.[27]

Agassiz cemented his reputation in the United States with his first American publication, *Principles of Zoology* (1848). Appearing in sixteen editions during Agassiz's lifetime, it became one of the most popular zoological treatises in the United States at the time. In this final decade before Darwin, Agassiz's textbook disseminated his idea of a fixed plan of creation operating on an enormously long timescale, from the Creation through the Ice Age to the Age of Man.[28]

The book's frontispiece encapsulated Agassiz's thesis of unchanging Idealism (fig. 6.8 and plate 15). Agassiz's circle diagram broke with the rigid and timeless linear hierarchy of the Great Chain of Being, the dominant metaphor for life since classical antiquity. Like a pie, his circle was divided into four sections, illustrating the four static zoological categories first proposed by Cuvier: radiates, mollusks, articulates, and vertebrates. Agassiz believed that each zoological category represented a general plan or Idea on which God created variations. The vertebrates, for example, represented the Idea of bilateral symmetry, while the circular radiates represented the Idea of radial symmetry. In his crowded public lectures, Agassiz liked to pose with his chalkboard drawings of creatures from each category, such as the radiates (fig. 6.9).[29]

The four categories flourished side by side while also progressing over "an immense period of time" from the Creation (at the center of the diagram) to the present (the outermost ring). No creatures adapted to a new environment, as insisted by transformists. Instead, all creatures were "autochthonoi," created where they now lived by a God who repeatedly repopulated Earth with successive orders of Creation. Unlike Darwin's branching tree metaphor, which showed all life emerging from a common ancestor through *continuous* change, Agassiz's circle metaphor showed life appearing through *discontinuous* change. Missing links were a non-issue in Agassiz's schema, since the governing metaphor was catastrophic breaking rather than fortuitous continuity.[30]

Applied to human beings, Agassiz's view of species fixity became known as *polygenesis*. According to Agassiz, God had created several human races, whose descendants remained unchanged over time. This view provided grist for the mill of advocates for the permanent biological inferiority of black people and embroiled Agassiz in sectionalist and religious controversy as the nation lurched toward Civil War. But while it tends to capture the limelight today, Agassiz's polygenism was just one aspect of his larger theory of separate creations and extinctions authored by a superintending God. The Ice Age was the most recent of those extinctions, in which God suddenly swept away

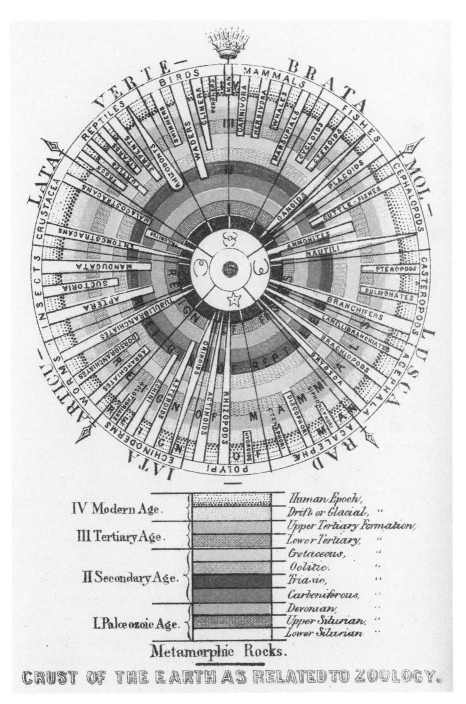

FIGURE 6.8. This colored circle in Louis Agassiz and Augustus Gould's popular *Principles of Zoology* (1848) offered a major alternative to the branching tree as a metaphor for the development of life on Earth. Agassiz denied that species changed over time; instead, he thought God repeatedly extinguished and then recreated life on Earth over millions of years according to his unchanging Idea for the body plan of each species. The circle is divided into the four major animal groups that represent God's unchanging Idea of body plans (Vertebrata, Mollusca, Articulata, and Radiata). The beginning of life on Earth is at the center of the circle, and life progresses until the Modern Age, which is crowned by Man. Agassiz showed how multiple special creations could coexist with the new idea of deep time.

FIGURE 6.9. A popular speaker on the national lecture circuit, Louis Agassiz stands contentedly with his chalkboard drawings of his beloved Radiata.

life, only to create it again later in time. The Ice Age was the doorway to the Reign of Man.[31]

A summer trip to Lake Superior in 1848 allowed Agassiz to collect data to support his theory of fixity of species and to sketch a picture of North America as a primordial land culminating in a continent-wide Ice Age. In June, he assembled a group of students and Boston colleagues for the two-month journey west. It was his first experience with the sheer vastness of North America. Although he had traveled from Switzerland to Scotland to view glaciers in 1840, that journey was only a thousand miles. From Boston to the westernmost point of Lake Superior at Thunder Bay was a few hundred more than that, and it got him only a third of the way west across the continent. The group took trains to Cleveland, then steamers to Lake Huron. From there they paddled in birchbark canoes to Lake Superior. They stuffed alcohol-filled barrels with reptiles, insects, birds, and fish to carry back to Agassiz's laboratory, where they would be unpacked and categorized. Along the way, the Agassiz group also benefited from the local knowledge of Indian women and men, from the Ojibwe who sewed, packed, and rowed the large canoes, to the system of temporary camps, rivers, and trails near the icy lakeshore that the Indians called the "neighborhood of rivers." A rock venerated by the Indians as the place where one of the gods rested after creating Earth was interpreted by the Agassiz group as extremely ancient basalts. Both Native peoples and white naturalists, that is, convened on Lake Superior's coastline as the vestige of the earliest Earth, though they did so from different traditions. Agassiz's memoir of his journey into deep time simultaneously—if perhaps inadvertently—captured the alternative temporal system of the Ojibwe. At camp and even in the canoe, Agassiz unfurled the stiff black linen canvas that served as his portable blackboard. Tracing thick lines with white chalk, he gave impromptu lectures on everything from sturgeons to glaciers, revealing at every stage his belief in a vast abyss of geological time.[32]

Two years after his return, all of this appeared in his illustrated book *Lake Superior* (1850), where according to his wife, Elizabeth Cary Agassiz, "The whole history reconstructed itself in his mind." By the whole history she meant the whole history of the North American continent. While nationalist US historians such as Francis Parkman and Henry Brooks Adams were writing large-scale national histories, Agassiz was embarking on something far bigger. *Lake Superior* unfolded the story of the continent's rise from the deepest antiquity—visible in the basalt columns at Thunder Bay that dated from a "very remote epoch"—to the glacial erratics scattered during the Ice Age (fig. 6.10). The gar pike swimming today in Lake Superior were modern versions of the ancient armored Ganoids from the primordial Reign of Fish. The fish had not migrated there, but "must have been created where they live" according to a "wise plan" so that they were "best suited for the country where they are now found."[33]

FIGURE 6.10. Glacial erratics and low, glaciated hills show the footprint of the Ice Age near Thunder Bay on Lake Superior. Louis Agassiz, *Lake Superior* (1850).

Extending his sights to the whole of North America, Agassiz speculated that only continental glaciation could explain the abrupt terminus of the boulders at around the thirty-fifth parallel north latitude (roughly the southern border of Tennessee and North Carolina). What other force besides ice had power sufficient to carry the boulders but then to stop suddenly, rather than sweeping over the whole globe as liquid water would? This phenomenon occurred "everywhere in the north," while in the tropics and temperate zone "we find no trace of these phenomena." All the glacial phenomena in northern Europe, Asia and America, moreover, were "of the same age" as the Alps. God had sent a great power to the North, one of the "successive changes" through which God acted in lieu of the "one creative act" advocated by the misguided followers of Lamarck.[34]

Agassiz's *Lake Superior* offered something extraordinary and new: North America as a time envelope, encasing the most ancient and the most recent phenomena, all evidence of God's repeated interventions in the continent's history. It was the first glimpse of the North American continental interior that was as temporally massive as it was spatially expansive. Even a thousand miles into the continental interior, one could find the oldest rocks in the world as well as evidence of a more recent divine freezing that prepared the land for the Age of Man. This was the final nail in the coffin of the flood theory, said Agassiz's friend Spencer Baird, curator of the Smithsonian Institution. Baird, too, was seeing scratches in the mountains near his home in Carlisle, Pennsylvania. "I do not see how the men of diluvial currents can stand after the push you give them with your glaciers," he told Agassiz.[35]

With its inviting prose style and high-quality pictures, Agassiz's *Lake Superior* reached an international audience, delivering news to Europe that America had been the site of the greatest dramas of the Ice Age. Agassiz sent it to Charles Darwin, who thanked him for giving him "such lively and sincere pleasure." The idea also moved westward, appearing in geology textbooks in Ohio within a decade. The journey helped Agassiz to write a new story of North America that reached deep into time and allowed him to venture a conclusion that he would repeat over the next twenty years to an American public increasingly willing to hear it: America—the ostensible New World—was "the oldest continental land known on earth."[36]

Glaciers of the North

By 1850, it had become plausible to imagine that the Ice Age had reached down the North American continent to about the thirty-fifth or fortieth parallel. Those lines had become filled with modern political meanings. The Missouri Compromise of 1820 had defined the 36°30' parallel as the northern limit of slavery in the western territories. Political maps of the United States traced a white gash across the land to mark the parallel, as though it were a natural feature. The idea of the Ice Age arrived to suggest that climate of the North and South were also natural features, ancient events that determined the politics of the modern United States.[37]

Sectional tensions, coupled with the rise of geographical science, furthered the idea that each half of the United States represented a

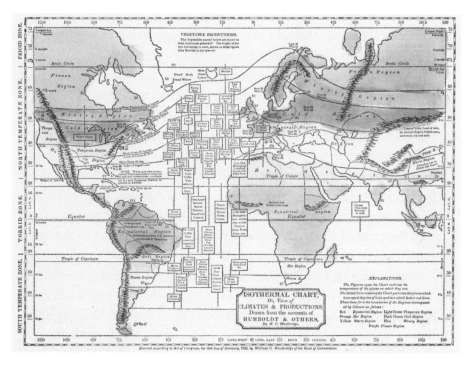

FIGURE 6.11. The grain-growing band of the Cold Region was the gift of the Ice Age to the North, according to American glacialists. Detail of Woodbridge, "Isothermal Chart" (1837).

fundamentally different kind of climate-based natural and social order that had been locked in long ago. The Prussian geographer Alexander von Humboldt's concept of isotherms—bands of equal temperature—circulated in popular American geography books. William Woodbridge's isothermal chart from 1837 showed the "Cold Region" of North America as a single grain-friendly band undulating between roughly the fortieth and fiftieth parallels. South of the fortieth parallel lay hotter climates suited for cotton and sugarcane (fig. 6.11).

The Swiss geographer and glacier expert Arnold Guyot built on this foundation to plump for the North. A friend of Agassiz who had explored the Alpine glaciers with him, Guyot had emigrated to the United States in middle age to escape the Revolutions of 1848. Settling first in Boston, Guyot explained the meaning of the cool North to receptive antislavery New England audiences in a series of lectures in 1849 that were reprinted in his book *The Earth and Man* (1849). "The people of the temperate continents will always be the men of intelligence, of activity, the brain of humanity, if I may venture to say so; the people of the tropical continents will always be the hands, the workmen, the sons of toil." Guyot's map of isotherms emphasized the band of temperate climates that united much of Europe and North America in a single band of "civilization."[38]

Climate determinism reached new heights during the Civil War. Mississippi's declaration of secession in 1861 claimed links between climate and society to be nothing less than a natural law. Cotton was "peculiar to the climate verging on the tropical regions, and by an imperious law of nature, none but the black race can bear exposure to the tropical sun." Northerners connected their cool climate and high latitude with civilization and the Union cause. In 1861, the American painter Frederic Edwin Church unveiled his new iceberg painting. He first called it *The North* in homage to the Union, whose antislavery cause he supported, and perched a glacial erratic near the icy archway to signal his growing interest in the science of geology (fig. 6.12 and plate 16).[39]

New York and Boston audiences saw the "Divine fingers" at work in Church's majestic scene of Northern beauty and power. When Church's friend, the polar explorer Isaac Israel Hayes, returned from the Arctic in 1861, he gave a stirring speech to New Yorkers about the role of the North in securing the future of the republic. "God willing, I trust yet to carry the flag of the great Republic, with not a single star erased from its glorious Union, to the extreme northern limits of the earth."[40]

For supporters of the Union cause, the North had become a total symbolic system. The region engrossed not just the terrain north of the Mason-Dixon line, but British Canada and the Arctic itself. When Church's painting went to London for exhibition in the summer of 1863, New Englanders cheered it as a victory for the cultural supremacy of the North. But in a nod to local public opinion, Church changed the name of his painting to *The Icebergs*. With the British public favoring the Confederacy, whose cotton exports supplied British factories, the

FIGURE 6.12. Frederic Edwin Church, *The Icebergs* (1861). Church originally called this painting *The North*, but changed the title to the more neutral *The Icebergs* so as not to inflame pro-Confederate opinion during the painting's London showing. Image Courtesy Dallas Museum of Art.

new title was deemed safely neutral. This was just a painting of icebergs, not a political statement.[41]

Northern naturalists now stepped in to argue that the Ice Age was the origin of this primordial difference between North and South. It was God's final preparation for a superior white civilization that would triumph in the post–Civil War era. Louis Agassiz became the major popularizer of this view during the Civil War and after. In a series of articles for Boston's pro-Union *Atlantic Monthly*, he announced that America was the oldest continent in the world, first to rise from Earth's infant seas and then given its final sculpting by the Ice Age. His prose was biblical in its majesty, befitting a story that was cosmic in its import. This was God's country, specially chosen, specially created. "First-born among the Continents," he began, "America, so far as her physical history is concerned, has been falsely denominated the *New World*." While most of Europe still languished beneath the primordial seas, America had already surfaced, stretching from Nova Scotia to the West. On those earliest dry lands on Earth, Agassiz painted an American Eden, where the first creatures sprang to life. A sequence of new Creations over eons had culminated in the Ice Age, when "God's great plough" had prepared America "for the hand of the husbandman."[42]

Although he opposed slavery, Agassiz saw the North as climatologically unsuited for black people, whom he regarded as an inferior race, the product of a separate creation from white people. He wrote privately to his friend, the abolitionist Samuel Gridley Howe, that he wanted to "accelerate their disappearance from the Northern States" after emancipation. Black people should go to the South, where the "climate is genial to them." Howe concurred, arguing that there were "irresistible natural tendencies to the growth of a persistent black race in the Gulf and river States." It was futile to resist nature. By contrast, in the North the "mulatto is unfertile." As national opinion moved to embrace transmutation doctrines in the wake of Charles Darwin's *Origin of Species* (1859), Agassiz continued to oppose them publicly and privately, a position that increasingly isolated him until his death in 1873.[43]

Influenced by Agassiz's ideas about the Ice Age, the painter William Haseltine produced several canvases of the coastal rocks at Nahant, a

FIGURE 6.13. Louis Agassiz pronounced these rocks in coastal Massachusetts to be the oldest on Earth. William Stanley Haseltine, *Rocks at Nahant* (1864).

popular Massachusetts vacation destination that showed these ancient igneous rocks scraped by later glaciation. Agassiz had a summer cottage at Nahant, with a laboratory attached. Nahant's smooth, striated rocks had been levelled by the same icy slab that had crushed much of North America during the "Ice-Period." A critic in the *Art Journal* commented on Haseltine's painting, *Rocks at Nahant* (1864) (fig. 6.13). "Agassiz pronounces the rocks of Nahant to be the oldest on the globe." Now Agassiz also showed that they had been glaciated in a much more recent era, for the benefit of Northerners.[44]

Agassiz was not the only Northern naturalist to make such claims. In Maine, the glacialist John DeLaski argued that New England's primordial rocks had been purposely scraped by more recent glaciers for the benefit of the North. A physician born in 1814 who now lived on the small island of Vinalhaven off the coast of Maine, DeLaski spent much of his spare time examining the glaciated landscapes around him. He developed professional relationships with the New England glacialists: Benjamin Silliman and James Dwight Dana of Yale, and Charles Hitchcock, son of Edward Hitchcock, who became the state geologist of Maine. A somewhat less fulfilling relationship was with Louis Agassiz, who joined DeLaski in Maine for excursions to view glaciation, only to publish their joint findings without mentioning DeLaski.[45]

DeLaski described the American Ice Age as the work of God. The glaciers in Penobscot Bay were part of a recent "universal glacier" that had been "directed by intelligence." Its brow must have reached the top of Mount Desert Island, which was 2,000 feet high. As though on cue, the landscape artist Jervis McEntee produced his *Mount Desert Island, Maine* (1864), with its tiny hiker resting atop a dome etched with "sacred glyphics" (fig. 6.14).[46]

DeLaski collected all these findings in a manuscript titled "Foot Steps of the Ancient Great Glacier of North America" (1869) with the intention of showing Americans the role of glaciers in the "past, the present and the future of the country." DeLaski argued that the glacier-prepared soil of the North—between 35 and 50 degrees north latitude—had fitted Northern white men for civilization and condemned people in the South to a backward role. Rich in minerals, glacial soil supported "invigorating" fruits and vegetables from which a whole society sprang:

> Where they thus flourish, there are the greatest population, the greatest men, the best system of learning, the most intuitive intelligence, the most wealth, the most commercial activity, the most profitable railroads, and the purest religion. The countries which grow them, are the countries of the best pork, the best beans, of the best butter and genuine beef and mutton, and consequently, the best developed human muscle and brain.

By contrast, the South languished with poorer soils that were merely a "glacial donation of the north," emptied of most of its nutritional content as it traveled downstream. The result was the shriveled animals, weak vegetation, and depleted human beings, the Southern whites and blacks who did not know how to labor for the future of civilization. The northern Indians who preceded modern whites also lacked civilization because they had no settled agriculture and had no concept of progress. DeLaski concluded that only "the white man of the North" could lead American civilization on its "divine missions" forward to "supremacy" after the Civil War to find a "unity of purpose."[47]

FIGURE 6.14. A tiny hiker sits atop a glaciated dome in Maine, an example of how American landscape paintings from this time often shrank humans while inflating the rocky signs of the continent's primordial past. Jervis McEntee, *Mount Desert Island, Maine* (1864).

The Ice Age of North America

This glacial vision was further developed in the decades after the Civil War, especially by scientists seeking to rebuild American nationalism along Northern ideals. Modernity and progress would be a North-led enterprise. The war had reshaped the worldview of many American scientists, encouraging a vision that was national rather than regional. The role of the federal government in science increased through centralized efforts such as the United States Geological Survey (f. 1879), national maps based on the federal census, and the land grant universities funded by the Morrill Land Grant Act (1862). Universities with scientists interested in glaciation spread from their old base in New England to the new universities of the upper Midwest and far West. Americans founded journals with a national readership such as *Popular Science Monthly* and the *Overland Monthly*, which consolidated local findings into a national vision of glaciation. Having favored the South during the Cretaceous, one Michigan geologist wrote in 1870, the climate of the Ice Age had benefited the "Northern States" by sprinkling fertile drift everywhere. This was nothing short of a "resurrection." Abroad, too, scientists endorsed the continental glaciation of North America. "It is no exaggeration to say," Scottish geologist James Geikie wrote in his book *The Great Ice Age* (1874), "that, the whole surface of North America, from the shores of the Arctic Ocean to the latitude of New York, and from the Pacific to the Atlantic, has been scarped, scraped, furrowed and scoured by the action of ice." There might even be not one ice age but two, his Scottish colleague James Croll chimed in, each draping moraines across the northern latitudes of the continent like garlands on a Christmas tree.[48]

Turning to the living glaciers of California's Sierra Nevada, American naturalists said these were orphans separated from the ancient parent glacier by "an immense period of time." The glaciers were not exactly sentient beings, as the Native Tlingit and Athapaskan peoples of the Pacific Northwest imagined them to be. To them, the "big ice" listened, smelled, made moral judgments about the world, and was responsive to humans. These Native views, conveyed through oral traditions and recorded by white naturalists and explorers, may also have conveyed ideas from the last few hundred years, a period of catastrophic glacial advance that was part of the Little Ice Age (c. 1400–1850). But the glaciers were not entirely lifeless for nineteenth-century white naturalists either, as their nomenclature of the glaciers suggested: the glacier had a face, a snout, and a tongue, it calved icebergs into the sea, and it left footprints on the landscape. Most importantly for them, its birth, growth, and death brought messages from the Christian God about who should rightfully inhabit and govern North America. These views came to dominate and finally conceal the Tlingit and Athapaskan perceptions of glaciers behind the language of empirical science, another example of how the transition to deep time was a form of imperialism, enacted through timekeeping rather than space marking.[49]

The meaningful glaciers of California were brought to public attention by two new transplants to the West. One was Joseph LeConte, a once-prosperous Georgia plantation slaveowner and naturalist who had lost everything in the Civil War. Now in his late forties and starting from scratch, he accepted a faculty position at the new University of California and was soon retraining himself as a glacialist in the Sierra Nevada Mountains 150 miles east of Berkeley. The second person was the young Scottish naturalist John Muir, who in the late 1860s was roughing it in Yosemite. It was here, in the cathedral of nature, that he would read the glacial manuscripts of God. Muir's vision hinged on the idea that the American landscape made visible the immaterial and unchanging thoughts of God.[50]

Clambering together among the granite peaks in the summer of 1870, Muir and LeConte saw the current shrunken glaciers as remnants of "their giant relatives now dead," as Muir put it. Muir, bearded and craggy-faced, was actively working to craft his public persona as America's wilderness prophet, the heir to the aging Ralph Waldo Emerson, who visited him in Yosemite. Muir explained to his readers that Yosemite was alive, its glaciers still pulsing with life. These icy giants were still cutting into its mountains "like endless saws." Like his companion LeConte, Muir disagreed with state geologist Josiah Whitney, who denied the action of glaciers in the Sierras. LeConte summoned his considerable scientific credentials—he had once worked with the great glacial theorist himself, Louis Agassiz—to refute Whitney. Yosemite Valley was "once filled to its *brim* by a great glacier," LeConte announced. A vast polar ice cap had smothered the region during the Ice Age, extending at least as far south as the fortieth parallel, and even farther south among the frigid peaks of the Sierra Nevada.[51]

The geologist Clarence King, surveying the fortieth parallel for the United States, sighed at how the glaciation of Yosemite dissolved all local views and gave Americans a broader sense of nation. "I think such vastness of prospect now and then extremely valuable in itself; it forcibly widens one's conception of country, driving away such false notion of extent or narrowing idea of limitation as we get in living on lower planes." Thirty years after the first glacial landscapes of New England had appeared, painters and photographers portrayed Yosemite as the result not just of glacial action, but divine action on behalf of the United States. Manifest Destiny, it seemed, extended not just forward to a democratic, messianic future, but backward in time to the Pleistocene. The painter Albert Bierstadt bathed these ideas in the glow of the divine. Bierstadt had visited the region in the 1860s and became a one-man factory of Yosemite paintings, each one emphasizing not just the glaciers but God's hand in carving the U-shaped valley for the glory of the United States (fig. 6.15 and plate 17).[52]

By the later nineteenth century, the divine glow of the Ice Age was dimming. Instead, American and European geologists were advocating multiple glacial epochs triggered by heat-transferring ocean currents. They gradually moved away from Agassiz's singular Ice Age, in which a great Agent, acting through a frozen Deluge, prepared Earth for the Age of Man. Now, "ordinary climatic agents" explained not just ice ages but all past geological ages, a triumph of uniformitarianism over Agassiz's Ice Age catastrophism.[53]

But even as the Ice Age theory shed its God-centered catastrophism in favor of ocean-current uniformitarianism, the idea of climate change itself did not. Our contemporary idea of climate change in fact retains the seeds of catastrophism, ready to bloom. From its birth in the nineteenth century, the idea of climate change has always required an agent (or Agent) to drive it, a force so powerful that it could shape global

FIGURE 6.15. The glaciated valley of Yosemite gleams with rays of divine light in Albert Bierstadt's *Valley of the Yosemite* (1864).

temperatures over the immensely long spans of time over which life on Earth was now believed to develop. Nineteenth-century Americans had assigned that agency to a benevolent God. In their view, climate change was essentially good, a sign that God still walked with his human children, freezing and warming Earth to prepare it for their prosperity. Inscribing his will in the glaciated valleys and flat prairies of North America, God used the ancient but ephemeral vehicles of rocks as signs to his children that he was timeless. "For the things which are seen are temporary, but the things which are not seen are eternal," the Bible assured Americans. The Ice Age was the visible and temporary cloak concealing the hidden and eternal God. As in so many other geological eras, nineteenth-century Americans found that deep time did not shrink their God but rather expanded his power and benevolence across millions of years.

Building on these nineteenth-century ideas, our own era has reassigned the agency of much recent climate change from God to human beings. The idea of the "Anthropocene" appeared around the turn of the twenty-first century.[54] Proponents of the Anthropocene see an era defined by human-caused climate change so significant that it deserves its own geological epoch. But in contrast to the Ice Age first proposed in the mid-nineteenth century, the Anthropocene of today is fundamentally dark and apocalyptic. Americans in the nineteenth century saw beauty, benevolence, and hope in an icy Earth shaped by an unseen Actor. The idea of the Ice Age, at its birth, was the visible, temporary sign of the invisible and eternal. In the Anthropocene of today, by contrast, we see only the wreckage of the world, and we read in the melting ice another sign that humans have failed, their deity nowhere in sight.

7

The Dinosaurs Go to College

AMONG THE MANY new faces arriving at Princeton University in the 1870s, one was not quite like the others. True, this individual came from an ancient family, like so many other Princetonians. And also walked on two legs. But the new arrival was quite a bit older—so old that its fossil bones had been found in a New Jersey marl pit.

The new face belonged to *Hadrosaurus foulkii*, a gentle, plant-eating dinosaur that had ambled through the forests of Cretaceous New Jersey (fig. 7.1). The reconstructed skeleton towered above the other creatures in Princeton's new natural history museum. Wrapping around the dinosaur was another marvel: one of the earliest and most complete prehistoric life painting sequences in the United States, showing life's evolution over millions of years from dark primordial seas to the modern Age of Man.[1]

As one of the first universities to display prehistoric fossils for the edification of its students, Princeton shows the process by which deep time became rapidly institutionalized in American higher education after the Civil War, spreading the idea to thousands of students. Not only did the size and number of colleges and universities grow, but their ethos transformed, as the denominational colleges of the previous 250 years gradually gave way to the established nonbelief of the new era. The natural sciences aspired to empirical and value-free inquiries; theology gave way to the social sciences and practical morality; religious sentiment transmuted into an ethos of public service; strident denominationalism became deliberate nonsectarianism; and the theologian-president

FIGURE 7.1. This stereograph of the E. M. Museum at Princeton in 1886 shows the upright Hadrosaurus skeleton and the sequence of seventeen landscape paintings depicting the history of life on Earth from its origins to the Age of Man.

metamorphosed into the fundraiser-in-chief. By the early twentieth century, the landscape of American higher education had been fundamentally transformed. The modern university had arrived.[2]

While religion was receding from the university, deep time took hold in many academic disciplines. Shattering the rigid classical curriculum of the previous two centuries, new disciplines appeared that probed the era stretching millions of years before the Greeks, Romans, and even the Egyptians and Chinese: archaeology, paleontology, geology, and biology. Other disciplines, such as those in the humanities and the new social sciences, launched their subject matter in the era after what was now called *prehistory*. This new word marked off *history*—the record of the human past inscribed in written documents—from prehistory, the vastly longer chunk of the past preceding written records. Princeton's future president Woodrow Wilson explained early in his career that his lectures opened in the prehistoric era, among the earliest human polities. By comparison, the time afterward seemed "short."[3]

And so it was that within a single decade between 1868 and 1878, Princeton's faculty transformed the Presbyterian campus into a temple of deep time. As part of their effort to convert the regional campus into a national powerhouse, the faculty incorporated the long chronology into the curriculum, opened the fossil museum, and trained a group of students who went on to become some of the most influential museum directors, scholars, and popularizers of deep time in early twentieth-century America. Yet far from signifying the secularization of the American university, Princeton shows that deep time did not eradicate Christianity in the academy. Instead, deep time cloaked divinity in new dress. Both deep time and Protestant Christianity were preoccupied with questions about unknowability and mystery, about agency in the natural and metaphysical realms, and about the link between knowledge and values. So while the visible face of the university changed, the invisible beyond it endured. Awe, wonder, and the sense of an ever-receding horizon of knowability: these, too, had to make their way into the modern university, to make scholarship and teaching theaters for imagination and not only for reason, an exaltation and not merely a chore. Deep time was essential to that process.

Reform Comes to Princeton

The Civil War years had been hard on Princeton. It lost not just students but its sense of national purpose. Founded in 1746, Princeton (formally known as the College of New Jersey until 1896) had trained future national leaders such as James Madison and Aaron Burr. The campus had also been a battleground in the American Revolution, with Nassau Hall itself briefly serving as the new nation's capitol in 1783. In the 1850s, the college still advertised its "national character." But at the outbreak of the Civil War, many southern students had decamped. Of the more than six hundred Princetonians who served in the Civil War, more than half

Plate 1 (Figure I.1). Gentle ruminants populate a segment of Rudolph Zallinger's mural, *The Age of Reptiles* (1947), at Yale University. Created before the asteroid extinction theory of the 1980s, the mural suggests gradual extinction through such agents as volcanos steaming ominously behind the oblivious T-Rex.

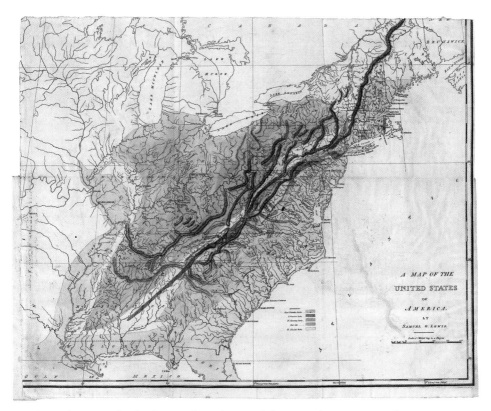

Plate 2 (Figure 1.3). The first geological map of the United States, created in 1809, uses colored bands of rocks to represent the ages of the strata. From "A Map of the United States of America" by Samuel G. Lewis. In William Maclure, "Observations on the Geology of the United States, Explanatory of a Geological Map."

Plate 3 (Figure 1.8). With undulating, colored bands representing rocks stretching to well over two billion years old, this geological map of North America from 1911 shows the transformation of the New World into a very old world in the span of a mere century.

Plate 4 (Figure 2.5). Three of the small, painted plaster trilobite casts made by Jacob Green and Joseph Brano in the 1830s. They accompanied Green's book on North American trilobites and were intended to give distant naturalists a sense for the great variety of trilobites and the endless generative powers of the deity.

Plate 5 (Figure 2.7). Jacob Green may have bought a plate much like this one when he visited the French trilobite expert Alexandre Brongniart, director of the Sèvres Porcelain Manufactory outside of Paris, in 1828.

Plate 6 (Figure 2.13). This magic lantern slide of the infant Earth, fresh and clean in the rosy dawn light of the very early Transition Period, must have enchanted the undergraduates at the University of Wooster in the decades after the Civil War.

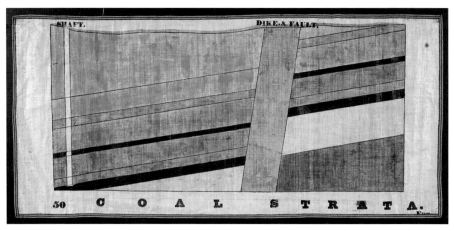

Plate 7 (Figure 3.8). Orra White Hitchcock's poster of coal strata, for use in Professor Edward Hitchcock's classes on geology and natural history at Amherst College.

Plate 8 (Figure 4.4). Robert B. Farren's *Life in the Jurassic Sea* (1850) is one of the many reproductions of Henry De la Beche's *Duria Antiquior* (A more ancient Dorset, 1830) that were circulated in both Great Britain and the United States. It is one of the earliest landscape representations of prehistoric life, showing recently discovered marine reptiles devouring each other with blood-thirsty abandon. The tropical landscape of palms and warm seas may recall De la Beche's time in Jamaica tending to his slave plantation and visiting sites of geological interest. © 2023 Sedgwick Museum of Earth Sciences, University of Cambridge. Reproduced with permission.

Plate 9 (Figure 4.7). In this scene of a Natchez plantation on the Mississippi River, workers directed by archaeologist Montroville Dickeson excavate mounds left by vanished Native peoples. John J. Egan, scene 18 from *Panorama of the Monumental Grandeur of the Mississippi Valley* (1850). Saint Louis Art Museum, Eliza McMillan Trust 34:1953.

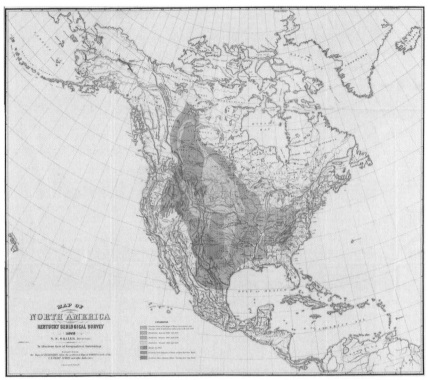

Plate 10 (Figure 5.1). In this map of the shrinking habitat of the American bison, the colored area at center is where the bison "still exists," while the largest area shows its range before 1800. J. A. Allen, "The American Bisons, Living and Extinct" (1876).

Plate 11 (Figure 5.8). Eohippus, the little dawn horse of North America, as painted by Charles R. Knight in 1905. American paleontologists liked to advertise the fact that the horse was native to North America, and therefore antedated Indian claims to being the true natives of the Americas.

Plate 12 (Figure 5.13). Two men overlook the Grand Canyon in William Henry Holmes's "Panorama from Point Sublime" (1882). With its strata piling upon strata, this image accentuates the new idea of the canyon's profound geological antiquity, asserted just a few decades before it was designated as a national park. The idea of deep time helped white Americans to justify Native exclusion from the new national parks. They argued that these cathedrals of nature were so ancient that they preceded the comparatively modern migration of the Indians into North America.

Plate 13 (Figure 6.1). Professor Benjamin Silliman of Yale overlooks his mineral collection as a red curtain unveils West Rock outside the window. Painted by Samuel Finley Breese Morse in 1825, the image shows how landscapes were often depicted before the idea of the Ice Age had been invented. Yale University Art Gallery. Gift of Bartlett Arkell, B.A. 1886, M.A. 1898, to Silliman College.

Plate 14 (Figure 6.6). The glaciated landscape near New Haven is captured in George H. Durrie's painting of low, smoothed hills and glacial erratics. *Summer Landscape Near New Haven* (c. 1849). Brooklyn Museum, Dick S. Ramsay Fund, 46.162.

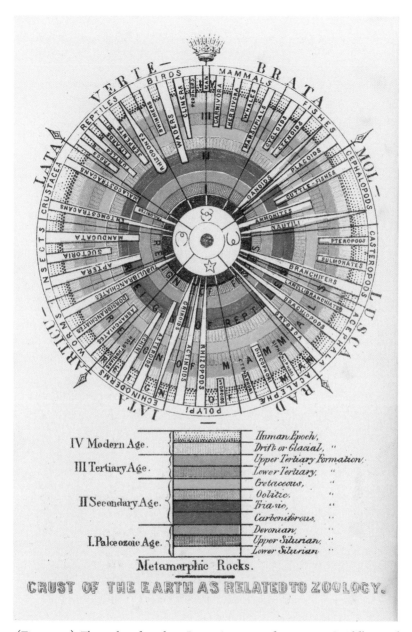

Plate 15 (Figure 6.8). This colored circle in Louis Agassiz and Augustus Gould's popular *Principles of Zoology* (1848) offered a major alternative to the branching tree as a metaphor for the development of life on Earth. Agassiz denied that species changed over time; instead, he thought God repeatedly extinguished and then re-created life on Earth over millions of years according to an unchanging Idea for the body plan of each species. The circle is divided into the four major animal groups that represent God's unchanging Idea of body plans (Vertebrata, Mollusca, Articulata, and Radiata). The beginning of life on Earth is at the center of the circle, and life progresses until the Modern Age, which is crowned by Man. Agassiz showed how multiple special creations could coexist with the new idea of deep time.

Plate 16 (Figure 6.12). Frederic Edwin Church, *The Icebergs* (1861). Church originally called this painting *The North*, but changed the title to the more neutral *The Icebergs* so as not to inflame pro-Confederate opinion during the painting's London showing. Image Courtesy Dallas Museum of Art.

Plate 17 (Figure 6.15). The glaciated valley of Yosemite gleams with rays of divine light in Albert Bierstadt's *Valley of the Yosemite* (1864).

Plate 18 (Figure 7.5). Trilobites and other creatures from Earth's first ocean have washed ashore, the better to be observed by Princeton students. Benjamin Waterhouse Hawkins, *Silurian Shore at Low Tide* (1875).

Plate 19 (Figure 7.7). Fossils from the Age of Fish (or Devonian) are visible in the rocks lining Niagara Falls. Benjamin Waterhouse Hawkins, *Devonian Life of the Old Red Sandstone* (1876).

Plate 20 (Figure 7.9). The torrid swamps of the Carboniferous produced the future coalbeds that powered American industry. Benjamin Waterhouse Hawkins, *Carboniferous Coal Swamp* (1875).

Plate 21 (Figure 7.10). Googly-eyed Labyrinthodons waddle ashore during the Triassic, a sign of vertebrate life's progressive move from sea to land. Benjamin Waterhouse Hawkins, *Triassic Life of Germany*.

Plate 22 (Figure 7.11). The famous cliffs of Dorset, England, are the stage for an iconic Jurassic battle between the Ichthyosaurus at center and the Plesiosaurus at left. Benjamin Waterhouse Hawkins, *Early Jurassic Marine Reptiles* (1876).

Plate 23 (Figure 7.13). The pose of the Jurassic monsters at center and left—legs tucked underneath rather than splayed outward like reptiles—was intended to resemble the gait of mammals, thereby showing the common plan on which God had created all vertebrates and refuting any Darwinian argument for a chain of progressive complexity through transmutation. Benjamin Waterhouse Hawkins, *Jurassic Life of Europe* (1877).

Plate 24 (Figure 7.14). New Jersey's internationally famous Cretaceous fossils come to life in this scene of carnage. Vicious red Laelaps attack a herd of green Hadrosaurus as three representative American species observe from the water. From left to right, the marine animals are Mososaurus and Elasmosaurus, with a pterodactyl perched at far right. Benjamin Waterhouse Hawkins, *Cretaceous Life of New Jersey* (1877).

Plate 25 (Figure 7.16). The Ice Age Himalayas are represented here, giving Princeton students a glimpse of the animals that had once wandered the land that was now part of Britain's empire in India. Benjamin Waterhouse Hawkins, *Pleistocene Fauna of Asia*.

Plate 26 (Figure 7.17). Two cavemen aim their weapons at a female Irish Elk, as the giant antlered paterfamilias looks on. Benjamin Waterhouse Hawkins, *Irish Elk and Palaeolithic Hunter*.

Plate 27 (Figure 8.2). Inspired by the paleontologists of his native Provence, Paul Cézanne tried to render landscapes of deep time by using squiggles and blobs of colorful paint rather than focused perspective. Paul Cézanne, *View of the Domaine Saint-Joseph [La Colinne des Pauvres]*, 1880s.

Plate 28 (Figure 8.4). Low-browed Neanderthal hunters spy a herd of rhinos in ancient France, as a disheveled mother and child cower in the cave. Charles R. Knight, *Neanderthal Flintworkers, Le Moustier Cavern, Dordogne, France* (1920).

Plate 29 (Figure 8.7). A white-skinned Cro-Magnon artist sweeps paint over the cave walls of a Renaissance-like atelier in Charles Robert Knight's *Cro-Magnon Artists of Southern France* (1920).

Plate 30 (Figure 8.14). The artist André Masson said the razor-toothed fish and "irrational" forms in his painting arose from his subconscious mind. *Battle of Fishes* (1926).

Plate 31 (Figure 9.1). A tiny ark floats in the distance as the waters of the biblical flood drain away. A human skull is visible in the foreground, the vestige of the sinful humanity eradicated by the deluge. Thomas Cole, *The Subsiding of the Waters of the Deluge* (1829).

Plate 32 (Figure E.1). This recent visualization of deep time shows the history of our planet as an uncurling ammonite. A surfer rides the edge of time into the future.

fought for the Confederacy. After the war, Princeton became a northern campus, drawing only 10 percent of students from the South.[4]

Intellectual changes were also afoot. Along with leaders of other post–Civil War American colleges, Princeton's faculty wondered how to adapt the traditional curriculum, centered on classics and moral philosophy, to the emerging disciplines in the sciences and social sciences. Yale and Harvard had added scientific schools in the late 1840s, while new universities such as the Massachusetts Institute of Technology (f. 1861) offered a glimpse of a scientific and utilitarian future. In 1832, Princeton hired Joseph Henry to the first Chair of Natural History, and he taught a variety of courses while also running a laboratory on campus in which he performed electromagnetic experiments. But in 1846 he left to become the first secretary of the new Smithsonian Institution. It was clear that more needed to be done if Princeton were to meet the demands of the Modern Age.

Help came in the person of theologian James McCosh, the new president who arrived fresh from Scotland in 1868. His thick brogue serving as a perpetual reminder that he was not from these parts, he was able to sidestep the sectional rivalries that still rankled. He immediately began to modernize the campus of ten full-time faculty by introducing a new scientific curriculum, adding some electives (though not as many as Harvard, with the brash young Charles W. Eliot at its helm), and hiring faculty who had not been trained at Princeton. Given his brusque and imperious manner, he also fared surprisingly well with the students, who called him "Jimmy" behind his back.[5] McCosh remained in the presidency for twenty years, overseeing an extraordinary transformation of the tradition-bound campus.

Among his projects was to show that faith need not be an impediment to studying the long history of life on Earth. Charles Darwin's *Origin of Species* (1859) had proposed that organisms changed over a long span of time through random variations in response to environmental pressure and toward no predetermined goal. By contrast, McCosh imagined that God foresaw all these changes, part of one mighty and unchanging Plan that culminated in the only creature with a soul:

Man. His *Typical Forms and Special Ends in Creation* (1856), published over a decade before he arrived at Princeton, showed his allegiance to respected like-minded naturalists, such as Louis Agassiz and Richard Owen. He allowed that God might clothe his workings in "natural laws or secondary causes." But Man was always the final outcome, with animals, plants, and climate somehow collaborating in each geological era to prepare Earth for the present "comfort of man."[6]

McCosh was pleased to find an ally for his views already on the Princeton campus: the Swiss naturalist Arnold Guyot. Like his show-boating friend and countryman Louis Agassiz, then basking in international stardom at Harvard, the docile and bespectacled Guyot was an expert on the Ice Age and the Alpine glaciers that he and Agassiz had explored together as young men. Also like Agassiz, Guyot was a Protestant, though a more orthodox one than the big-picture Agassiz. He had prepared for the ministry as a young man before dedicating himself to natural history. The Revolutions of 1848 sent him fleeing to the United States, and thanks to Agassiz's help, he found a temporary home in Cambridge, Massachusetts, where he was soon invited to give a series of public lectures on natural history.

Guyot's Boston lectures attracted large audiences eager to hear the cutting edge of science: a unified history of the planet, based on the latest scientific findings from European and American naturalists, that culminated in the present moment. The lectures were published as a popular book, *The Earth and Man* (1849). The title announced the link Guyot made between geography and modern human beings. His ideas were inspired by the German geographer Carl Ritter, who held that the configuration of Earth's continents and their associated climates were an important influence on human society, designed by God to ensure the march of progress. "*Excellent, excellent, excellent,*" Ritter wrote to Guyot of *The Earth and Man*. Guyot flattered his American audiences with his own twist on Ritter's tale: humankind's destiny culminated in the United States, where Old and New World ways mingled to form a new civilization. At its head would march the white race, symbolized by the classical perfections of the ideal man, the Apollo Belvedere. To Ritter's thesis, Guyot added another personal insight: that geological eras could be harmonized with the record of Genesis. Guyot's ideas about reconciling geology with Genesis were cited approvingly in orthodox theological journals.[7]

In his role as a clearinghouse of Continental thought, Guyot also helped to introduce into the United States the educational ideas of Swiss educational reformer Johann Heinrich Pestalozzi. Building on Jean-Jacques Rousseau's work, Pestalozzi had taught that the child's innate abilities would unfold not with harsh discipline and rote learning but through a curriculum that enlisted the head, the hand, and the heart. At the core of this child-centered pedagogy was nature, believed to lie outside corrupting society and therefore suited to nurturing virtuous people who could in turn reform institutions and individuals. "All instruction of man is then only the Art of helping Nature to develop in her own way," Pestalozzi had announced. Guyot, teaching geography and pedagogy in the Massachusetts normal schools, helped this wave of Pestalozzian teachings to sweep across American schools. Soon nature became the backbone of early childhood education, exemplified by the kindergartens that took off after the Civil War.[8]

In 1854, Guyot was appointed professor of physical geography and geology at Princeton. His religious convictions and object-centered nature pedagogy promised a smooth landing on the rural Presbyterian campus. Instead, he was greeted by chaos. The natural history specimens, accumulated willy-nilly over the last thirty years, were scattered everywhere. The students later recalled, "Our stuffed crocodiles lay during last year high and dry on the tops of the book shelves: with tails erect, and mouths wide open, and glittering teeth, as if ready to swallow alive the first rash individual who should give the slightest intimation of thieving designs on the books."[9]

Guyot rolled up his sleeves. On a trip to Europe, he collected around six thousand glacial erratics from Switzerland and representative fossils from all periods of Earth's history. Back at Princeton, he made geology and geography into required courses. The library acquired a handsome collection of leading books on Earth's antiquity by Charles Lyell, Louis Agassiz, Roderick Murchison, James Geikie—even Charles Darwin's *Origin of Species* stood in a special section on "evolution of species" that also included proponents of non-Darwinian species change such as President McCosh.[10] The students, in short, beheld a rainbow of opinion on the topic of evolution.

In the classroom, Guyot lectured in his thick French accent to the upperclassmen about geology and geography, bellowing over the continual buzz of the students with senioritis lounging in the back of the lecture hall. There was no laboratory or fieldwork: only a textbook, lectures explaining prehistory as *"prophetic"* of the Age of Man, as well an extensive new collection of cloth wall hangings to which Guyot gestured during the lectures. One featured his native Switzerland long ago when low, snowless Alps framed a swamp grazed by pachyderms and ungulates eventually felled by the Ice Age to come (fig. 7.2).[11]

By 1874, the campus newspaper declared that Guyot had accomplished nothing less than a "revolution" at Princeton. His many specimens of rocks and fossils had allowed the campus to open a new geological museum in Nassau Hall. "The leading idea in the arrangement of fossils," Guyot explained, "is that they should strike the eye as an open book in which the student can read, at a glance, and in real forms, the history of the Creation from the dawn of life to the appearance of man." Named the E. M. Museum in honor of Elizabeth Marsh Libbey, the wife of a Princeton donor, the geological museum was fitted with the latest modern convenience of steam-heated pipes.[12]

Passing through an entrance hall lined with portraits of Princeton presidents and professors, students emerged into sunlit splendor: the geological hall—sixty-six feet long by forty-eight feet wide—illuminated by a new skylight. Rows of glass-fronted fossil specimen cases filled the main floor as well as an upstairs gallery reached by a spiral staircase. A giant extinct sloth stood amid other prehistoric reconstructions made of plaster and contributed by entrepreneur Henry Augustus Ward. A former Agassiz student, Ward now owned an outfit in Rochester that supplied real and plaster natural history specimens to over a hundred American classrooms and museums, as well as down-market commissions like Buffalo Bill's Wild West show and P. T. Barnum's circus. Eager to keep his museum up to date, Guyot wrote to paleontologists working in the American West, pleading for them to send bones. Western fossils

FIGURE 7.2. Princeton professor Arnold Guyot pointed to wall hangings such as this one during his geology lectures at Princeton. The poster depicts his native Switzerland long ago in the "Middle Diluvial Age," when pachyderms and hippos grazed in a lush meadow at the foot of snowless Alps.

would "open to us a long vista into the mysteries of the ancient life of our continent," Guyot wrote to the famous paleontologist Edward Drinker Cope.[13]

Guyot also created an archaeological museum in an adjacent space. Among other human artifacts, he insisted on displaying a model of the recently discovered Stone and Bronze Age lake dwellings from his native Switzerland, discovered by chance in 1854. Believed to float above the water on stilts, these huts were a subject of particular national pride, nursing hopes not only that northern Europe developed inspiring architectural forms beyond the pale of Roman civilization, but that a common Swiss identity somehow united a new federal republic comprising twenty-six cantons that had been established only in 1848. Under Guyot's watchful eye, American students might take away a lesson about the place of human prehistory in shaping the patriotic identity of new republican confederations, in addition to savoring the sheer absurdity of living daily life on stilts. The reconstructed Swiss lake dwelling joined a mishmash of Old Master paintings and fake classical statuary awaiting transfer to a projected new art history museum, which was not completed until 1890. For nearly two decades, Princeton students encountered extinct creatures along with Rembrandt paintings, Venuses, and Swiss lake dwellings, not to mention busts of their professors and even President McCosh. "Temple" was the word the students used to describe this jumble. And a temple it was—a temple to science, art, and the new idea of deep time, perfumed by varnished mahogany cabinetry.[14]

The museum was not a clean victory, however. Its opening in 1874 brought a severe rebuke to the idea of deep time it so brazenly paraded. From just down the lane at the Princeton Theological Seminary, the Presbyterian theologian Charles Hodge published a pamphlet called *What Is Darwinism?* His answer: "Atheism." His pen dipped in indignation, Hodge insisted that Darwin's theory of species transmutation by natural selection substituted beastly ancestors for the noble creation of man by God. The pamphlet was small, but Hodge's reach was immense. As one of the most eminent American theologians of the century, Hodge trained around three thousand clergymen during his long career, more than any other theologian in the United States during the nineteenth century.[15]

Hodge did not deny the reality of the long chronology but rather pointed out some of its logical implications to believing Christians. He was cautious about whether the days of Genesis were literal or figurative, as were many theologians of this time, and reminded readers that scientists had yet to offer any "sure data for the calculation of time." But more important for Hodge was the significance of the long chronology: it was the stage over which Darwin's "virtually atheistical" theory of natural selection operated. Those millions of years seemed to Hodge empty and "abandoned" by God. The universe wheeled along, controlled only by "chance and necessity." Hodge insisted that God was constantly present everywhere, working through and with the physical laws he had created to uphold the universe. He balked at the long epochs demanded by geology, saying they "have no meaning." The tiny, "imperceptible" variations that appeared over millions and millions of years were likewise unimaginable, a "demand on our credulity." Containing only the random process of natural selection, time became hollow and empty, echoing with God's absence. The conclusion was obvious. "This banishing God from the world is simply intolerable."[16]

Hodge was not alone in his skepticism about deep time. First proposed in the seventeenth century, Archbishop James Ussher's short biblical chronology of a six-thousand-year Earth was now dusted off for modern combat. In 1871, Presbyterian missionary Sebastian Adams published a twenty-three-foot-long panorama titled the *Synchronological Chart of Universal History*. Timelines and colorful cartoons ensured that its facts would be "indelibly photographed upon the brain of the young." History flowed from the Garden of Eden in 4004 BC to the present day. The opening cartouche recasts the Garden of Eden for the Victorian Age (fig. 7.3). A strategic hairstyle and cooperative plants clothe the first couple in modesty. Eve stares mutely downward as Adam explains Creation to her.[17]

In the face of critics like Charles Hodge and the resurrectors of Ussher, Guyot and McCosh had to show that deep time was just as filled with God as shallow time. In the millions of years since the first life forms appeared, Guyot and McCosh saw God creating life again and again, directing it toward the perfection of man. Deep time in fact was *more* filled with God than Ussher's six thousand years, requiring a deity who created and felled life again and again over millennia. And what better instrument to show the deity's omnipresence in deep time than their new museum? McCosh's lectures on psychology to Princeton students emphasized the various powers or faculties of the mind, one of which was "Time." Everybody had a "*natural* tendency to see how events are related to one another in time," McCosh explained. Here was a chance to show the relationship of events in prehistoric time by adding the *pièce de résistance* to Princeton's new museum: a sequence of prehistoric paintings showing the history of life on Earth.[18]

There was only one man for the job: the English artist Benjamin Waterhouse Hawkins.

FIGURE 7.3. Sebastian Adams's very long *Synchronological Chart of Universal History* (1871) begins with this small scene of Adam and Eve in the Garden of Eden in precisely 4004 B.C.

Benjamin Waterhouse Hawkins Comes to Princeton

By the time he was hired to produce prehistoric paintings at Princeton, Benjamin Waterhouse Hawkins was famous in Britain and the United States. He was especially known for the thirty-three life-sized cement sculptures of prehistoric animals that he had made for the Crystal Palace Park in south London in 1852, where much of the Crystal Palace Exhibition was rebuilt following the closure of the original one in Hyde Park. Here Hawkins fashioned the world's first prehistoric park. In addition to the primordial beasts, it also included tilted artificial rock formations showing geological change through time and fake tides alternately submerging and then revealing the prehistoric monsters (fig. 7.4).[19]

London's prehistoric park was also an ideological program. It was the brainchild of the renowned English naturalist Richard Owen, archopponent of transformist doctrines. The poses of the prehistoric reptiles he directed Hawkins to make—large legs tucked underneath rather than small legs splayed outward like lizards—were intended to resemble the gait of mammals. This would reveal the common Plan on which

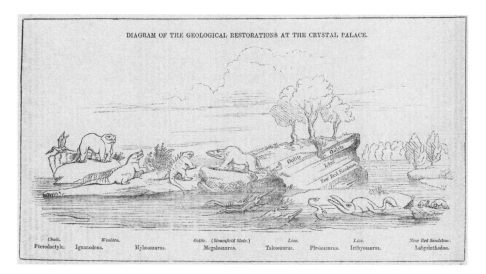

FIGURE 7.4. Dinosaurs standing scarily at the Crystal Palace exhibition known as Prehistoric Park, located in London. It also featured tilted artificial rock formations showing geological change through time, as well as fake tides alternately submerging and then revealing the prehistoric marine monsters. From Benjamin Waterhouse Hawkins, "On Visual Education as Applied to Geology" (1854).

God had created all vertebrates and refute any Lamarckian argument for a chain of progressive complexity through transmutation. Hawkins's book, *A Comparative View of the Human and Animal Frame* (1860), used bipedal animals and humans to show the "oneness of plan upon all animals are constructed," a pattern fixed by God at the beginning of time. Hawkins endorsed Pestalozzi's principles of teaching nature *"through the eye"* rather than with dry lectures. His enormous sculptures joined his other public-facing output—posters, diagrams, small dinosaur models—as a way of *"educating the masses."* In 1853, Hawkins honored Owen and about twenty other men of science by hosting an elaborate eight-course New Year's Eve dinner inside an enlarged and hollowed-out cast of the Iguanodon.[20]

Opinionated and temperamental, Hawkins was also a bigamist. With two families to support in England, he was constantly scrambling for commissions. In 1868 he gladly accepted the invitation to come to New York, extended by the science popularizer E. L. Youmans, founder of

Popular Science Monthly. After creating more drawings and life-sized prehistoric models for a museum in Central Park (destroyed by vandals under "Boss" Tweed before they could be displayed), Hawkins went to Philadelphia to "undertake the resuscitation of a group of animals of the former periods of the American continent" at the Academy of Natural Sciences in Philadelphia. He assembled the bones of the New Jersey dinosaur *Hadrosaurus foulkii*, the first reasonably intact dinosaur skeleton ever found. The creature had been named for the Haddonfield marl pit on which its thirty-five bones had been unearthed, and the lawyer-naturalist, William Parker Foulke, who had brought them to scientific attention in 1858. Hawkins assembled the bones (and plaster models of missing bones) into the first full dinosaur reconstruction in the world. Lacking the skull, he fashioned a scaled-up version using an iguana cranium that he painted green to signal the creature's reptilian nature.[21]

Amid all these other firsts, Hawkins achieved a final first in Philadelphia: he assembled *Hadrosaurus foulkii* in a standing posture. This idea had first been proposed by American paleontologist Joseph Leidy after the discovery of more Hadrosaurus bones. "A remarkable reptile, of huge proportions, has been proved to have existed during the Cretaceous period of the Western Continent," Leidy had announced in 1865. Based on the disproportion between front and hind limbs, he concluded that it could raise itself into a bipedal position and browse "kangaroo-like" in the foliage. To this suggestion, Hawkins added his personal views about the significance of bipedalism. Repulsed by Darwin's "absurd" suggestion that humans were descended from the animals, Hawkins thought the upright posture had anticipated the "dignified" bipedalism of man. The fossil record showed not the transformation of life forms through time, but their repeated appearance over time, as God continually repopulated the Earth with higher life forms using a few basic patterns. The public that viewed bipedal dinosaurs would see God's hand in the progressive ascent of life over long periods of time. The dignified bipedal Hadrosaurus prophesied the dignified bipedal Victorian gentleman.[22]

These views endeared Hawkins to President McCosh, and an invitation to become a visiting professor at Princeton followed forthwith. Guyot also knew Hawkins. The two had met in 1873 at Louis Agassiz's summer school at Pekinese Island off the coast of Massachusetts, where they had absorbed Pestalozzian hands-on techniques for teaching nature to young people. Hawkins seemed like a good fit for the intellectual climate at Princeton, able to popularize the anti-Darwinian idea that successive regimes of life ruled Earth over long expanses of time, with God heavily involved at each juncture.[23]

The commission: produce seventeen paintings illustrating the progression of life over a vast period time, and erect a bipedal dinosaur. The goal: unveil for Princeton students a "tangible history of creation ... representing, as far as science can define them, the scenery, vegetation, and animals of each age of the world, terminating with the creation of man."[24] Hawkins got to work.

A Tangible History of Creation

Created between 1875 and 1877, Hawkins's prehistoric paintings for Princeton are best understood as two-dimensional versions of the concrete dinosaur models he had created over twenty years earlier for London's prehistoric park.[25] For Hawkins was not an *artist* in the Victorian sense of the term, a visionary interpreter whose forms gestured toward loftier meanings. Instead, he was a literal transcriber of nature and an effective popularizer. His Princeton paintings put prehistoric animals into prehistoric landscapes. But they are not *landscape paintings* in the expansive sense being pioneered by American painters of the Hudson River school, such as Thomas Cole, Asher Durand, and others. While those artists cast American landscapes as species of the sublime, Hawkins painted prehistoric landscapes as didactic models. In fact, they are best understood as specimen display cases set against a habitat diorama that replicates the London park's ponds and islands.[26] The main action in the paintings is pushed to the foreground, where the prehistoric creatures line up in an orderly fashion for inspection by undergraduates. All avians are rooted to the land, just as their concrete counterparts were in the London park.

Hung side by side so that they criss-crossed the gallery from the Creation to Man, the Princeton paintings functioned like a Gothic cathedral's stations of the cross, revamped for the age of science. Students were to stroll through the whole history of life on Earth, aided by the guidebook Guyot published to help them decode the gallery's religious and scientific meanings. The guidebook's title, *Creation; or, The Biblical Cosmogeny in the Light of Modern Science*, left no doubt as to the goal. Guyot refuted the results of "so-called modern, higher criticism" of the Bible while also offering "modern science." The guidebook reproduced some of the paintings and added labels with the scientific names of the animals and plants, all the while commenting on the religious meanings to be extracted. Guyot's tangible history of Creation deployed prehistory to make the students better Christians.[27]

The heavy didacticism of the whole venture—the slow, numbered walk through the paintings, guidebook in hand—is replicated in the rest of this section of the chapter, which marches steadily, even metronomically, through the gallery as Princeton students would have done in Guyot's era. The point is to reveal the heady ambition and meticulous science that went into making the E. M. Museum the first great temple of deep time for American college students. Pedantry in some sense was the purpose. But you can skip ahead to the next section, "The Age of Man," if the details of the paintings don't interest you.

Students would begin at the beginning, at the first painting (fig. 7.5 and plate 18). It depicted the moment of Creation: a Silurian beach on an infant planet almost entirely engulfed by oceans. The Silurian marked the dawn of Creation and of America, according to McCosh and Agassiz. Low tide has exposed the first living organisms washed up from their ocean nursery. Some frantically wave their tentacles as they die on the inhospitable shore. In contrast to Darwin, who would have seen these early forms evolving into later ones, Hawkins presents them as three of the four unchanging archetypes of life first proposed

FIGURE 7.5. Trilobites and other creatures from Earth's first ocean have washed ashore, the better to be observed by Princeton students. Benjamin Waterhouse Hawkins, *Silurian Shore at Low Tide* (1875).

by Georges Cuvier and affirmed by Louis Agassiz: Radiata (the flower-like creatures at left), Articulata (the trilobites at the center) and the squid-like Mollusca (at right, awaiting fossilization in the nearby strata). The second floor of Agassiz's Harvard Museum of Comparative Zoology had also arranged the creatures this way. These basic types, according to Guyot's textbook explanation, reappeared throughout the "untold ages" of Earth's history. The vaguely ark-like shape of the central island suggests the divine. So do the sun's feeble rays, which penetrate the heavy mist still shrouding the primeval planet. A photograph of the Silurian painting in place captured how the gallery's skylight would have enhanced the sensation of heavenly light (fig. 7.6).[28]

The second painting brought students forward in time to the Age of Fish, an era dating to millions of years ago, according to Hawkins (fig. 7.7 and plate 19). Its formal name was the Devonian, after the English formations in Devon that entombed its characteristic fossils. But here Hawkins made an American move, setting the Devonian at Niagara Falls because he had seen a painting of the mighty cataract at the Academy of Natural Sciences in Philadelphia. Niagara Falls had by now become an American national icon and tourist attraction. It symbolized American democracy and territorial grandeur while dwarfing European nature.

FIGURE 7.6. The E. M. Museum's skylight illuminates the painting of the Silurian shore at left, magnifying the effect of heavenly light on the infant planet's beached Silurian creatures.

FIGURE 7.7. Fossils from the Age of Fish (or Devonian) are visible in the rocks lining Niagara Falls. Benjamin Waterhouse Hawkins, *Devonian Life of the Old Red Sandstone* (1876).

Niagara also seemed tailor-made for a time period before terrestrial life, showing "only rocks sky and water," according to Hawkins. "I want a copy of that waterfall," he continued, since he had never been. "I have fixed upon it for the foundation of my next picture, 'Age of Fishes' for the Geological Museum at Princeton."[29]

Hawkins redirected the spatial immensity of the cataract into the temporal vastness of the Devonian. Rushing water has eroded the rocks over a long period of time, as the stratified cliffs reveal. Draped helpfully over the dark foreground rocks are the fossil fish that Agassiz had said represented the first era of vertebrate life on Earth. This was the first appearance of ideal types that recurred, unchanging, throughout the fossil record according to a predetermined plan of the Creator. Hawkins copied the fish fossils from a popular book that also denied transmutation in favor of separate creations by God: Hugh Miller's *The Old Red Sandstone*. Miller's book contained numerous illustrations of primordial fish, along with the names by which Agassiz had classed them as exempla of his unchanging types. The bizarre shovel-headed creature from Miller's book is identifiable in Hawkins's bottom-most rocks (fig. 7.8).[30]

Third in the sequence of paintings came the tropical Carboniferous, where life at last crept onto land (fig. 7.9 and plate 20). In the sunless murk, plants rot into future coal seams, the economic engine of Britain and the United States. No animate life is present, reflecting not just the scarcity of animal and insect fossils, but the reality that the drama of the Carboniferous was in the plants, which could reach heights of a hundred feet or more. These giants were "the material for the vast beds of coal so precious to civilized man," Guyot explained to his Princeton students as he emphasized the immense length of this geological period.[31]

Hawkins arrays characteristic plants from left to right in the immediate foreground for ease of teaching: the Lepidodendron (with its crowd of branches and mottled, scarred trunk, all conspicuous in fossil remains), Sigillaria (with its crown of long, grass-like fronds), and Calamites (the ancestor of modern horse-tails). Yale geologist James Dwight Dana had taught that Earth "dragged slowly" in its infancy, speeding up

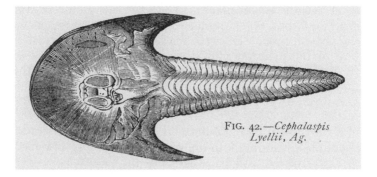

FIG. 42.—*Cephalaspis Lyellii*, Ag.

FIGURE 7.8. Hawkins's painting *Devonian Life* included fossils such as this shovel-headed fish, copied from Hugh Miller's popular book from this time, *The Old Red Sandstone*.

FIGURE 7.9. The torrid swamps of the Carboniferous produced the future coal-beds that powered American industry. Benjamin Waterhouse Hawkins, *Carboniferous Coal Swamp* (1875).

with "dynamical agencies" as it approached the Age of Man. Hawkins's static composition reflects this. Nothing is happening. We await the industrial age to unlock the energy in the picture.[32]

Having inspected the swampy Carboniferous, the students would press onward in time, arriving at two paintings of the Triassic. Both paintings were set in Europe, probably because until the 1870s these Triassic formations had been better exposed there than in the United States. Triassic strata contained some of the earliest major fossil finds for what seemed to be numerous amphibious—as opposed to totally aquatic—creatures, able to breath the still-noxious air that lingered from the Carboniferous and the Permian. This was seen as an important step for Earth as it progressed toward the Age of Man.[33]

One of the two paintings, *Triassic Life of Germany*, nodded toward the anti-Darwinian German naturalists who had fleshed out the fossil record for this period (fig. 7.10 and plate 21). The Triassic had been named in 1834 by the German paleontologist Friedrich August von Alberti for a sequence of three (hence "triassic") distinctive rock layers in central Germany. Richard Owen was among the major British naturalists who drew parallels between the German Trias and equivalent British formations. The painting's formal structure emphasizes life's progressive move from sea to land. Guided by a pathway of moonlight, a group of frog-like Labyrinthodons paddle toward shore. Their grinning leader is identical to the cement version in Crystal Palace Park. The animal's faint footprints are visible on the beach, a nod to the fame of the Labyrinthodon footprints found in in England and Germany.[34]

The renowned fossil-rich cliffs of Dorset in southwest England are the setting for the next geological era, *Early Jurassic Marine Reptiles* (fig. 7.11 and plate 22). Hawkins stages an iconic meeting of two of the earliest and now most famous prehistoric reptiles discovered in southwest England in the 1820s: the googly-eyed Ichthyosaurus at center, which roars at the sinuous Plesiosaurus at left. The painting reproduces the drama of the exhibit at Crystal Palace Park, where these two dinosaurs faced off in the Secondary (Mesozoic) area. Even the grayish-blue limestone ("lias") cliffs at Dorset are faithfully reproduced here, as they were in the London park. Hawkins draws attention to the color by setting the cliffs against a red horizon. The behavior of the animals suggests the best science of the time. Plesiosaurus was thought to paddle among coastal seaweeds, where it could nibble on fish at a safe distance

FIGURE 7.10. Googly-eyed Labyrinthodons waddle ashore during the Triassic, a sign of vertebrate life's progressive move from sea to land. Benjamin Waterhouse Hawkins, *Triassic Life of Germany*.

FIGURE 7.11. The famous cliffs of Dorset, England, are the stage for an iconic Jurassic battle between the Ichthyosaurus at center and the Plesiosaurus at left. Benjamin Waterhouse Hawkins, *Early Jurassic Marine Reptiles* (1876).

FIGURE 7.12. First published in the late 1830s, these are some of the first dinosaur images directed at American children. Clearly labeled as extinct animals, the enormous reptiles show the popularity of the new idea of extinction, even for young people. Although the scene is a gentle one, with reptiles paddling in a tropical seashore as a bat-like pterodactyl swoops from the clouds to greet them, the author—and possibly the readers he called "my young friends"— delighted in descriptions of the animals' "ruffian sort of life." Samuel Griswold Goodrich, *Peter Parley's Wonders of the Earth, Sea and Sky* (1840).

from deep-water predators. One has even hauled itself onto land. More Plesiosauruses appear in the distance, like ducks bobbing for bread crusts.[35]

This ancient English scene may already have been familiar to Princeton undergraduates. A similar one had appeared in an American children's book in 1840, informing them that these extinct animals dated from "a great many years ago" (fig. 7.12).[36]

Staying in the Jurassic, Hawkins's next painting moves to land—specifically, an island in a certain London pond (fig. 7.13 and plate 23). The scene highlights the intellectual contributions of Richard Owen. In 1841 he had erected the new category *Dinosauria* and had used three of the genera pictured in this painting to define those terrible lizards: the carnivorous Megalosaurus (the menacing carnivores at center), the plant-eating Iguanodons (fleeing to the left), and the spiny-backed Hyaelosaurus (waddling in from the right), all fossils uncovered in southern England. As in the London park, both the Megalosaurus and Iguanodon are represented as lumbering, mammal-like quadrupeds, as Owen had decreed. More Crystal Palace Park borrowings appear in the crocodilian Teleosaurs making landfall at left. Six leathery pterodactyls lurk on the rocks and splash in the shallows, reflecting the belief that they were able both to swim and to fly. They are grounded in the painting, though, just as the cement ones were in the London park, as though Hawkins could not summon the imagination to paint them flying above.[37]

The moonlit setting reflected the view that the Age of Reptiles was one of lawless violence, a dark chapter to be endured as Earth awaited the civilized Age of Man. Blood-red clouds add to the lurid, gothic atmosphere of a Victorian novel. This may have been a nod to Charles Dickens's *Bleak House* (1853), whose first paragraph featured a Megalosaurus in the city of London. "Implacable November weather. As much mud in the streets as if the waters had but newly retired from the face of the earth, and it would not be wonderful to meet a Megalosaurus, forty feet long or so, waddling like an elephantine lizard up Holborn Hill." By the time Hawkins painted the scene for Princeton in the mid-1870s, Richard Owen—long skeptical of Lamarckian transmutation—had become a severe critic of Darwin's *Origin of Species*. The painting broadcast Owen's anti-Darwinian views to American college students.[38]

Before leaving the Age of Reptiles, Hawkins offered a final scene of blood and gore. This one was set in America—and, in a triumphant move for the aspirational university—in New Jersey (fig. 7.14 and plate 24).

FIGURE 7.13. The pose of the Jurassic monsters at center and left—legs tucked underneath rather than splayed outward like reptiles—was intended to resemble the gait of mammals, thereby showing the common plan on which God had created all vertebrates and refuting any Darwinian argument for a chain of progressive complexity through transmutation. Benjamin Waterhouse Hawkins, *Jurassic Life of Europe* (1877).

Today the painting goes by the name *Cretaceous Life of New Jersey*, but stereographs from the time show the painting with a different label that does not specify a location: *Scenery and Vegetation in Cretaceous Period*. Still, students probably would have understood the local reference to the famous fossil Hadrosaurus dug up from a nearby New Jersey farm. Guyot featured this example of "the mastery of reptiles" in his guidebook for the students.[39]

The foreground of this ancient New Jersey scene was, as usual, an island in a pond. A gang of vicious red Laelaps attacks a herd of Hadrosaurus, colored green to suggest that they are peaceful vegetarians. Two

of the enemies are locked in a death waltz, a fate that sends most of the Hadrosaurus herd fleeing to the safety of a distant island. From the water, three representative American species look on. At left, two chunky Mososaurus (whose fossils had also been found in New Jersey) have beached themselves close to the battle. Four serpentine Elasmosaurus ride the breakers; their fossils had been unearthed in Kansas in 1867. At right, a lone pterodactyl rests on the chalky strata that gave the Cretaceous its name.[40]

If the scene resembled the Jurassic fight of the previous painting, it was intentional. The point was to show that American Cretaceous reptiles were just as vital, numerous, and fierce as their European counterparts. American Cretaceous reptiles were "no less brilliant and no less marvelous than those of Mantell and Owen in the Old World," declared University of Michigan geologist Alexander Winchell.

FIGURE 7.14. New Jersey's internationally famous Cretaceous fossils come to life in this scene of carnage. Vicious red Laelaps attack a herd of green Hadrosauruses as three representative American species observe from the water. From left to right, the marine animals are Mososaurus and Elasmosaurus, with a pterodactyl perched at far right. Benjamin Waterhouse Hawkins, *Cretaceous Life of New Jersey* (1877).

Like Hadrosaurus, the thuggish Laelaps had also been discovered in a New Jersey marl pit, in 1866. Edward Drinker Cope described it as a fierce carnivore eighteen feet long and able to raise itself to a height of six feet. Its claws were "like instruments for holding living prey." Elasmosaurus helped Americans to establish the viability in America of a sea animal "far exceeding" the neck length of the English Plesiosaurus. Compared to Plesiosaurus, announced Leidy, the Kansas beast "possessed a much longer neck,—one indeed that exceeded that of all known animals."[41]

Lacking a Crystal Park precedent, Hawkins may have turned to some of the first American images of the Cretaceous landscape to craft this particular painting of what amounted to mascots for Princeton. Perhaps he had seen Cope's image titled "Fossil Reptiles of New Jersey," featuring the ferocious combat between these Cretaceous American reptiles in the *American Naturalist* (1869) (fig. 7.15). Standing on a small island in the foreground, a clawed Laelaps bares its teeth to the Elasmosaurus in the water, while a fork-tongued Mososaurus looks on. A Hadrosaurus snacks on ferns in the forest behind.[42]

Forward the students would march through time until they arrived at six paintings of the recent Pliocene and Pleistocene. The significant representation of modern scenes—forming roughly a third of the total in the gallery—reflected not just the numerous fossils from these more contemporary eras but Hawkins's awareness of dazzling new findings

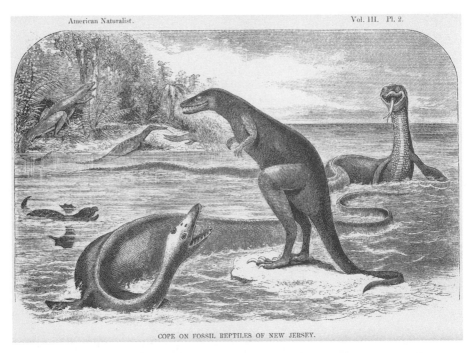

FIGURE 7.15. Among the earliest illustrations of American dinosaurs in a prehistoric American landscape, this image from 1869 that accompanies an article by Edward Drinker Cope may have inspired the Princeton painting in Figure 7.14 and plate 24.

from Britain's colonies in New Zealand and especially India. These Indian fossils he now presented to American students in *Pleistocene Fauna of Asia* (fig. 7.16 and plate 25). Deep in India's Ice Age, the four-horned mammal *Sivatherium giganteum* trails two pachyderms and tortoises against the backdrop of the snowcapped Himalayan foothills. The fossils of the odd Sivatherium had been found several decades before in the Sivalik, a region of northern India named by the British to link India's geological and ethnological past. The name literally means "Shiva-like," for the Hindu deity. Its discovery meant that "the severe investigations of modern science" could finally expel the superstitions of "the people of India even now," who insisted that the fossils were remains of the ancient Titans celebrated in their "ancient writings." British naturalists in India traded modern "Hindoo" crania and ancient

FIGURE 7.16. The Ice Age Himalayas are represented here, giving Princeton students a glimpse of the animals that had once wandered the land that was now part of Britain's empire in India. Benjamin Waterhouse Hawkins, *Pleistocene Fauna of Asia*.

Sivalik animal fossils with their colleagues in Britain and the United States, such as the Philadelphia physician Samuel George Morton. The enormous Sivatherium, wrote Hugh Falconer, superintendent of India's Saharanpur Botanical Garden, was even larger than a buffalo and a rhinoceros. Hawkins's painting shows the Sivatherium with the curious skull as it appeared in Falconer's scientific illustrations. Guyot did not mention this painting in his guide for the students, but attentive students may have discerned the larger truth, neatly encapsulated by Louis Agassiz: that fossils were found "wherever the civilization of the white race has extended." Fossil hunting not only uncovered the vestiges of civilization, Victorians proposed, but was part of the civilizing mission itself.[43]

The Age of Man

At the far end of the museum, after their long journey through time, weary students at last arrived at the Age of Man. There began "the moral world," Guyot's guidebook informed them. Details in these paintings once again helped students to extract moral messages from the science.[44]

The painting *Irish Elk and Palaeolithic Hunter*, for example, shows two men rushing from a cave beneath a giant Megaceros (or Irish elk) strutting its magnificent antlers (fig. 7.17 and plate 26). Clad in skins but clean-shaven, the hunters aim their spear and stone tool at the dainty doe and her young. The scene is set in Europe, where items of human manufacture had recently been found alongside the remains of extinct animals. At Brixham Cave in England in 1858 and at Amiens and Abbeville in France in 1859, archaeological finds suggested that humankind was far older than the six thousand years proposed long ago by Archbishop Ussher. Although the antiquity of all other life forms on Earth had been asserted over the previous half century, humanity itself had been cordoned off, seen as the recent and separate crown of God's Creation, uniquely endowed with a soul. Now, in the wake of these revolutionary finds, human beings were inserted into deep time. It was in this context that the new words *prehistory* and *prehistoric* were coined, since new terminology was needed to discuss the extremely long era before the dawn of the writing that formed the subject of history.

A series of landmark publications in the 1860s and early 1870s brought the antiquity of humanity to popular attention: Charles Lyell's *Geological Evidences of the Antiquity of Man* (1863), Thomas Huxley's *Evidence as to Man's Place in Nature* (1863), John Lubbock's *Pre-historic Times* (1865), E. B. Tylor's *Primitive Culture* (1871), and Darwin's *The Descent of Man* (1871), among others. "[W]e must extend by long epochs the most liberal estimate that has yet been made of the antiquity of Man," Huxley wrote in 1863. It was noteworthy that Lyell, long a proponent of Earth's antiquity but a skeptic of the antiquity of man, now warmed to the idea. The Egyptian pyramids were "things of yesterday" compared to these human relics, he wrote to George Ticknor in the United States.[45]

The Age of Man was pushed back at least into the Ice Age, now named the "Palæolithic" era (also Lubbock's coinage). The new term emphasized the stone tools found in many sites. In the Paleolithic dwelled the "cave-men," yet another Lubbock coinage from this time. Some of those cave dwellers were perhaps even a separate human species, the heavy-browed Neanderthals who had been named for the German river valley in which their fossil remains had been found in the 1850s. Others wondered whether primeval man dwelled even lon-

FIGURE 7.17. Two cavemen aim their weapons at a female Irish Elk as the giant antlered paterfamilias looks on. Benjamin Waterhouse Hawkins, *Irish Elk and Palaeolithic Hunter*.

ger before the Ice Age, perhaps in the Pliocene or Miocene or even earlier. "Where, then, must we look for primæval Man?" wondered Huxley in 1863.[46]

Where indeed? Princeton students could look for primeval man in Hawkins's elk painting, a compendium of the iconography of prehistoric human origins established over the previous decade. It mixed religious tradition (the Adam-like naked or partially clothed men) with new archaeological finds (the cave and stone tools). The full flower of Victorian gender binaries was also on display, as the conquering heroes flex their muscles for the mincing doe. As with his other Princeton paintings, Hawkins recycled the giant elk from his Crystal Park and New York projects, updating the scene with a dawn sky promising humanity's bright future.[47]

Other parts of the museum also revealed Arnold Guyot's fascination with the recent discovery of prehistoric humans in Europe. Atop the giant elk painting Guyot hung the bust of the "perfect man," Apollo Belvedere, suggesting a lineage of European achievement from cavemen to the ancient Greeks. He added human skulls and implements of "prehistoric man" from Europe to the new museum. Eventually he hoped to supplement these with implements and pottery from the archaeology department, representing a linear sequence from the paleolithic forward. Soon the new art museum would open and siphon off the classical statuary and Old Masters, now considered to be "historic art."[48]

The Princeton natural history museum put into sharp relief the ways in which the antiquity of man challenged traditional religious belief. Guyot and Hawkins both accepted the deep antiquity of man, filling the museum with paleolithic hunters and primordial spears. But they categorically refused the descent of man from animal ancestors, the argument pursued by Darwin in his *Descent of Man*. Darwin had systematically approached nearly every aspect of human behavior that had allegedly separated humans from the animals—reason, language, emotions—and argued that they had evolved from animal antecedents. The E. M. Museum would have none of it. Guyot maintained an "impassible abyss" between animals and the spiritual life of man and mocked paleontologists seeking a "missing link" between man and ape. Hawkins likewise mocked the "degrading" hypothesis that humans descended from the beasts, calling upon artists to illustrate Darwin's fallacy.[49]

Their rebuttal to Darwin was the giant dinosaur we met at the beginning of this chapter, the local New Jersey plant-eater *Hadrosaurus foulkii*. Hawkins assembled the beast on a raised dais in the middle of the room, surrounded by his paintings. A replica of the one he had erected in Philadelphia, it telegraphed the same message. God had created bipedalism repeatedly over time, beginning among primitive creatures such as the Hadrosaurus and culminating in Man today. Just like the Princeton undergraduates, the dinosaur walked upright, showing God's hand in nature. Across the abyss of time, students would see that the museum's bipedal Hadrosaurus—not an ape—was preparing the way for bipedal man. The difference between humans and other animals was not to be found in their anatomy, but in their moral nature. "Any length of time" proposed by Darwin was irrelevant, Guyot told the students, for it would never make a monkey into a man.[50]

Princeton in the Nation's Service

What a feast of deep time this was. The Princeton E. M. Museum unfolded a spectacular vision of the long history of life on Earth, fulfilling Guyot's dream of creating a tangible history of Creation. The array of fossils and paintings mixed religion and science into a seamless new package that filled deep time with God's creative energies. Far from being empty of God, as charged by more conservative Christians such as Charles Hodge, the millions and millions of years of Earth history were full of God. From the earliest life forms, bathed in divine rays as they washed from the primordial seas, life progressed until it reached modern Man—the Princeton students now bathed in the light from the skylight of the museum.

Could the E. M. Museum help to restore Princeton to its antebellum glory and once again become the steppingstone for America's national leaders? It was hard to say, but a final gesture seemed to forge a new link between prehistory and national history. As photographs from the time reveal, the Hadrosaurus faced a full-length portrait of George Washington (figs. 7.18 and 7.19). Guided by a white statue of the winged Roman

FIGURE 7.18. In this photograph of the E. M. Museum from the late nineteenth century, the gigantic New Jersey Hadrosaurus looks at a small statue of winged Mercury, who points to a portrait of General George Washington, victorious after the Battle of Princeton. Students might take away the message that New Jersey had a glorious past, whether in the age of dinosaurs or the age of revolution.

messenger Mercury, the dinosaur gazes across the abyss of time directly at the general, who stands victorious after the Battle of Princeton. Whether accidental or purposeful, the message was powerful. To a college seeking its place in a nation rocked by the twin calamities of civil war and Darwinian evolution, the meeting of these two giants from deep time and national time offered hope that Princeton could organize these new and unsettling ideas into a coherent package.

One of the students who had enrolled in those geology classes with Professor Guyot put this hope into words in a speech in 1896. Now a professor of political science at his alma mater, Woodrow Wilson delivered a rousing speech titled "Princeton in the Nation's Service." It was

FIGURE 7.19. Charles Willson Peale's *George Washington at the Battle of Princeton* (1783–84) was positioned in the museum such that the victorious general stared back at the small statue of Mercury and the Hadrosaurus skeleton.

Princeton's duty, set in a free population amid roiling social change, to implant a sense of obligation through lessons drawn from the past, said Wilson. Students should ignore the scientific spirit of the age and its contempt for the past, for even science with all its stupendous achievements "has not freed us from ourselves." The past was the great teacher. Anchored not just in national history but in the very Creation itself, perhaps Princeton could once again serve the nation.[51]

Princeton was just one site for deep time's institutionalization in colleges and universities across the United States. From sporadic mentions before the Civil War, deep time became systemic in higher education afterward, touching thousands of American college students. Courses presumed a new link between prehistory and history; campus museums acquired collections of prehistoric animal and plant fossils; and field trips ferried students to ancient American sites. Through this quiet but steady institutionalization, deep time moved from a private hunch to an everyday reality for millions of Americans. College-educated Americans learned that they could blend Christian belief with an antiquity so incomprehensibly old that it touched the divine mystery. To such a public, prehistory could now be deployed for an even grander agenda: the reform of modern American society itself.

8

The Caveman within Us

BY THE 1920S, deep time had shaped not just the geological age of North America, but what it meant to be a modern American. Humanity's descent from ape-like ancestors, a radical suggestion in Darwin's *Descent of Man* (1871), had gained widespread acceptance in the United States a half century later. Human evolution could still spark controversy, as demonstrated by the Scopes Monkey Trial of 1925. But as the remains of prehistoric humans continued to be unearthed the world over, many Americans pondered not the fact of human evolution, but what it implied for the United States in the Modern Age.

Enter the caveman. Although the term had been coined in the 1860s, it was not until the 1920s that the caveman saturated American popular culture. The caveman (and a growing cast of other stock cave people) became the common ancestor of a nation that lacked clear roots in a primordial era. The discovery of Peking Man in the late 1920s had given the Chinese a remote progenitor who fit the needs of a rising socialist power, just as the discovery of prehistoric paintings in European grottos suggested Continental roots in artistic genius. Americans, however, believed they lacked a suitable precursor. Having rejected the Native peoples of North America as candidates for that role, intellectuals and artists cast about in numerous directions. They seized on the caveman as the public face of scientific discoveries and theories drawn from the fields of archaeology, paleontology, anthropology, sociology, and psychiatry. The caveman offered lessons on everything from ethics to handshakes to treating neurasthenia. Novels, etiquette manuals, magazine articles, paintings, and films converged on the idea that the caveman had something to teach modern Americans because the caveman lurked somewhere in all of them. "The caveman within us," as the title of a popular 1922 self-help book put it, would guide modern Americans to health and happiness.[1]

The American caveman emerged from the nation's fractious past and troubled present. While the half century after the Civil War had ushered in scientific and technological wonders, it had also yielded some bitter fruits: urban jungles teeming with millions of impoverished immigrants from Asia and southern and eastern Europe, skies blackened by soot, a swelling proletariat crushed by greedy plutocrats, rampant consumerism, and a vague sense that the soul itself was in peril, the candle within slowly guttering in the airless void. The drift of the Modern Age called for new anchors. Some found stability in a past civilization: the Renaissance, the Middle Ages, and the imagined genealogy of a "Western Civilization" rooted in ancient Greece. For others, deep time's longer angle of vision—a human prehistory so old that it predated the earliest civilizations by hundreds of thousands and perhaps even millions of years—promised the sturdiest anchor of all. With Christianity's certainties eroded by natural selection and the Eden story displaced by the growing consensus that apes lurked in the distant branches of the human family tree, the caveman led Americans far backward to a time of primal rituals, customs, and dreams. From this new common ancestor, they drew new connections to their own modern bodies and minds, listening closely to the faint whispers from the dawn of humanity.[2]

The Americans who sought the caveman within did not form a unified group. Some were scientists and social reformers; others were writers, painters, and poets; still others were museum directors and patrician observers. Many drew from European ideas, reshaping them to suit the American scene. Sometimes their caveman was a strapping, fur-clad brute. At other times, he was a concept, the symbol of humanity's earliest rituals and myths. With the caveman, they forged a rich conceptual vocabulary flexible enough to address a rainbow of fears and aspirations. Nor was there consensus about what the caveman boded for America. The Enlightenment had invented the idea of a secular future, but without the longer backstory now inhabited by the caveman. The hope now was that amid the anxieties of the Modern Age, the caveman would reveal an essential, unchanging human nature that would surely light the way forward. We all carried within us some primal truths, common to our nature, that if revealed would set modernity on the proper path. That hope, too, proved illusory. The caveman sparked controversy rather than settling it. For some Americans, the caveman promised mastery, a superior ancestor guiding the selection of an improved future population. Others enlisted the cave ancestors as forces of liberation who would heal neurasthenic moderns by unleashing their authentic inner self. A few Americans took a stranger trip, seeing irrationalism in the bizarre squiggles left on cave walls by the earliest humans. The cavemen, it seemed, might be the first Surrealists.

This chapter explores the first half century in which Americans grappled with the new idea that humans, too, might have emerged from deep time. So far in this book, we have seen how nineteenth-century Americans added trilobites, fossil forests, dinosaurs, and even the plodding and dim-witted Uintatherium to their national history. But adding prehistoric humans tested their imaginations further still. The US Constitution had enshrined "We the People" as the nation's supreme law of the land, anterior and superior to all other legislation. But what of "We the *Homo sapiens*," a species whose distant ancestors dissolved into a fog stretching hundreds of thousands of years before the founding document? After 150 years of nationhood, Americans had become accustomed to debating who precisely "We the People" were in a political sense; a series of amendments to the Constitution after the Civil War expanded political belonging more and more. But the preamble to the Constitution said nothing about the preamble to the human species. By the first decades of the twentieth century, Americans were confronting a much wider circle of "We," as humanity's ancestors appeared to extend backward in time to a tree full of ape-like ancestors. Who were we, Americans wondered, now that the nation floated atop a nearly bottomless sea of time?

The Plant within Us

Before the caveman could speak to Americans, prehistory more generally had to be made relevant to modern American society. In the wake of Darwin, thinkers around the Atlantic began to debate whether human society was governed by the same process of natural selection that drove animal and plant evolution. Social scientists cast back into deep time, linking modern society to what they dubbed *ancient society*. American anthropologist Lewis Henry Morgan explained the new mission in his influential book, *Ancient Society* (1877). "Whatever doubts may attend any estimate of a period, the actual duration of which is unknown, the existence of mankind extends backward immeasurably, and loses itself in a vast and profound antiquity. . . . It is both natural and proper to learn, if possible, how all these ages upon ages of past time have been expended by mankind."[3]

Lester Frank Ward was among the first to take up the challenge, linking deep time not only with modern society but specifically with modern American society. Born in Illinois in 1841, Ward is remembered today as the father of sociology in the United States. The first president of the American Sociological Association, Ward embodied the Gilded Age's faith in science and progress. Society was governed by immutable laws akin to the physical laws of nature. Once understood, these laws could be enlisted to guide reforms that would usher the nation into a happier future. This future, like the secular past offered by deep time, was also imagined in secular terms. It was not a millennial second coming, as it had been for so long in the United States under the influence of ideologies such as Manifest Destiny. Instead, the new sociological future was a human-engineered and human-directed project.

In 1883, Ward published *Dynamic Sociology*, one of the first major texts on modern sociological thinking. Against the ruthless "survival of the fittest" policies promoted by British sociologist Herbert Spencer and his American disciple William Graham Sumner, Ward thought people should actively intervene to improve society. He called this *telesis*, human progress through "foresight and intelligent direction." Sociology would be a science of human liberation.[4]

Sociology was just half of Ward's self-imposed brief. He had an equally distinguished second career studying fossil plants. As Ward was unrolling his early sociological theories in the 1880s, he was also quarrying fossilized plants in the American West as an employee of the United States Geological Survey. He traveled across the Missouri River and through Yellowstone under its influential director, John Wesley Powell. Sorting through the stony remains of ancient forests and fields, Ward became convinced that plants were the link to life's common past. Plants could show how "how existing forms have come to be what they are," he explained in his *Sketch of Paleobotany* (1885), based on his fieldwork in the West.[5] Ward gave that title to the whole new field of what he termed *paleobotany*: the study of primordial plants. Unlike animals, which could flee unfavorable climatic conditions, plants had to evolve in place or perish. In a single region, plants revealed a great continuity of life as animals could not. Plants might even be older than animals, Ward proposed. There was much that plants could teach us.[6]

Crucially, Ward linked his two fields of expertise to pose a new question: what was the remote origin of modern American society? Like his hero Charles Darwin, Ward held that all life—including humans—descended from a common ancestor deep in the mists of time. But how deep? While some conservative theologians clung to the older idea of an "impassable" gulf between the plant and animal kingdoms, Darwinian logic held that both kingdoms should evolve from an even more primitive common ancestor.[7] Darwin imagined humans descending like all mammals from some "fish-like animal." Following that logic, perhaps human origins lay even earlier, among the plants. As though to illustrate his own kinship with plants, Ward posed for a photograph in Yellowstone, leaning against a fossil tree (fig. 8.1).[8]

Ward found an ally for his Darwinian views in Gaston de Saporta, an elegant French nobleman and fossil plant expert who lived in the Mediterranean splendor of Aix-en-Provence. Provence functioned in France as the West did in the United States: as a national and cultural periphery with an uneasy relationship to the metropolitan center. Paris's exclusive and hierarchical scientific circles could be hostile to foreign theories, including those of Darwin.[9] Nor were they always welcoming

FIGURE 8.1. As though communing with prehistoric plants, the paleobotanist and sociologist Lester Frank Ward leans against a fossil tree trunk in Yellowstone, c. 1887.

to naturalists from other regions of France, and the marquis de Saporta was part of a circle of naturalists, writers, and avant-garde artists who were creating a regional identity for Provence that was distinct from Paris. Saporta crafted a geological *terroir* for Aix, while others in Saporta's circle pursued local paleontology. Antoine-Fortuné Marion, director of Marseille's natural history museum, found Neolithic human bones and tools in the hills east of Aix, a discovery that fed hopes that a stable Provençal racial type linked the modern-day inhabitants to the prehistory of the region. Local paleontologist Philippe Matheron unearthed Provençal dinosaur bones and even eggs. He christened a large plant-eating dinosaur *Rhabododon priscum* ("ancient fluted-tooth"), seeing the creature as a serious regional rival to the celebrated English Iguanodon, the way Americans were trying to dethrone Iguanodon with New Jersey and Kansas dinosaurs.[10]

As it did elsewhere, the new idea of deep time also provoked visual experiments in Provence—experiments that, as we will see, would be welcomed several decades later in the United States as "modern art." The artist Paul Cézanne, also a member of Saporta's circle, was inspired by the strata-lined cliffs and dinosaur fossils of Provence to revolutionize his painting. Born in Aix in 1839, Cézanne had also been snubbed by Paris and abandoned it in his mid-twenties for home. He was friends with Marion, and together the two young men mixed art and science into a thrilling whole. They even doodled together on the same page, Marion drawing and labeling geological strata, while Cézanne drew human figures.[11]

Inspired by the remote Provençal antiquity that Marion and other paleontologists and geologists were unearthing, Cézanne sought to convey that past with new painting techniques. He turned away from traditional landscape painting, based on focused perspective, instead layering colorful blobs of thick paint to suggest the deep time of the earth in quarries and cliffsides. His painting *La Colinne des Pauvres*, completed in the 1880s, depicted a famous hillside near Aix where Marion had found fossils. Quivering green and blue squiggles were meant to bypass rational thought, provoking instead what Cézanne called "color sensibility," a feeling for time and space (fig. 8.2 and plate 27). The painting would eventually find its way to the first American exhibit of "modern art" at the Armory Show in New York City in 1913. It was then purchased by the Metropolitan Museum of Art at the highest price of any painting in the show. All that lay decades in the future, when Americans had begun to work out for themselves the possibilities for social reform through deep time. But even in the 1880s, at the moment of its creation, the painting and many others like it illustrated the creative and novel interweaving of forms of knowledge and expression that deep time opened.[12]

It was no wonder that amid this scientific and artistic ferment, Lester Frank Ward found such a congenial correspondent in the marquis de

FIGURE 8.2. Inspired by the paleontologists of his native Provence, Paul Cézanne tried to render landscapes of deep time by using squiggles and blobs of colorful paint rather than focused perspective. Paul Cézanne, *View of the Domaine Saint-Joseph [La Colinne des Pauvres]*, 1880s.

Saporta. As the working-class Ward was toiling at two jobs, the marquis sat quietly in the Château de Fonscolombe, his family's noble seat. There he accumulated local evidence to support Darwin's thesis. Darwin murmured encouragements from England. "I am glad to hear that you are at work on your fossil plants," he wrote to Saporta in 1872, "which of late years have afforded so rich a field for discovery." In the 1870s, Ward also began to write to Saporta, launching a transatlantic exchange of fossil plants, letters, photographs, proof plates, and growing esteem that culminated in Ward's visit to Fonscolombe two decades later, in 1894. The fossils of the American West were like the ones Saporta was uncovering in France, Ward explained. He asked the marquis to help him identify "problematic organisms" from the American West that seemed to be neither plants nor animals. Perhaps they hailed from an even earlier world, before the two kingdoms had separated. The fossil exchange bore fruit in Saporta's book, *Le Monde des plantes avant l'apparition de l'homme* (The world of plants before the appearance of man, 1879). Saporta used fossil plants—including plants from North America—to argue for the primordial interdependence of plants and animals over an incalculably long period of time. Life was a grand gradation, Saporta wrote, revealing all the links between the explicit "me" of the human personality and the absolute insensibility of a lichen clinging to a rock.[13]

Ward and Saporta elevated plants to a starring role in the long story of life. They showed that plants lurked in humanity's deep ancestry. Many paleontological finds, wrote Ward in 1883, "establish man as a product of geologic time," no doubt stretching further back than the 200,000 years of the last glacial maximum to possibly millions of years before that. The man of the glacial era was already "developed man," anatomically modern, standing upright with a large brain and "with the power of oral intercommunication—articulate speech," the ligaments of human society. Human society must have its origins in this "dim and hoary antiquity."[14]

Paleobotany had met sociology. Ward linked the evolutionary progress of plants and animals to the social progress of humanity: the two cases were comparable. "Many of them are still in a progressive state," he wrote of nonhuman beings. "As the vegetable kingdom has for millions of years been slowly rising from sea-weed to moss, from moss to fern, from fern to cycad, from that to pine, and so on to the oak and the apple, and as the animal kingdom has, in like manner, continued to exhibit forms of a higher and higher organization from the monad to the man, so it may be expected that both these kingdoms will, for perhaps an indefinite period of time, continue this progressive differentiation." Human beings progressed by working with this natural progress. By selecting which plants and animals would work best, "human agency" had created nourishing maize, wheat, and apples. What if the same human agency were put to work to "improve society"? Working in sociology and paleobotany side by side in the 1880s, Ward had seen that the entire story of life could be "arranged in some sort of connected and ascending series." Darwin and others had supplied a "universal theory of development or evolution." And now, adding his own "high utilitarian motive, focalizing all considerations in the good of man," Ward linked the enormous antiquity of life on Earth and the lowliest mosses and seaweeds to the modern project of improving his own society. Far from being a "sterile" and "dead" system of thought like so many before it, sociology would be dynamic by drawing on the same theory of evolution that united all the other sciences of Ward's era. It would do what all sciences should do: "benefit man." Deep time might be deployed to reform modern society: this was the promise of Ward's "dynamic sociology."[15]

Ward went to work. His first reform target was women, whose subordination had long been justified by hoary traditions such as patriarchal marriage and religion. In his 1888 article, "Our Better Halves," published in the popular magazine *Forum*, and in his later book *Psychic Factors of Civilization* (1893), Ward looked to a future of greater equality for women in American society. He promoted his *gynaecocentric* theory (another of his coinages), locating women's complementarity to man "far back in prehistoric, presocial, and even prehuman times." Ward also offered his own idiosyncratic interpretation of Darwin's branching tree. "Woman is the unchanging trunk of the great genealogic tree," whose qualities were passed on to the future, whereas men were merely a branch whose qualities died with the individual. "Woman *is* the race," he concluded, and women the key to progress.[16]

The American feminist Charlotte Perkins Gilman seized on Ward's vision of woman as the racial future. Calling Ward "the greatest man I have ever known" and his *Forum* article the most important theory since Darwin's, Gilman used Ward's ideas to frame her calls for women's liberation. She kept up to date on new works on prehistory.[17]

Born in Connecticut in 1860, Gilman saw how women's lives were chained by tradition. She launched her long public career in 1890 with a poem about prehistory. Titled "Similar Cases," it was first published in the *Nationalist*. She began with America's primordial horse, little

Eohippus, and brought her readers all the way to the "Anthropoidal Ape"—human beings in the modern United States. Gilman used Neolithic man to satirize the idea that nothing could be done to improve society without altering a basic and unchanging human nature. She unveiled "human nature" as a patriarchal myth that ensured the retrograde status quo:

> There was once the Neolithic Man, an enterprising wight,
> Who made his simple implements unusually bright.
>
> Unusually clever he, unusually brave,
> And he sketched delightful mammoths on the borders of his cave.
>
> To his Neolithic neighbors, who were startled and surprised,
> Said he: "My friends, in course of time, we shall be civilized!
>
> We are going to live in Cities and build churches and make laws!
> We are going to eat three times a day without the natural cause!
>
> We're going to turn life upside-down about a thing called Gold!
> We're going to want the earth and take as much as we can hold!
>
> We're going to wear a pile of stuff outside our proper skins;
> We're going to have Diseases! and Accomplishments!! and Sins!!
>
> Then they all rose up in fury against their boastful friend;
> For prehistoric patience comes quickly to an end.
>
> Said one: "This is chimerical! Utopian! Absurd!"
> Said another: "What a stupid life! Too dull upon my word!"
>
> Cried all: "Before such things can come, you idiotic child,
> *You must alter Human Nature!*" and they all set back and smiled,
>
> Thought they: 'An answer to that last it will be hard to find.'
> It was a clinching argument—to the Neolitic [*sic*] Mind![18]

The poem brought her national recognition. William Dean Howells congratulated her, and Lester Frank Ward called it a "remarkable poem" and "the most telling answer that has ever been made" to those Darwinists who claimed that traditional gender roles were rooted in a natural order.[19]

Gilman also deployed prehistory for reformist ends in her landmark treatise, *Women and Economics* (1898). She diagnosed "Americanitis": the nervous depletion of American women caused by the mismatch between modern industrial life and primitive institutions like the family. Prehistory showed that women's subjection was not natural but arbitrary. A poem opening the book set the stage for a moment of prehistoric equality between men and women:

> In the dark and early ages, through the primal forests faring,
>> Ere the soul came shining into prehistoric night,
> Twofold man was equal; they were comrades dear and daring.

Equality vanished when humans ate the fruit of the Tree of Knowledge. The original sin of humankind was not Eve's transgression, but men's knowledge that they could dominate and enslave women. While poets, priests, artists and novelists might sing the praises of this arrangement, it was up to the sociologist, "from a biological point of view," to show its true perversion for the development of civilization. Adopting Ward's framework—that women embodied the constancy of the race and men its temporary aspects—Gilman urged a realignment of duties for men and women along the more egalitarian principles of primordial times. For Gilman, the deep past opened a promising new vista for a future of social equality.

Prehistoric Art: Men without Chairs

Museums also became didactic temples of deep time. None were more influential than the American Museum of Natural History (AMNH) in New York City under its imperious president, Henry Fairfield Osborn. From his lofty perch, which he held from 1908 until 1935, Osborn created a cultural juggernaut that did more than any other institution of that era to shape the American vision of what prehistory looked like and how it could set a changing nation on the correct path.

Born to the Gilded Age purple in 1857—his father was a railroad tycoon—Osborn despaired of the decadence of modern America. City life, indoor education, feminism, jazz, abstract art, and even women's frivolous hats: all these had sapped the vital energies of the American male. Immigrants from eastern and southern Europe, Chinese workers, and blacks fleeing the Jim Crow South were all diluting the white race.

Cultural institutions such as the AMNH offered hope for the endangered species of Osborn's class. He would bring the wilderness indoors and tame it into instructive tableaux. Entering the hushed galleries, the unruly and untutored masses of the industrial age would be transformed into the educated and discerning citizens of the modern republic. A multimedia blitz of exhibitions, books, magazines, and pamphlets would create "not more but better and finer representatives of every race," Osborn declared. One museum trustee predicted that undesirable European immigrants would be "transformed into Americans."[20]

How to achieve this grand vision for America? A three-week car trip through the Dordogne region of southwestern France in August 1912 inspired Osborn. Coming early in his term as museum director, the tour brought him face to face not just with early man but with the cradle of art itself: France.[21]

This, at least, was the claim of French paleontologists. Beginning in the 1850s, France had begun to generate intriguing findings in human paleontology. Britain, too, was producing spectacular finds, such as Cheddar Man and Piltdown Man (later unveiled as a hoax). But the French situation had an element of political urgency that energized its scientific corps on the global stage. Eager to help restore French national pride after their humiliating defeat in the Franco-Prussian War (1870–1871), French scientists exalted their nation as the nursery of human genius. They pointed to the ochre and red paintings lining limestone caves around the Dordogne River and to the stone tools surfacing in sun-drenched quarries of Provence. "Do not be surprised that the word *art* is used in this circumstance," advised Marion, the Provençal paleontologist. They christened France's prehistoric peoples with the names of the sites in which their remains were found: Cro-Magnon, Magdalenian, Solutrean, Aurignacian. These French terms were soon applied universally to human prehistoric eras. "From the start the study of Prehistory was essentially French," one French paleontologist later observed.[22]

To Osborn, no early humans were more exalted than the Cro-Magnons of the Dordogne region. The Cro-Magnons had been named for a grotto called the Abri de Cro-Magnon (Shelter of Cro-Magnon) near Les Eyzies, a hamlet tucked near the winding Dordogne River. After some workmen in the 1860s had stumbled upon a human skull, flint tools, and the bones of extinct animals, the French geologist and prehistorian Louis Lartet followed up with excavations. In 1868, he discovered skeletons of five prehistoric human beings at the rocky shelter. The rush was on to find some artistic merit in the tools. The Cro-Magnon cave dwellers, according to one of the first English-language publications on the subject, revealed art superior to the "grotesque" artifacts fashioned by North American Indians.[23]

Osborn's journey to the Dordogne convinced him that he had found the prehistoric key to reforming American civilization. Unlike the native peoples of North America, who were mere hunter-gatherers, he told Americans, Cro-Magnons were "the great race of art-loving hunters." Europe's "art of the cave man" was accomplished "just as we would" in the present day, concurred AMNH anthropologist Clark Wissler.[24]

Osborn expanded on Cro-Magnon's meanings for modern America in his book *Men of the Old Stone Age* (1915). Sitting in his French-style château overlooking the Hudson River after his return from France, Osborn made prehistory accessible to Everyman. Although the book was based largely on the findings of others, especially French paleontologists Marcellin Boule and Henri Breuil, its thirteen printings established Osborn as the American authority on Cro-Magnons and other early humans. Teachers around America wrote to the museum asking Osborn to clarify the pronunciation of this strange French term, *Crô-Magnon*.[25]

Osborn's book left them with no confusion about what Cro-Magnons embodied. They were the ancient Greeks of the Paleolithic—actually superior to the Greeks—capable of correct representation, beauty, truth, and a creative spirit. They were the perfect variety of men, rugged artist-hunters energized by the bitter Pleistocene climate. One stared

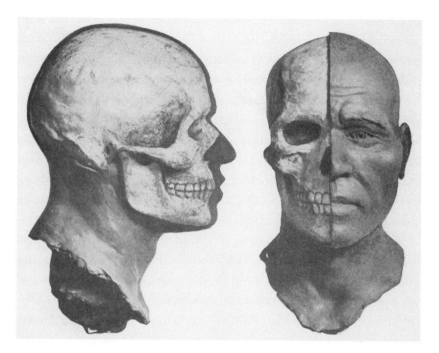

FIGURE 8.3. The head of Cro-Magnon Man, a type of *Homo sapiens* superior in artistic genius even to the ancient Greeks, according to museum director Henry Fairfield Osborn. J. H. McGregor, "Head of the 'Old Man of Cro-Magnon'" (1916).

out from the book, his skull reconstructed by one of Osborn's team (fig. 8.3). It was perfect, exactly like the "highest types" of *Homo sapiens*. The Cro-Magnons' fate held lessons for modern Americans. The decline and fall of Cro-Magnons came as they abandoned nomadism in favor of sedentary agriculture.[26]

Conceived as the art-producing Greeks of the paleolithic, Cro-Magnon Man soon entered the arsenal of the eugenicists. Eugenics—the effort to shape ideal humans through correct breeding—had by now gained a prominent following in American society, especially among elites who feared contamination of a white, Protestant nation by immigrants, blacks, and others. Madison Grant, a trustee of the American Museum of Natural History, unrolled these views in his book, *The Passing of the Great Race* (1916). Patrician in upbringing, bearing, and outlook, Grant was obsessed with purity of lineage. He mourned the numerical decline of the great race of Americans descended from Nordic stock and complained about the increasing number of Jews in the United States. Grant held up Cro-Magnon Man as an example of how a higher race responsible for "the birth of art" could be replaced by an inferior race. Inspired by Cro-Magnons, white Americans should use their superior intelligence to block competition from the lowest races of eastern Europe and western Asia. Osborn's preface to the book urged the preservation of the race that had given the nation "the true spirit of Americanism." The greatest threat to the American republic, Osborn explained, was the dying out of "our religious, political, and social foundations . . . and their insidious replacement by traits of less noble character."[27]

The opening of the museum's new Hall of the Age of Man clarified these connections between Cro-Magnon Man and modern Americans. Funded by Osborn's uncle, the industrial magnate J. P. Morgan, the hall was filled with specimens of early humanity, from mandibles to sculptured busts. With the "Age of Man" stretching to half a million years before, according to Osborn, the museum could showcase the progressive series by physical appearance and "relative intelligence." Osborn advertised the Hall of the Age of Man as the climax of the museum's sequence of life, estimating that the museum had over a million visitors annually.[28]

The hall also featured three murals painted by the artist Charles Knight, produced under Osborn's watchful eye. Working at the museum since the 1890s on paintings of prehistoric reptiles and mammals, the shy and testy painter now turned his attention to prehistoric human beings. Knight did more than anybody else at the time to shape the American public's perception of the appearance, behavior, and environment of prehistoric humans. Osborn's involvement ensured that there was also a message for moderns.

Born in 1874, Knight grew up middle-class Brooklyn, but he too was touched by the titans of industry who dominated the Gilded Age: his father was the private secretary of J. P. Morgan. As a boy, Knight had loved to draw and observe animals at the Central Park Zoo, and then later at the new Bronx Zoo, another cultural jewel built with Gilded Age money. He often visited the American Museum of Natural History.[29]

Knight admired realism in art, the almost literal transcription of the natural world onto canvas. His taste had been shaped by a stay in Paris in 1895 and 1896. He sketched animals at the Jardin des Plantes, roamed the Louvre, the Cluny, and the Luxembourg galleries, and visited old citadels and churches. While in Paris, Knight also met two of his heroes, the sculptor Emmanuel Frémiet, whose Paris zoo statue of a bear fighting a "primitive" man greatly impressed him, and Jean-Léon Gérôme, whose animal paintings and realistic, classicizing style Knight strove to emulate. He was also influenced by one of Gérôme's American pupils, George de Forest Brush, who painted Native Americans in the tradition of the noble savage, and who was Knight's teacher at the Art Students League in New York City.[30]

Knight's realist style was on grand display in his three murals for the Hall of the Age of Man. His paintings of Neanderthals, Cro-Magnons, and Neolithic stag hungers merged realism with the latest science from such luminaries as Henri Breuil of the Institut de Paléontologie Humaine in Paris. Arrayed from most primitive to most advanced, Knight's murals suggested the progressive chain of development through which the public should comprehend prehistoric humans.[31]

In the first mural in the sequence, *Neanderthal Flintworkers* (1920), fur-clad Neanderthals spy a herd of rhinos drinking in the river below (fig. 8.4 and plate 28). Slinking out of his rocky shelter, explained Knight, the "gorrillalike" Neanderthal hunter needed "to kill for his woman and his newborn babe," who cower under the overhang. Knight had followed Osborn's instruction to give them olive skin, like an "Italian laborer." Avoiding the German connection rendered problematic for Americans by World War I, Knight situated the Neanderthals (originally found in Germany) in France's fossil-rich Dordogne River region. Stirred by his passions and determination to fight for life, Knight gave his Neanderthal man rippling biceps, telling readers of *Popular Science* that "This is our ancestor." Modern girls might be taller, Knight went on, but this manly hunter "thrills us," we weaklings of the city.[32]

Most elite Americans rejected any hint of Neanderthal ancestry. Osborn's wife, Lucretia Perry Osborn, assured readers of her popular book *The Chain of Life* (1925) that the Neanderthals had disappeared in the

FIGURE 8.4. Low-browed Neanderthal hunters spy a herd of rhinos in ancient France, as a disheveled mother and child cower in the cave. Charles R. Knight, *Neanderthal Flintworkers, Le Moustier Cavern, Dordogne, France* (1920).

face of superior civilizations, "like some aboriginal races to-day." Sociologist E. A. Ross saw prehistoric traits in the European immigrants arriving in America, whom he described as "hirsute, low-browed, big-faced persons of obviously low mentality." To Ross, a eugenicist, these immigrants clearly "belong in skins, in wattled huts at the close of the Great Ice Age." Some paleontologists hoped to excise Neanderthals altogether from the modern human family. British archaeologist Arthur Keith placed them on an extinct side branch of the human family tree in the frontispiece illustration of his *Antiquity of Man* (1915) (fig. 8.5).[33]

Knight's next canvas in the series, *Cro-Magnon Artists of Southern France* (1920), unveiled humanity's ascent from the ape-like Neanderthals. It hung above a doorway, flanked by Knight's paintings of woolly mammoths (fig. 8.6). Dwarfed by the giant reconstructed mammoth skeletons, visitors were plunged into the frigid depths of Ice Age Europe. Knight's painting contrasted the "highly evolved" race of Cro-Magnons with the brutish Neanderthals (fig. 8.7 and plate 29). The Cro-Magnons had migrated west from Asia—the prevailing theory of human origins—and driven out the inferior Neanderthals. Here was "a race of warriors, of hunters, of painters and sculptors far superior to any of their predecessors," Osborn explained, whose contrast with the preceding Neanderthals was "as wide as it possibly could be." The crania of Cro-Magnons reflected the fact that they were "people like ourselves in point of evolution." Cro-Magnon Man had "a modern look," one of his museum colleagues concurred, and "could walk our streets unremarked."[34]

Knight's Cro-Magnons line up in a flat pyramid, as though on the Parthenon frieze. A "magnificent" man, tall and muscular, anchors the center of the painting, his hairless skin a marmoreal white. He sweeps colors and shapes over the cave wall, displaying the artist's gift of insight into the world. Knight yearned for connection with these noble ancestors, and his own painterly gestures recapitulated those of the Cro-Magnon painter. This ancient human was not just a technician: he was an artist, seeing into the very soul of things. He had a personality, a whole psychology, that lifted his woolly pachyderms into the realm of art. It seemed to Knight that there was a "magical something" in man's

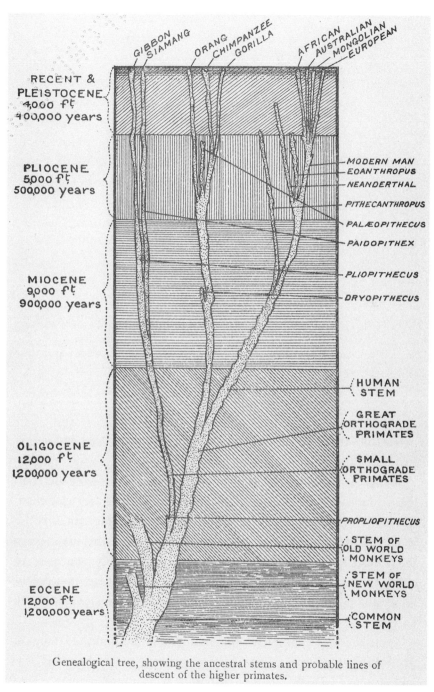

Genealogical tree, showing the ancestral stems and probable lines of descent of the higher primates.

FIGURE 8.5. British anthropologist Arthur Keith excised Neanderthals from the branch leading to modern humans in the frontispiece to his book *The Antiquity of Man* (1915).

The sketch by W. M. Berger shows a section of the Pleistocene Hall with the actual skeletons of mammoths and mastodons; and above them a glimpse of two of the large murals in position on the wall.

FIGURE 8.6. Knight's painting of Cro-Magnon hangs above the doorway in this sketch of Pleistocene Hall at the American Museum of Natural History, published in the popular *Scribner's Magazine* in 1922.

personality that transcended the abyss of time separating prehistoric Cro-Magnon from modern man.[35]

In this Cro-Magnon atelier lay the infancy of art-valorizing societies. Crouching assistants mix colors and hold lamps aloft as a fur-clad chieftain looks on, clutching his bony scepter as the sign of his rank. Like the Renaissance patrons who commanded their artists to include them in the tableau, Osborn himself instructed Knight, embodying a vision of elite patronage for the arts in America. The candlelit cave was a paleolithic Athens, a primeval Florence that could reform decadent New York City.[36]

Though artistic, these Cro-Magnon men embodied the strapping manhood lacking among America's wan urbanites. Osborn issued clear instructions to Knight. "[E]ach man must be in a pose natural to wild men, without chairs, who are accustomed to damp or stony ground, and therefore squat, or kneel on a rough piece of skin." Knight should "Omit women from the Cro-Magnon picture entirely, and do not make the poses too classical or artistic."[37]

The links between Cro-Magnons and the American ideal were reinforced a few years later, when the Hall of Man began to display Osborn's diagram of the "Family of Man" (fig. 8.8). The tree placed Cro-Magnons and white people on a separate evolutionary branch from all other humans.[38] According to this representation, both human types had emerged comparatively recently, splitting long ago from other lineages such as the distant "Negro." Osborn's printed guide to the Hall of Man called Cro-Magnon Man the direct ancestor of "An American, Representing the Caucasian group." Cro-Magnons, Osborn argued, were "the educated men of their day." His wife, Lucretia Osborn, drew out the implications for her own era. "The modern world existed potentially in the development of that brain."[39]

Absent from Henry Fairfield Osborn's family tree of man were American Indians, and he discounted any relationship they had to the modern United States. His response to the discovery of "Nebraska Man" in 1922 was indicative of his stance. That year, Osborn received a single, fossilized tooth from a colleague working in Nebraska. Turning it over in his sunlit window in New York, Osborn dubbed it *Hesperopithecus*: the ape of the western world. In the journal *Science* he declared it "the *first anthropoid ape of America*," nearer to men than the apes. Darwin had held that humans evolved from Old World apes. But Hesperopithecus suggested that the New World might also have a role to play in the story of human evolution. It, too, had a large, upright ancient ape, perhaps one that had wandered to America from Asia along a belt of open grasslands and forests during the Pliocene. In other words, Hesperopithecus might be ancestral to mankind, with the New World playing a starring role in the long drama of human evolution.

News of this "Earliest Man" soon crossed the Atlantic. The "Prehistoric Columbus" appeared on the front page of the *Illustrated London News* in 1922 as a club-wielding brute whose ape-like head sat atop a modern human body (fig. 8.9). Nebraska Man was so ancient that he shared North America's windswept plains with the native rhinos, camels and horses of the Pliocene period that had since gone extinct there.

FIGURE 8.7. A white-skinned Cro-Magnon artist sweeps paint over the cave walls of a Renaissance-like atelier in Charles Robert Knight's *Cro-Magnon Artists of Southern France* (1920).

"The antiquity of this earlier genus [of Pliocene humanity] must have been enormous," the article announced.[40]

Based on this single tooth, Osborn declared that Nebraska Man had evolved into North American Indians, who had failed to improve on their ancestors' primitive dentition and probably much else. They were

living fossils. Lucretia Osborn's popular book confirmed that "no aboriginal American tribe ever reached the iron stage," the last stage in the sequence of prehistoric human development. In this way, the Osborns severed any physical or cultural connection between Nebraska Man and modern American society. The ancestry question seemed settled, at least to the Osborns. By the end of the 1920s, the tooth had been identified as belonging to a pig-like ungulate, and the matter of Hesperopithecus faded away.[41]

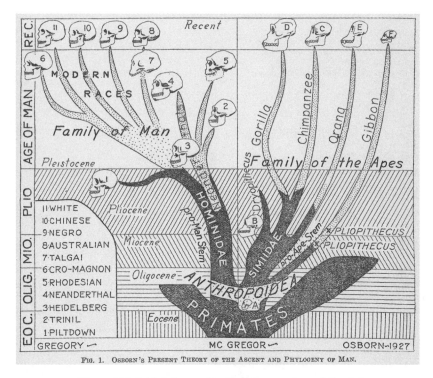

FIG. 1. OSBORN'S PRESENT THEORY OF THE ASCENT AND PHYLOGENY OF MAN.

FIGURE 8.8. Henry Fairfield Osborn placed white people and Cro-Magnon man (#6 and #11) on the same branch of his human family tree to highlight their imagined kinship, relegating other "modern races" to different branches.

By the 1920s, the cultural elites at the American Museum of Natural History had connected a select group of prehistoric humans to a reforming vision for the United States. Fleshing out the paltry fossil record with wishful thinking, patricians such as Henry Fairfield Osborn linked themselves to a new vision of what they now termed "Old World culture." This was not Old World culture in the sense of ancient Greeks or the cathedral at Chartres. It was far older than that, stretching to the damp grottoes of Cro-Magnon Man, lit by fires long since dead. "Old World chronology," one museum employee wrote of this prehistoric past, was "now fully recognized as the foundation to history and the comprehension of civilization." Knowledge of their artistically gifted Old World ancestors, the Cro-Magnons, would rescue white Americans from cultural decadence and spur them to a higher path.[42]

FIGURE 8.9. Nebraska Man, or Hesperopithecus—the "ape-man of the western world"—was depicted in the 1922 *Illustrated London News* as a club-wielding human with an ape-like head. The artist, Amédée Forestier, imagined Nebraska Man living on the wind-swept plains of North America when native rhinos, horses, and camels still wandered the land.

All Paths Lead into Darkness

As the cavemen settled into museum displays, they also infiltrated the American psyche. Riding on the rising authority of psychiatry in the early twentieth century, the technique of psychoanalysis was popularized in the United States chiefly through the theories of Austrian psychoanalyst Dr. Sigmund Freud and Swiss psychoanalyst Dr. Carl Jung. Both visited the United States several times, and their works, translated into English and widely reviewed in the press, forever changed American society. By the time Freud died in 1939, the poet W. H. Auden had declared the doctor no longer a person but "a whole climate of opinion."[43]

We are strangers to ourselves, announced the psychoanalysts. Psychoanalysts scrapped the knowable, rational mind of the Victorians, its surfaces precisely mapped with moral codes such as duty and amativeness. In its place they erected the stratigraphic mind of id, ego, and super-ego. These were not divisions of the brain but rather aspects of a single mind at war with itself. The primitive id harbored forbidden urges and desires, safely hidden in the unconscious mind by various mechanisms such as repression. This unconscious mind was not some lesser version of the "real" outside world or the "real" conscious mind, they cautioned. Freud in fact declared the unconscious mind to be the "real psychic," the larger circle that included within itself the conscious. But sometimes the thoughts and drives of the unconscious mind intruded into the conscious mind, disguised as baffling dreams, inexplicable conflicts, destructive habits, and unsatisfying goals. Untreated, these could lead to neuroticism, the mismatch between the true inner self and the demands of modern civilization, which sickened moderns by inhibiting the instinctual drives that had guided primal man. The role of the analyst was to unveil these dark longings and urges to the conscious mind, where they were woven into a tapestry of meanings. Insight and self-awareness—not the renunciation of sin—would be modernity's guides. This was the gift of primordial humans to their anxious children.[44]

Psychoanalysts used archaeological metaphors to describe their new mind and their new science. Freud's "Aetiology of Hysteria" (1896) compared the psychoanalytic process to unearthing the fragments of a ruined ancient city whose silent stones were made to speak. A few years later, his *Interpretation of Dreams* (1899) likened modern dreams to the bow and arrow, "the discarded primitive weapons of grown-up humanity." Freud's theory of the Oedipus complex—the allegedly universal desire of men to kill their father—also drew on the primordial world. When modern men recognized that their destructive habits and behaviors were rooted in this ancient story, they could heal their minds. The turn of psychoanalysis was not just inward from society to self, but backward from modernity to antiquity. "All paths lead into darkness," Freud decreed.[45]

Freud's archaeological metaphors were not unique to psychoanalysts. During the late nineteenth century, many physicians in America and

Europe described the brain and nervous system as the product of Darwinian evolution. They had evolved during "the abyss of planetary time," according to physiologist George Romanes in his *Mental Evolution in Man* (1888). British neurologist J. Hughlings Jackson concurred that the nervous system formed an "evolutionary hierarchy." The brain was made of nervous strata that ascended from the lowest and least complex (such as the spinal cord) to the highest and most complex regions where consciousness reigned: "the anatomical substrata of consciousness." These views were warmly received by American psychologists eager to eradicate metaphysical speculation from psychology and to put this new science on an empirical footing. Like Jackson, whose nervous-stratum model of the brain he cited approvingly in his *Principles of Psychology* (1890), Harvard psychologist William James lobbied for a "strictly positivistic" rather than metaphysical view of the brain, calling the brain the "mechanical substratum" of thought.[46]

In contrast with the somatic emphasis of many physicians, Freud was convinced that the mind—unseen, internal, fugitive—was "the true psychic reality," as he put it in *The Interpretation of Dreams*. This liberated Freud and his followers to apply stratigraphic metaphors to the mind rather than only the brain. Like the brain, announced Freud, the mind was the product of evolution, with unconscious forces contributing to the survival of the human species. Freud's insistence on the role of deep time's evolutionary processes in the mind—and not just the brain—broadened the interpretive field to allow for fundamental questions of self, identity, reason, and morals. All paths might lead into darkness, Freud had cautioned. But the reward for the intrepid traveler on those dark roads leading to the inner caveman was psychic healing. An understanding of the dreams of the caveman within us freed anguished modern people to forge new paths into the light of self-understanding.[47]

Among the first major popularizers of psychoanalysis in the United States was Dr. Beatrice Hinkle. Just as Charlotte Perkins Gilman had turned to the Neolithic past to criticize modern patriarchy in 1890, now over two decades later, Hinkle took up Gilman's unfinished business, using psychoanalysis to condemn the failures of the Modern Age and especially women's subjection since "primitive times." Born in San

Francisco in 1874, Hinkle trained as a physician, then moved to New York City in 1905, where she joined the Heterodoxy Club, a feminist debating group founded in 1912. Hinkle was attracted to the many new ideas swirling at this time, including Christian Science, hypnotism, New Thought, and the sociological theories of Lester Frank Ward. This rainbow of interests was typical for this era, when new medical conceptions of the human mind mingled with the traditional Christian emphasis on the mind as the primary site where God met man. Hinkle joined the staff of leading neurologist Charles Dana at Cornell Medical College, and they opened one of America's first psychotherapy clinics. By 1909 she had become aware that "several German physicians have devised methods for bringing up from the depths of the patients' mind circumstances and incidents forgotten by them at the present time." She left for Vienna that year to study psychoanalysis, meeting both Sigmund Freud and Carl Jung, who was only a year her junior. During Jung's trip to the United States in 1912–1913, Hinkle introduced the Swiss analyst to her progressive New York City reformist clubs, such as the Liberal Club. She accompanied him to the Armory Show, the first showing of so-called modern art in the United States—the very exhibition that displayed Cézanne's painting of the primordial hills of France.[48]

In 1916, her English translation of Carl Jung's *Wandlungen und Symbole der Libido* (1912) appeared as *Psychology of the Unconscious*. The work in which Jung definitively broke from Freud, it was also the first translation of Jung's theories for American readers. Hinkle's long introduction, over fifty pages in length, doubled as a primer on general psychoanalysis for Americans. She explained the distinctive terminology that eventually permeated Americans' everyday speech, such as the unconscious, the ego, repression, transference, fixation, and the libido. She offered psychoanalysis as a scientific way to heal the soul, previously the province of religion and metaphysics. "Psychoanalysis is the name given to the method developed for reaching down into the hidden depths of the individual to bring to light the underlying motive and determinants" of symptoms, behaviors, and attitudes, she wrote. Hinkle's *Psychology of the Unconscious* launched her long career of disseminating psychoanalytic theories in books such as *The Re-Creating of the*

Individual (1923) and articles for general-interest magazines such as the *Nation* and *Harper's* in the 1920s and 1930s.[49]

As her translation of Jung revealed, Hinkle had rapidly become disillusioned with Freud's male-centered perspective and terminology. She was swayed instead by Jung's focus on the mother and his more generous attitude toward women. "Jung's development of this point of view shows very clearly that, just as the problem of the father is the great fact of Freud's psychology, the problem of the mother is the essence of Jung's, with the struggle carried on between the two great forces of love and power." The Oedipus complex, penis envy, the allegedly "dark continent" of women's sexuality: these Freudian ideas convinced Hinkle that "even the scientific man is under the sway of the same old wish which conceived of woman as created out of Adam's rib." By contrast, Jung's ideas applied more readily to everybody. Everyone shared "a common denominator of humanity" in the dark abyss of time, she explained. She rejected Freud's sex-centered libido theory in favor of Jung's view that the libido was energy for life, comparable to French philosopher Henri Bergson's *élan vital*. Nor did the Oedipus complex involve any real sexual desire by a boy for his mother, Hinkle explained. Instead, the child's attachment to the mother had to do with her role as protector and feeder. Finally, Hinkle unveiled Jung's concept of the "archetypes," the primordial myths common to all humans since the dawn of the species. Troubled moderns could gain access to the archetypes through the "collective unconscious," which was "the common matrix belonging to all human beings at birth," a "primitive, unorganized" psychological inheritance of "instincts and impulses."[50]

In the *Nation* and *Harper's*, Hinkle used Jungian psychoanalysis to criticize Progressive-era feminism. Political and economic equality offered outward freedom, but left the inner person shackled. Modern American women clung to an "unthinking acceptance of the masculine point of view, even about themselves." Nothing was more tragic than college-educated women who, suffering from neurotic ailments, allowed their husbands and fathers to explain to them why they were restless and miserable. What was needed in modern society was a version of feminism that recognized that the inner creation of a new person was as important as the outer struggle against society and politics. Feminism was not criticism but creativity, she argued. "For behind their stridency and revolt lies the great inner meaning of woman's struggle with the forces of convention and inertia. This is nothing less than the psychological development of themselves as individuals, in contradistinction to the collective destiny that has exclusively dominated their lives." Citing Lester Frank Ward, she argued that women were originators of "all the primitive arts." Their greatest gift to humanity was maternal love, "the basis of love for others and of altruism." Everything that men knew of love had been awakened in them by women.[51]

Both Jung and Hinkle used archaeological and geological imagery to explain the primordial past of mythical archetypes from which moderns drew. In the journal he freely shared with his students, Jung painted a huge eye buried under rocks at the bottom of the ocean. Many myths, he believed, involved a single-eyed creature, which symbolized anything from the all-seeing to a lack of perspective (fig. 8.10).[52] Hinkle illustrated the collective maternal power with a drawing made by a woman deploying "that oldest form of expression, picture drawing" to channel the "archaic" ideas bubbling up from her unconscious. The woman sketched a many-breasted, many-tentacled octopus, a symbolic representation of the mother-power, according to Hinkle (fig. 8.11). Deep in the bowels of Earth lay a female creature whose nourishing breasts and umbilical cords reach to the modern surface of the planet. Hinkle explained: "It is easy to read from this the story of the most ancient struggle—that of the bondage to the mother who is here represented in impersonal and powerful form with tentacles like umbilical cords binding the children of men to her."[53]

Hinkle's popularization of psychoanalysis was far-reaching—and for some Americans, transformative. Jack London recalled reading Hinkle's translation of Jung in the summer of 1916. It left him "standing on the edge of a world so new, so terrible, so wonderful, that I am almost afraid to look over into it." Why, London thought, "if he could learn to analyze the secret soul-stuff of the individual and bring it up into the light of foreconsciousness, could not he analyze the soul of the race, back and back, ever farther into the shadows, to its murky beginnings?"[54]

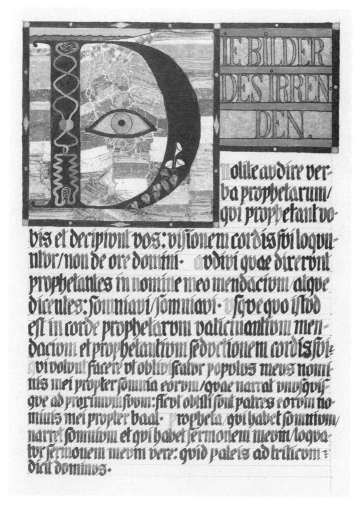

FIGURE 8.10. The Swiss psychoanalyst Dr. Carl Jung drew a large eye buried deep underground, symbolizing a primordial human myth to which introspection would give us access.

Other Americans encountered the primordial abstractions of the psychoanalysts through popular books, magazines, and films. Books for grade school teachers such as Wilfrid Lay's *The Child's Unconscious Mind* (1919) unveiled the "hinterland of the mind, whose geological strata have been laid down during . . . eons of time." Sexologist William J. Fielding's *The Caveman within Us* (1922) deployed Jung and Freud to

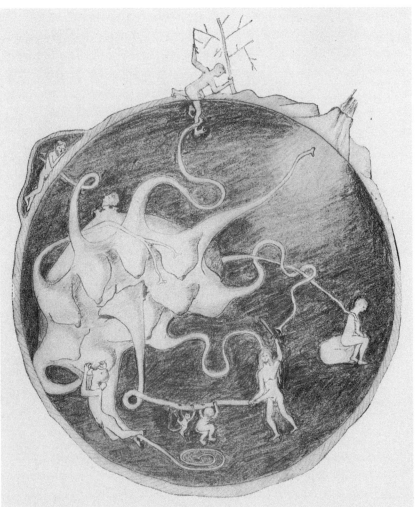

FIG. 1. Representation of collective maternal power: merely breasts and umbilical cords. Note the cord on which the dæmons representing the pair of opposites, good and evil, disport themselves.

FIGURE 8.11. An unnamed woman sketched this many-breasted and many-tentacled underground octopus, which the American psychoanalyst Dr. Beatrice Hinkle interpreted as the symbolic representation of the primordial mother power in her book, *The Re-Creating of the Individual* (1923).

argue that nervous ailments were caused by "a struggle between the Caveman and the Socialized Being." Although a hundred thousand years separated their existence, humanity's "great primitive heritage" was "a dominant influence on the present." Americans also learned that modern customs such as courtship and marriage could be traced to their caveman ancestors. The silent film *Three Ages* (1923), starring Buster Keaton, jumped forward through time from the Stone Age to ancient Rome to the Modern Age of the 1920s. In the Stone Age, a fur-clad cave maiden is claimed by rival suitors, one who arrives on the back of a brontosaurus, the other astride a mammoth. "From the beginning of time parents have found different methods of choosing their daughter's mate," a placard explains. The advice columnist Lillian Eichler's *The Customs of Mankind* (1924) extended the caveman into the modern etiquette manual. From half a million years ago came the modern wedding dress and table manners, she explained. Visiting cards were the modern descendants of carved rocks placed at cave entrances. The handshake developed from a caveman's dropped club, his hand extended as a symbol of peace (fig. 8.12). Eichler summed up the ethos of her era: "Thus do our savage ancestors slumber within us."[55]

Slumbering within us, the caveman in his murky strangeness became crucial to artistic modernism as it developed in the first three decades of twentieth-century America. Despite the labels "modern" and "contemporary," such art was not severed from the past. Instead, many modernist painters and sculptors responded to earlier eras in fresh and challenging ways. Some were stirred by the psychoanalytic project of excavating humanity's primordial rituals and ideas from the unconscious mind. Modern art museums in turn enlisted the caveman for new roles. Shunning the didactic, linear evolutionary sequences that now dominated natural history museums like Osborn's AMNH, they used the caveman to liberate Americans for lives of moral autonomy. Like the psychoanalysts, they imagined the caveman speaking to moderns through abstract symbols that left interpretation to each individual. In 1935, Auden clarified the distinction between the older didactic mode and the new psychoanalytic view of art. "The task of psychology, or art for that matter, is not to tell people how to behave, but by drawing their

In primitive life it was a symbol of peace to drop the weapon and extend the hand—unarmed. From this crude beginning developed our handclasp.

FIGURE 8.12. According to author Lillian Eichler, the handshake originated with the caveman, who dropped his club to shake hands with another caveman. *The Customs of Mankind* (1924).

attention to what the impersonal unconscious is trying to tell them, and by increasing their knowledge of good and evil, to render them better able to choose, to become increasingly morally responsible for their destiny."[56]

New York's Museum of Modern Art (MoMA), founded in 1929, pioneered the use of human prehistory as modern art. The exhibition "Prehistoric Rock Pictures in Europe and Africa" (1937) offered disjuncture, mystery, abstraction, and fragmentation in place of the linear progression and visual realism of Osborn's diagrams and Knight's paintings. Many of the MoMA's photographs of rock art from across Europe and Africa showed humanoid figures, animal shapes, and geometric designs scattered in abstract space. A photograph of a rock painting from Southern Rhodesia (now Zimbabwe) typified the abstraction of the MoMA exhibition (fig. 8.13). The organizers insisted on the value of this prehistoric art in evoking "antediluvian first things." These first things functioned at a spiritual level hovering beyond the mere social and economic utility of Depression-era government art projects. MoMA director Alfred H. Barr Jr. explained. "We can, as modern men, no longer believe in the magic efficacy of these rock paintings; but there is about them a deeper and more general magic quite beyond their beauty as works of art or their value as anthropological documents." The encounter with prehistoric rock pictures was emotional rather than intellectual, reviving spirits depleted by the banality of modern life. "We modern Europeans," wrote Leo Frobenius, the original German excavator of the artifacts, "concentrating on the newspaper and on that which happens from one day to the next, have lost the ability to think in large dimensions. We need a change in *Lebensgefühl*, or feeling for life."[57]

The museum announced what viewers of the exhibition had probably already noticed: that the pictures bore a striking resemblance to works by contemporary painters such as Paul Klee, Jean Arp, and Joan Miró. Surrealists, Dadaists, and other Modernist painters had abandoned the positive, integrating artistic vision of the turn of the twentieth century. The squiggles and mysterious fragments of prehistoric art entered the Modernist canon, aided by museum officials. Initiated by Cézanne half a century ago in the hills of southern France, the links

FIGURE 8.13. A reproduction of this ancient rock painting from Southern Rhodesia (modern-day Zimbabwe) appeared in the Museum of Modern Art exhibition *Prehistoric Rock Pictures in Europe and Africa* (1937).

FIGURE 8.14. The artist André Masson said the razor-toothed fish and "irrational" forms in his painting *Battle of Fishes* (1926) arose from his subconscious mind.

between deep time and Modern art were becoming normalized for American audiences. MoMA director Alfred Barr connected prehistoric art and Modernism by reusing some the same paintings by Wassily Kandinsky, André Masson, Jean Arp, and Paul Klee for its 1936 exhibition "Fantastic Art, Dada, Surrealism" as he did for the prehistoric rock painting exhibition of the following year. The thematic intersections between prehistory and Modernism were explored in Masson's *Battle of Fishes* (1926) (fig. 8.14 and plate 30). The razor-toothed fish first popularized by Louis Agassiz a century before in his primordial Age of Fish were once again locked in battle. Masson threw sand randomly onto to the canvas to suggest "irrational" forms channeled from his unconscious mind. Tongue in cheek, one newspaper challenged readers to tell the difference between Paul Klee's *Small Experimental Machine* (1921) and prehistoric rock paintings from the African desert. "First Surrealists were Cavemen," announced a review of MoMA's exhibit.[58]

This was all quite dizzying. By the third decade of the twentieth century, deep time had become normalized in American society. For some Americans, deep time opened a new future of freedom and possibility; for others, it offered the comfort of a pedigreed ancestor; for still others, deep time flung reality itself out the window. Any hope that the caveman might yield consensus on how to live in modern society by unveiling a stable human nature was gone. The caveman pointed in so many directions, from feminism to eugenics to surrealism, that the hunt for human nature remained one of the enduring quests of twentieth-century science.[59] Humanity's deep time created more uncertainty rather than less.

Some Americans found this situation intolerable. The rise of deep time had destabilized so many traditional certainties—too many, in their view. The origin of the Earth, the creation of life, and the place of humanity in the cosmic order: all these had been swept away by the rise of deep time and its triumph in the early twentieth century. As we will see in the next chapter, refugees from modernism took shelter in a new Eden, one they created specially to oppose deep time and everything it stood for. Actually, not quite everything: as enchanted by the dinosaurs as the most hardcore Darwinians, the Young Earth Creationists filled their new Eden with everything from T-Rexes to pterodactyls.

9

Pterodactyls in Eden

"WE CANNOT BE certain that Adam regularly interacted with dinosaurs or pterodactyls in Eden, but it is not an unreasonable conclusion in a biblical worldview."[1] So writes a supporter of the Creation Museum near Cincinnati, built in 2007. Hewing to a roughly six-thousand-year-old Earth, the museum shows dinosaurs inhabiting the Garden of Eden alongside Adam and Eve (and not eating the First Parents because the first dinosaurs, being innocent of sin, were vegetarians). Outside, an enormous model of Noah's ark houses dinosaurs.

The Creation Museum is one product of a century of evangelical Protestant institution-building that has erected colleges and universities, Bible institutes, websites, publishing houses, books, journals, and museums. Even as geology and paleontology flourished as mainstream secular science in the twentieth century, adding further eons to the history of the planet and life upon it, some evangelical Christians denied Darwinian evolution and the billion-years timespan over which it operated. They pushed the short chronology as a central axiom of their belief system and rejected the higher biblical criticism that had emerged in the nineteenth century, which in treating the Bible as a historical text had exposed inconsistencies, redundancies, obscurities, and numerous linguistic complexities.

One group of Protestants, known since the 1980s as Young Earth Creationists, urge a "biblical worldview" and "literal" interpretation of the Bible that sets the date of Earth's creation around six thousand years ago. The New World having become old, they assert the newness not

FIGURE 9.1. A tiny ark floats in the distance as the waters of the biblical flood drain away. A human skull is visible in the foreground, the vestige of the sinful humanity eradicated by the deluge. Thomas Cole, *The Subsiding of the Waters of the Deluge* (1829).

just of the New World but of the whole planet. God created Earth around six thousand years ago, they contend. Then came a globe-drowning flood, with a tiny band of survivors huddling for safety aboard Noah's ark until finally the waters receded to reveal a cleansed Earth. The story was familiar not only from the Bible but from nineteenth-century American paintings, such as Thomas Cole's *Subsiding of the Waters of the Deluge* (1829) and Edward Hicks's *Noah's Ark* (1846) (fig. 9.1 and plate 31; fig. 9.2).

Much has been written on American Protestantism and its relationship to Creationist viewpoints of various stripes. Here, we need only look briefly at how Young Earth Creationism collides with the story of America's deep time we have been telling here. Although it has found a

FIGURE 9.2. As storm clouds gather, animals familiar to modern people board Noah's ark. Edward Hicks, *Noah's Ark* (1846).

foothold abroad, Young Earthism is a homegrown product of the United States, seeded in the hothouse of conservative Protestantism in the late nineteenth century and growing ever since. The Young Earth viewpoint remains more popular in the United States than elsewhere. A Gallup poll in 2012 estimated that 46 percent of Americans believe that God created Earth and humans in their present form around six to ten thousand years ago, with that view being stable since the early 1980s.[2]

Denials of evolution by natural selection did not have to accompany a denial of deep time, as we have seen. Those who created Princeton's natural history museum exhibits in the 1870s told a glorious story of an ancient Earth progressing from trilobites to cavemen over a span of millions of years—all superintended not by natural selection but by God himself reaching into Earth's history with successive acts of creation and destruction.

But for some Americans, denying evolution also meant denying deep time. As deep time ripened into a mainstream idea among Americans in the late nineteenth and early twentieth centuries, short chronologists found their moment. Inventing a kaleidoscopic blend of bizarre geology and shallow history, they burst into real visibility in the early twentieth century. Never a single outlook, Young Earthism is a collage of shifting positions resulting from a close monitoring and mirroring of mainstream scientific, secular culture in the United States. Young Earthism has also gained an institutional foothold through the shrewd deployment of tax breaks and age-specific marketing plans.

At bottom, deep time denial arose from a crisis in authority. By the early twentieth century, deep time implied to some Americans that truth itself was not an absolute. Did we simply float in a directionless ether of naturalism, bereft of meaning and purpose? Did we even create ourselves out of our own internal psychological resources, as some of the psychoanalysts seemed to be suggesting? To this crisis of authority, deniers of deep time responded by asserting a short chronology as a bulwark against modernity's uncertainty, subjectivity, and moral relativism. Defense of the short chronology was not a field of inquiry but an epistemological precondition. Against the limitless flux and open-endedness implied by deep time, short chronologists urged limits, fixity, and enclosure. While deep timers confronted modernity by seeking new methods and guides in the murky abyss of time, deniers took shelter from modernity in the four corners of the Bible.[3]

The short chronologists began by joining some late-nineteenth century conservative Protestants in rejecting aspects of the German higher biblical criticism, which acknowledged the Bible as a historical text, created by many human hands over hundreds of years. Instead, they insisted on what they called an inspired or literal reading of the Bible that steered clear of what they scorned as interpretation. But the short chronologists took an extreme view. By the 1890s, even some conservative theologians were calling for a repudiation of James Ussher's 4004 BC Creation date (first proposed in 1650), since it could not be extracted from the data of the Bible, which was vague and incomplete. "The Scriptures furnish no data for a chronological computation prior to the life of Abraham," wrote William H. Green of the Princeton Theological Seminary in 1890, as a colleague condemned the ongoing printing of Ussher's chronology by Bible houses and tract societies. In the face of such doubts about Ussher's chronology, some fundamentalists gripped the 4004 date ever more tightly, bringing it into the twentieth century. The new *Scofield Reference Bible*, first published in 1909, nipped temporal interpretation in the bud by hanging the famous Ussher Creation date of "B.C. 4004" above the first chapter of Genesis. At the very moment that Yale scientist Bertram Boltwood was finding American rocks more than two billion years old, the Scofield Bible offered a tiny fraction of that new number. The events of Genesis covered a period of 2,315 years, Scofield assured his readers.[4]

The short chronologists' departure from deep time is clearly seen in the ways they imagined the human mind. For it was here that advocates of deep time had made some of their most unconventional proposals. Twentieth-century geologists continued to build on the idea of earlier naturalists such as Charles Lyell and Charles Darwin, who had bundled deep time with cognition, marveling at the way long time spans of millions of years challenged the human mind. They found deep time fascinating not just as the enormous stage for Earth's long history, but for what it suggested about thinking itself. Why were our brains, which had evolved over deep time, unable to grasp this large quantity of time? "The imagination in vain endeavours to grasp" it, Lyell said of the age of the planet. Darwin agreed that Earth's history was of "a length quite incomprehensible by us." Lyell's geology in fact hinged on human cognition. What was known about geology, he argued, paled in comparison to what remained to be learned. Knower and known were linked. The present—the topmost stratum of the geological column—was the thinking human agent, who retrieved the fossil record for analysis. As that record receded in time, it became less and less known to human observers. More and more fossils were lost as the eons rolled by, such that modern people could know almost nothing of the earliest moments of Earth's creation. The intriguing unknowability of a near-infinitude of time transposed onto the natural world the mystery and infinitude traditionally associated with the deity. As American naturalist David Dale Owen wrote in 1846, geology was the most "sublime" of the natural sciences because it made people aware of "remote periods" and "by-gone existences" in the "dark abyss" of Earth's antiquity.[5] This transference of awe-inspiring unknowability from God to the ancient lands of the United States had helped to justify America's national parks. These secular Edens cleansed the human mind and spirit with their billion-year-old landscapes.

For short chronologists, all this was a bridge too far. The knower and the known as overlapping entities? This was a nonstarter. For short chronologists, Lyell's view that the powers of the mind somehow mapped onto the very objects perceived by the mind—the strata revealing Earth's long history—failed to acknowledge that all of Earth's history was plainly revealed to the mind in the Bible. Lyell's bizarre cognitive proposals were just as intolerable as Darwin's theory that human beings had shared a common ancestor with the apes. And Jung's psychoanalytic theories about retrieving our common caveman unconscious through dreams about monsters and sex? These were dead on arrival in Young Earth circles.

Short chronologists departed from these disturbing ideas about cognition in several ways. First, they denied that the Bible was a historical text. In their view, the Bible did not record the idiosyncratic stamp of human minds from long ago, minds that bore the imprint of their historical time and place. Instead, the Bible was the product of a single, unchanging, unified Intelligence. George McCready Price, a Canadian-born, self-styled geologist, was the first great popularizer of a Young Earth manifesto in the early twentieth century. He followed the dictates of Seventh-day Adventist founder Ellen White, who rejected the "indefinite periods" of "infidel geologists" in her book *Spiritual Gifts* (1864). The first of Price's many books, titled *Outlines of Modern Christianity and Modern Science* (1902), joined White in opposing the long chronology. Price also railed against the materialist view of the mind that infected modern science. The proper use of the human mind was as a path to "reverence" of an "intelligent Designer," an "*order-loving Mind*," and an "eternal Mind."[6]

Short chronologists also parted ways with deep timers in their view of the knowability of Earth's history. They argued that the origins of Earth and life were knowable through the creation account in Genesis, which was given to humans by God in a series of dispensations that ordered the biblical narrative. While aspects of the biblical narrative might remain mysterious, Earth's infancy was no more mysterious by virtue of its earliness than the later parts. These early days did not recede from human cognition in the dark abyss of time, the way Lyell had claimed.

The cognitive availability of Earth's history remains a Young Earther preoccupation today. Some have recently proposed "catastrophic plate tectonics." Responding to the theory of plate tectonics developed in the 1960s, in which great slabs of Earth's crust collide and submerge in mostly invisible processes, they suggest the reality of the recent Genesis flood in part because it is more imaginable. In mainstream geology, Earth's crustal movements proceed at pace of less than six inches a year, far slower even than a snail. By contrast, crustal movements become lightning-speed events in catastrophic place tectonics. While the long chronology proposed by Lyell and others "overwhelmed the human mind," one Creationist has written, catastrophic plate tectonics proves the reality of the flood described in the Bible by positing a mechanism of rapid recycling of the ocean lithosphere into the mantle rocks. The mind can "imagine" through a "reasonable inference" that this is the flood described in Genesis.[7]

What is not knowable, short chronologists claim, is the billions of years that they argue deep timers claim to know in their totality. Short chronologists see this as treading on the omniscience of the deity. First articulated by George McCready Price in the early twentieth century, the denial of the knowability of long time spans continues to be articulated by Young Earth Creationists today. As one put it recently: "For a person to make the claim that humans and dinosaurs did not coexist, they would have to be able to see all history at exactly the same time, which would make that person omniscient and omnipresent, qualities of God. So, when someone says emphatically that humans and dinosaurs did not exist together in the past, that person is claiming to be a god, while calling God Himself a liar, or, at best, deceptive."[8] While this quotation misconstrues the deep time position—which as we saw with Lyell and Darwin is one of awe at the unknowability of deep time—the speaker reveals that the debate is not entirely about whether Earth is young or old. It is about what humans can and cannot know.

Furthermore, in contrast to Lyell, short chronologists do not believe that the deep time visible in the landscape challenges the human mind. On the contrary, they believe that the landscape comforts the mind by knitting land, Bible, and deity into a seamless whole. The landscape transcribes the Bible into rock; call them Jurassic or Paleozoic, the rocks were laid down rapidly in Noah's flood just as the Bible describes it. These views were widely propagated by George McCready Price in his textbook, *The New Geology* (1923). It was illustrated with hundreds of photographs of the North American landscape, many used without crediting the United States Geological Survey. Price slashed through a thicket of geological terminology to reach a sacred conclusion: the origins of life were to be determined not by science but by theology. "The Bible record of a universal Deluge is so wonderfully confirmed by these contemporary documents in stone, that inexorable logic compels us to go back of all this and face the great problem of creation itself," Price wrote. The truth of Creation was "the most sublime and august thought which can occupy the mind."[9] *The New Geology* is illustrated with a potpourri of iconic dinosaurs that claim the creatures for a young Earth (fig. 9.3). This was a stroke of marketing genius that continues to pay dividends to Young Earthers today, who can sell dinosaurs in every medium imaginable while still hewing to the short chronology. Did Price's readers recognize these iconic dinosaurs as the very ones painted by Charles Knight to illustrate deep time and evolution, the very concepts Price opposed?

After a few decades of marginalization, Young Earth Creationism resurged with the publication in 1961 of *The Genesis Flood*, by John C. Whitcomb and Henry M. Morris. This book enshrined the six literal days of Creation as Young Earth orthodoxy. It flatly rejected the re-

FIG. 350. DINOSAURS
Restorations of what scientists think some of these animals looked like. 1. Skeleton of Diplodocus (order Sauropoda) ; 2. Ceratosaurus (order Theropoda) ; 3. Stegosaurus (order Orthopoda) ; 4. Hadrosaurus (suborder Ornithopoda) ; 5. Triceratops (order Ceratopsia). (2-5, Restorations, after Osborn and Knight.) (From "The New International Encyclopædia," by permission.)

FIGURE 9.3. George E. McCready Price's *The New Geology* (1923) denied the antiquity of the Earth, but nevertheless copied these dinosaurs from the paintings of the famous prehistory illustrator Charles Knight.

peated capitulation of Christian scholarship to the scientific establishment's beliefs in deep time and evolution by natural selection. Refusing any compromise, its authors advocated biblical inerrancy in all domains. Ignored outside fundamentalist circles, the book's reach within them was enormous, with numerous reprintings over the decades. Poking holes in all kinds of geological dating methods from radiometric dating to tree ring counting, the authors declared that "the modern mind" had been seduced by science and materialism and that it was dogmatic in its claims to finality in the metaphysical and epistemological realms. These modern minds were naturally impatient with those who found "ultimate truth" in the "supernaturally-inspired Book."[10]

The political edge of Young Earth Creationism had also sharpened by midcentury. Against fears of communism, materialism, and psychoanalysis, Young Earth Creationists made their creed nationalistically

American. The Sputnik-era push for the proper teaching of evolution in American schools renewed the Creationist project of resisting naturalistic accounts of life on Earth along with their heretical container, deep time. Fundamentalists shifted the debate from a scientific war—which they had a hard time winning on facts—to a war on culture.

This was ironic. The first proponents of deep time in the United States had seen the long chronology as a form of United States nationalism, liberating the New World from the contempt of the Old World by virtue of its greater antiquity. Deep timers established the national parks as the cathedrals of US nature, the patriotic shrines to the continent's unimaginable antiquity, gifts from the federal government to its democratic people. By the mid-twentieth century, Young Earth Creationists had asserted something close to the opposite. Identifying the United States as God's favored nation, they deployed the young Earth as a cultural weapon. The United States and the rest of the world were equally young, they said, all washed clean by the worldwide Genesis flood a few thousand years ago. But only the United States, if it embraced the tenets of fundamentalism, would be spared the wrath of God at the inevitable Armageddon. To consider the alternative was un-American. Although Young Earth Creationists were born on the margins of American religious life, their evangelical descendants now oversee what is arguably the most powerful religious movement in the United States and one of the most powerful around the globe. Their alliance with the Republican Party and conservatism more generally is one of the political and cultural success stories of the late twentieth and early twenty-first centuries.[11]

The institutional and media footprint of Young Earth Creationism is now vast. It replicates the virtual reality of deep time objects first developed in the early nineteenth century in order to illustrate the very opposite of deep time: shallow time. Many of these shallow time objects are drawn from the United States, reflecting the demographics of the Creationist base. Young Earth Creationists can visit the sites they read about and correct the deep time dogmas insidiously disseminated by national and state park employees. At the Grand Canyon, they offer alternative tours intended to assign new meanings to the mile-deep strata. An archipelago of shallow time institutions mirrors the mainstream institutions that propagate deep time and Darwinian evolution. Colleges and universities such as Cedarville University in Ohio employ geologists to teach Young Earth Creationism and biblical flood geology. There are think tanks, such as the Institute for Creation Research in Dallas, Texas, and conferences, such as the International Conference on Creationism. Museums that look like deep time natural history museums—complete with fossil displays and habitat dioramas—in fact argue that Earth is a few thousand years old. A video produced by the Creation Research Society shows dinosaurs thundering across an arid landscape, as a narrator describes how soft tissue remains prove the creatures' recent origin. Since the 1960s, the society has also published the *Creation Research Society Quarterly*, which mimics the visual and scientific program of mainstream geology to argue that Earth is around six thousand years old. "But whose problem is time?" one writer asks while erasing twenty million years of history from the Texas Panhandle. From Utah comes a trilobite fossil nestled in a human footprint apparently wearing a sandal. "[T]he Bible alone provides a possible explanation of this remarkable occurrence of trilobites and humans obviously alive and together at the same time," according to the author, who clarifies (with a photograph proving the coexistence of humans and trilobites) that this is "in some way related to Noah's Flood." The biblical flood's spread over Utah is illustrated by a chart of the Creation event, complete with strata that show the stages of the flood.[12]

And of course, there are dinosaurs—many, many dinosaurs in all shapes and colors and sizes. Recognizing that dinosaurs are big business, deep time deniers have found a way to market dinosaurs within the Young Earth paradigm. Ken Ham, the public face of Answers in Genesis, and Steve Green, the billionaire president of the craft store chain Hobby Lobby, promote a young Earth by capitalizing on the public fascination with dinosaurs. Ken Ham's Creation Museum near Cincinnati displays huge dinosaurs galore, while Hobby Lobby hawks dinosaur merchandise of all kinds, from tiny plastic dinosaurs to signs advising that you should always be yourself, unless you can be a dinosaur, in which case you should always be a dinosaur. Along with the

Green-initiated Bible Museum in Washington, DC, the Creation Museum can file for nonprofit status with the Internal Revenue Service—not as a religious organization but as a museum.

Today we are surrounded by dinosaurs: beautiful, transporting visions of dinosaurs pioneered by the advocates of deep time in the nineteenth century, and today also claimed by those who deny their vast antiquity. Young Earth Creationists now claim patriotism by utterly repudiating the American nationalist victory of the nineteenth century, which was to assert that the United States sat atop the oldest lands in the world, a sign of God's favor. Marching into an alternative reality of their own making, the Young Earthers have created a land of enchantment teeming with dinosaurs and Bible stories. In a glorious color panorama on the Answers in Genesis website, mighty brontosauruses and stegosauruses are escorted onto the ark by Noah as parrots and pterodactyls swoop overhead. Dark clouds gather in the distance. The storm is coming.[13]

Epilogue

IN *THE BOOK OF SAND* (1975), the Argentinian writer Jorge Luis Borges tells the story of a man who buys an infinitely long book. At first the man is fascinated as the pages multiply between the covers. But finally he is horrified that he can find no beginning and no end. "If space is infinite," he despairs, "we may be at any point in space. If time is infinite, we may be at any point in time."[1] Declaring the book monstrous, he hides it in the national library.

We might say that the book of sand was first written in the nineteenth century. Released from the constraints of the biblical chronology, Earth's past now stretched into a time so remote as to be incomprehensible. Here was a world without knowable beginning, without prospect of an end. But instead of hiding from deep time's existential horrors, the scientists, artists, and others who attended its birth stepped forward, striving to understand the thing that they did not understand. In the process, they created a haunting gallery of images that speak to the urgency of their desire to comprehend the incomprehensible. When comprehension ended, they summoned their imaginations. With ink and paint, they crafted beautiful paper worlds filled with curious creatures and strange viewpoints, many rendered in glorious color.

We still have a hard time grasping deep time. John McPhee, who coined the term in 1981, said that the years of deep time "awe the imagination to the point of paralysis."[2] But deep time also fills us with wonder, our minds straining to touch eternity. We, too, mobilize our imaginations when comprehension ends, giddy with anticipation at the marvels

FIGURE E.1. This recent visualization of deep time shows the history of our planet as an uncurling ammonite. A surfer rides the edge of time into the future.

our mind will bring into being. That the nineteenth century bequeathed to us a dilemma rather than a solution is therefore perhaps cause for celebration rather than despair.

The wheels of deep time keep turning. In the United States, the long chronology continues to generate endless forms. A spiral jetty in the Great Salt Lake. A giant clock destined for burial in Nevada that might keep time for ten thousand years. And on a ranch in North Dakota, perhaps even traces of the first hours after the impact of the great dinosaur-annihilating asteroid. All sorts of time schemes continue to proliferate. Physicists and philosophers speculate about nonlinear times: perhaps time is a crumpled handkerchief.[3] Others ignore deep time in favor of a thin present. Behaviorism and its therapeutic descendant, cognitive behavioral therapy, dispense with the deep temporal narrative driving Jungian and Freudian psychoanalysis. Instead, they offer behavioral adjustments suspended in the eternal present of neural networks. In precise doses, the pharmaceutical industry adjusts our brain chemistry, claiming the irrelevance and even futility of the introspection that led to the caveman within. Virtual reality gives us a simulation of veracity, a thinned experience leaving no traces. In short, deep time is continually generative. Having broken through the biblical short chronology, it has opened to us an endless vista of intriguing possibilities for how to think and how to live.

The nature of time continues to elude us. We wonder whether it lies outside us or within us, in the very structure of our minds. But it most assuredly lies between us, a shared language for talking about the most consequential matters of our common humanity. Perhaps we are like the surfer at the edge of this deep time spiral (fig. E.1 and plate 32). Looking backward as billions of years of Earth's history uncoil in stately majesty, we ride a wave toward the undiscovered horizon.

ACKNOWLEDGMENTS

MY FIRST THANKS go to Michael Gordin of Princeton University. Several years ago, he reached out to me about including this manuscript in the Modern Knowledge Series he edits for Princeton University Press. I'm grateful for the opportunity to be included in this series, and for his wise and astute comments on the manuscript.

Many friends and colleagues read chapters and made helpful comments along the way. Special thanks go to Suzanne Marchand, Mark Peterson, Jennifer Ratner-Rosenhagen, Daniel Rodgers, and James Turner for crucial insights at decisive junctures in my thinking. Many others offered comments that were immensely helpful. In alphabetical order they are: David Bell, Thomas Broman, Austin Clements, Jamie Cohen-Cole, Henry Cowles, Andrea Davies, John Dixon, Jared Farmer, Paula Findlen, Anthony Grafton, Sherril Green, Emily Bradley Greenfield, Fiona Griffiths, Kimberly Hamlin, Andrew Hartmann, Charlotte Hull, Sarah Igo, Theresa Iker, Sveinn Jóhannesson, Richard John, Jamie Kreiner, Karen Kupperman, Robert Lee, Emily Levine, Joan Malczewski, Adrienne Mayor, John McNeill, Stephen Mihm, Jonathan Morton, Kathryn Olivarius, Brad Pasanek, Lukas Rieppel, Jessica Riskin, Kevin Schultz, Silvia Sebastiani, Serena Shah, John Sime, Jeffrey Sklansky, Matthew Sommer, Scott Spillman, Mitchell Stevens, James Stoner, Cameron Strang, Laura Dassow Walls, Kären Wigen, Malin Wilckens, Juliette Winterer, Peter Wirzbicki, and Rebecca Woods.

Several current and former Stanford students have been thorough and careful research assistants. I owe special thanks to Dr. Charlotte Hull, who not only did archival research but whose precise, efficient work securing the rights and permissions for the numerous illustrations in this book was nothing short of heroic. Charlotte: thank you! Warm thanks also to Nabila Akthar, Claire Rydell Arcenas, Glory Liu, and Erin Wenokur, who did some archival research as well over the years. Digitization for all images held in Stanford's libraries was done by the Stanford Libraries Digital Production Group (with imaging by Kylee Diedrich, Cheryll Go, Chris Hacker, Trieu Nguyen, and Wayne Vanderkuil), and production coordination by Chris Hacker.

My editor at Princeton University Press, Eric Crahan, ably shepherded the book through its various stages. He recruited three exceptionally shrewd anonymous reviewers, whose long and detailed reports improved the manuscript immeasurably. He also shared in my enthusiasm for the earth sciences in all their visual glory. At the Press, Whitney Rauenhorst also worked expeditiously behind the scenes on many matters large and small. I thank my copyeditor, Molan Goldstein, for her judicious corrections and recommendations.

The following conferences and seminars offered forums to test and improve my ideas over the years, and I'm grateful to the organizers and directors for the invitation to present chapters in progress: the Shelby Cullom Davis Center Seminar at Princeton University; the Conference on Knowledge and the Science of Improving Minds at the Max Planck Institute for the History of Science in Berlin; the Early American History and Culture Seminar at Columbia University; the Intellectual History Reading Group; the Eric Voegelin Institute for American Renaissance Studies at Louisiana State University; the David Rumsey Map Center at Stanford; the Long Now Foundation in San Francisco; the New Directions in the History of Science Conference at Stanford (co-organized with my colleague Kathryn Olivarius); the Apes & Us Workshop at Stanford (co-organized with my colleague Jessica Riskin), the US History Workshop in the Department of History at Stanford; and Il Gruppo. Much of this book was written during my two terms as chair of the Department of History at Stanford. I'm grateful to the excellent staff members, especially Burçak Keskin-Kozat, who ensure the smooth and efficient functioning of the department.

My family was always game for visits to dinosaur tracks, petroglyphs, cave paintings, Neolithic lake dwellings, extinct giant sloth footprints, Ötzi the Iceman, megaliths, extinct and live volcanoes, glaciers, Neanderthal grottos, and anything calling itself a dinosaur museum. This book is for them, and for the other comrades I call my dinosaur friends. You know who you are.

NOTES

Introduction

1. McPhee, *Basin and Range*, 21.
2. Boltwood, "Ultimate Disintegration Products," 87; Dalrymple, *Ancient Earth*.
3. Samuel George Morton to Timothy Abbott Conrad, 11 May 1833, LSGM.
4. Elias, *Time*.
5. Gibbes, *The Present Earth*, 6; A. Winchell, *Sketches of Creation*, 201.
6. Siegfried, *America Comes of Age*, 348–49, 353; Nieuwland, *American Dinosaur Abroad*; Scully, "The Age of Reptiles."
7. S. Turner et al., "Forgotten Women"; Mayor, *Fossil Legends*; Strang, *Frontiers of Science*; Schuller, "The Fossil and the Photograph."
8. Samuel Hildreth to Samuel George Morton, 22 November 1834, PSGM; Gideon Mantell to Samuel George Morton, 4 December 1834, PSGM.
9. Perrin, "The Chemical Revolution"; Goodman, *The Domestic Revolution*; Socolow, ed., *The State Geological Surveys*, unpaginated preface.
10. "Synchronous" is from Timothy Abbott Conrad, "Observations on the Eocene deposit of Jackson, Miss, with descriptions of 34 new species of shells and corals," c. 1831–33, PTAC; on other aspects of timekeeping, see Ogle, *The Global Transformation of Time*.
11. Owen, "Scientific Pursuits," 44; J. Green, *The Inferior Surface of the Trilobite*, 10.
12. Draper, *History of the American Civil War*, 1: 65; Dickinson, *Poems*, 1211; Thrailkill, "Fables of Extinction," 225.
13. Buckland, "Eyes of Trilobites," 300; J. Green, *The Inferior Surface of the Trilobite*, 10; P. Smith, *From Lived Experience*.
14. Winchell, *Sketches of Creation*, 175.
15. Bjornerud, *Timefulness*.

Chapter One: Why the New World Was New

1. Vespucci, *Alberic[us] Vespucci[us]*; *Encyclopedia Virginia*, https://www.encyclopediavirginia.org/mundus_novus_1503.
2. Rossi, *The Dark Abyss of Time*, 132–52; Grafton, *Joseph Scaliger*, 4; Sheehan, "The Stamp of Time Elapsed"; Dal Prete, *On the Edge of Eternity*.
3. Ussher, *The Annals of the World*, 1; J. Barr, "Why the World Was Created in 4004 BC"; Haber, *The Age of the World*; Cohn; *Noah's Flood*.

4. Winterer, *American Enlightenments*, chap. 2; Gerbi, *The Dispute of the New World*.

5. Buffon, *Natural History*, 124–52, esp. 136–38.

6. Crèvecoeur, *Letters*, 42; C. Peale, *A Scientific and Descriptive Catalogue*, viii; Webster, *An Address*; 70; Hattem, *Past and Prologue*, chap. 6.

7. J. Q. Adams, *Memoirs*, 433; W. Wilson, "Princeton in the Nation's Service," 459.

8. Benjamin Silliman, "Journal of Travels in England, Holland and Scotland, Vol. III," entry for 19 November 1805, PBS.

9. Barton, "Journals and Notebooks, 1785–1806," PBSB; Barton, *New Views*, cviii.

10. Dunbar, "Description," 169; Hutton, "Theory of the Earth," 215, 288; E. James, *Account*, 2: 394.

11. Maclure, "Observations," 427; Winterer, "The First American Maps of Deep Time."

12. Dugatkin, *Mr. Jefferson and the Giant Moose*; Barrow, *Nature's Ghosts*, chap. 1.

13. Barton, "Fossils," n.d. [c. 1809?], PBSB; R. Peale, *An Historical Disquisition on the Mammoth*; Cohen, *The Fate of the Mammoth*.

14. W. Smith, *Strata Identified by Organized Fossils*; Oxford University Museum of Natural History, ed., *Strata*.

15. Wilcox, *The Measure of Times Past*, 2–8; Rosenberg and Grafton, *Cartographies of Time*.

16. E. Hitchcock, "The Connection," 262; Amos Eaton, Journals A & B, Box 2, folder 1, entry for December 14, 1827, PAE; Samuel George Morton to Isaac Lea, 21 September 1833, LSGM; Bozeman, *Protestants in an Age of Science*.

17. Schulten, *Mapping the Nation*.

18. Lyell, *Principles of Geology*, 1: 166.

19. Willis and Stose, "Geologic Map of North America."

Chapter Two: Beginnings

1. Amos Eaton, Journal E, Box 2, Folder 3, entry for June 1826, PAE; Conrad, "Notes," 243; Sheriff, *The Artificial River*; Spanagel, *DeWitt Clinton*.

2. J. Green, *A Monograph*, 13; E. Hitchcock, *Elementary Geology* (1841), 126.

3. On the skeletal nature of Genesis, see Hendel, *The Book of Genesis*; Auerbach, *Mimesis*.

4. Eaton, "The Globe Had a Beginning"; Hutton, "Theory of the Earth," 304. On the nebular hypothesis, see Numbers, *Creation by Natural Law*; Brush, "The Nebular Hypothesis." On the early period in American geology, see Corgan, ed., *The Geological Sciences*; Greene, *Geology*.

5. John Gebhard to Amos Eaton, 22 September 1834, sketch titled "Section of the Strata 1/4 Mile East of Schoharie," Box 1-f5, PAE; Conrad, "Notes," 243.

6. Cuvier and Brongniart, *Essai*; Lajoix, "Alexandre Brongniart"; Paredes, *Sèvres Then and Now*; A. Brongniart, *Traité des Arts Céramiques*.

7. Brongniart and Desmarest, *Histoire Naturelle*, 45, 50, 64.

8. Brongniart and Desmarest, *Histoire Naturelle*, 56–58; Benjamin Silliman to Alexandre Brongniart, 15 January 1822, PBS.

9. Brongniart and Desmarest, *Histoire Naturelle*, 58–59, 62.

10. George Featherstonhaugh to James Madison, 28 February 1828, in Stagg, ed., *The Papers of James Madison Digital Edition*; Amos Eaton journal, June[?] 1829, Box 2-f4_pt1., PAE; Dekay,

"Observations," 188; Benjamin Silliman to Edward Hitchcock, 27 December 1824, EOWHP; James de Carle Sowerby to Thomas Hodgkin, 28 February 1828, PSGM.

11. The other trilobite publications include J. Green, *The Inferior Surface*, and journal articles published in the 1830s. On Green's casts, see Sime, "The Illustration of Nature Recast"; Charwat, "Treasures in Our Collections."

12. Blum, "'A Better Style of Art,'" 72.

13. J. Green, *Notes of a Traveller*, 1: 18; A. Green, "Biographical Sketch," 31; Noll, *Princeton and the Republic*, 277.

14. J. Green, *Notes of a Traveller*, 1: 28, 224, 225, 244.

15. J. Green, *Notes of a Traveller*, 1: 163, 244; 2: 5.

16. Rebekah Pollock, "Like Father Like Son," https://www.cooperhewitt.org/2014/09/29/like-father-like-son/; J. Green, *Notes of a Traveller*, 3: 6; Marchand, *Porcelain*; Wise, *Aesthetics*; Tresch, *The Romantic Machine*; Wosk, *Breaking Frame*.

17. J. Green, *Notes of a Traveller*, 3: 6–7.

18. J. Green, *Notes of a Traveller*, 1: 257.

19. Comstock, *Outlines of Geology*, v.

20. J. Green, *A Monograph*, 8–9, 29; Samuel George Morton to Samuel Hildreth, 10 July 1837, PSH; L. Agassiz, *Monographie*, xix.

21. J. Green, *A Monograph*, 95 (unpaginated); Sime, "The Illustration of Nature Recast," 1; "Stated Meeting, February 1, 1842," 147–48; Benjamin Silliman, "Lecture," 18 March 1835, PBS; Jacob Green to Philip Torrey, 12 February 1832, PJT; Chadarevian and Hopwood, eds., *Models*.

22. Amos Eaton, Journal E, Box 2, Folder 3, entry for 7 June 1826, PAE; Bigsby, "Description," 368.

23. J. Green, *A Monograph*, 14, 52; Jacob Green to Benjamin Silliman, 1 October 1833, PBS; Silliman, "Professor Jacob Green's Monograph," 396; Conrad, "Ode to a Trilobite" (1840): https://www.palaeopoems.com/palaeopoems/ode-to-a-trilobite#:~:text=Early%20American%20palaeontologist%20Timothy%20Abbott,in%20literature%20and%20his%20profession; E. Hitchcock, *Elementary Geology* (1841), 127.

24. Buckland, "Eyes of Trilobites," 299–300; quoted verbatim by Green in *The Inferior Surface of the Trilobite Discovered*, 10–11. On Buckland, see Robson, "The Fiat and Finger of God."

25. Willard, *Universal History in Perspective*; Benjamin Silliman, "Lecture on Geology, 11 March 1835," PBS. The extensive scholarship on scriptural geology includes C. Wright, "The Religion of Geology"; Gillispie, *Genesis and Geology*; Guralnick, "Geology and Religion"; Bozeman, *Protestants in an Age of Science*; Moore, "Geologists and Interpreters of Genesis"; Stiling, "Scriptural Geology in America"; Numbers, *The Creationists*; Hendel, *The Book of Genesis*, 179–82; M. Roberts, "Genesis Chapter 1 and Geological Time."

26. *Albany Evening Journal*, 14 October 1833. See also Rudwick, *Scenes from Deep Time*, chap. 1.

27. Conrad, "Notes," 243.

28. Murchison, *The Silurian System*, 1: 3, 11.

29. Murchison, *The Silurian System*, 1: 10; Rudwick, *Worlds before Adam*, 451. On the controversy over the naming of the Silurian and closely related deposits, see Rudwick, *The Great Devonian Controversy*; Secord, *Controversy in Victorian Geology*.

30. Murchison, *The Silurian System*, 1: xxxii, n.1; Roderick Impey Murchison to William Venables Vernon Harcourt, 15 August 1831, Electronic Enlightenment Project, https://doi-org.stanford.idm.oclc.org/10.13051/ee:doc/baascoRH0010044b1c; William Venables Vernon Harcourt to Roderick Impey Murchison, 29 August 1831, Electronic Enlightenment Project, https://doi-org.stanford.idm.oclc.org/10.13051/ee:doc/baascoRH0010048a1c. Studies of Murchison include Secord, "King of Siluria"; Stafford, *Scientist of Empire*; Morton, *King of Siluria*; Collie and Dimer, eds., *Murchison's Wanderings in Russia*; Rudwick, *Worlds before Adam*, 444–49; Collie, *Science on Four Wheels*.

31. Murchison, *The Silurian System*, 1: 8; E. Hitchcock, *Elementary Geology* (1841), 64; Benjamin Silliman, Lecture, 26 January 1840, PBS; Roderick Impey Murchison to Benjamin Silliman, 4 April 1841, PBS; Schneer, "Ebenezer Emmons," 447. The first use of *Taconic* for the rocks in question was Dewey, "Sketch of the Mineralogy." On the controversy over the usage of Taconic, see Schneer, "The Great Taconic Controversy."

32. Conrad, "Third Annual Report," 200; Conrad, "Observations on the Silurian," 228; Conrad, "Notes," 247; E. Hitchcock, *Elementary Geology* (1841), 64.

33. Murchison, *Siluria*, 422, 416, 418; Thackray, "R. I. Murchison's *Siluria*."

34. Castelnau, *Essai sur le Système Silurien*.

35. Professor Lippitt, "A Trip to Siluria"; Murphy, "The Ancient History of Plants," 593. On the Eve myth at this time, see Hamlin, *From Eve to Evolution*, chap. 1. The magic lantern slide at the College of Wooster was taken from Franz Unger, "Überangs-Periode" (Transition Period), which was the first illustration in his *Die Urwelt in ihren verschiedenen Bildungsperioden*; for its context, see Rudwick, *Scenes from Deep Time*, 102–3.

36. C. Hitchcock, "The Earlier Forms of Life," 257; J. W. Dawson, *The Story*, 18, 19, 20, 23, 24, 27, 32; Sheets-Pyenson, *John William Dawson*; Kaalund, "Of Rocks and 'Men'"; Zeller, *Inventing Canada*, 102–3.

37. Chauncey Wright, "Sir Charles Lyell."

38. LeConte, *Elements of Geology*, 273.

39. H. Adams, *The Education of Henry Adams*, 230, 400.

40. LeConte, *Elements of Geology*, 274.

Chapter Three: Fossil Futures

1. Lyell, *Travels*, 1: 3; K. Lyell, ed., *Life*, 2: 64–65. On Lyell's American visit, see Dott, "Lyell in America"; Dott, "Charles Lyell's Debt"; L. G. Wilson, *Lyell in America*; Weeks, *The Lowells*.

2. Jevons, *The Coal Question*, 2, 289; S. P. Adams, ed., *The American Coal Industry*, vii; S. P. Adams, "Promotion."

3. Holland, *The History*; Benjamin Silliman, diary entry for 14 May 1830, PBS.

4. J. Beckert, *Imagined Futures*; Malm, *Fossil Capital*; MacDuffie, *Victorian Literature*.

5. Lyell, *Travels*, 2: 22.

6. Witham, *Observations*, 13.

7. Witham, *Observations*, 13.

8. Lesley, *Manual*, 25. The various viewpoints are summarized in A. Scott, "The Legacy."

9. E. Hitchcock, *Elementary Geology* (1841), 99; Taylor, *Statistics*, 25; Rogers, *The Geology*, 2: 1018.

10. Curwen, ed., *The Journal*, 210; Lyell, *Principles*, 1: 76; Rudwick, "Charles Lyell Speaks," 147.

11. Lyell, *Principles*, 1: 2, 76, 146; Rudwick, "Charles Lyell Speaks," 150; more generally on Lyell, see Rudwick, *Worlds before Adam*.

12. Lyell, *Principles*, 1: 102.

13. K. Lyell, ed., *Life*, 2: 60; Lyell, *Travels*, 2: 22.

14. John Locke to Samuel Hildreth, 29 December 1836, PSH; Benjamin Silliman to Alexandre Brongniart, 7 August 1821, PBS.

15. Jared P. Kirtland to Samuel Hildreth, 3 September 1834, PSH; Edward Hitchcock to Samuel Hildreth, 20 April 1832, PSH.

16. O. M. Herron to Samuel Hildreth, 25 August 1832, PSH; George Featherstonhaugh, "Journals," vol. 7, entry for 13 August 1835, PGWF; Silliman, "Obituary," 313.

17. Benjamin Silliman, "Journal of Travels in England, Holland and Scotland, Vol. III," entry for 21 November 1805, PBS; Parker Cleaveland to Benjamin Silliman, 16 May 1815, PBS.

18. Benjamin Silliman to Alexandre Brongniart, 7 August 1821, PBS; Benjamin Silliman to Samuel Hildreth, 22 February 1840, PSH; Benjamin Silliman to Samuel Hildreth, 26 June 1827, PSH; L. Wilson, *Lyell in America*, 104; Benjamin Silliman, "Lecture VI" (18 March 1835) and "Lecture VII" (27 March 1835), PBS.

19. Brongniart and Desmarest, *Histoire Naturelle*, 4, 8, x–xi.

20. Witham, *The Internal Structure*, 49, 3, 50, 4; Witham, "On the Vegetation," 116.

21. Witham, "On the Vegetation," 117; Nutt, "Geological remarks upon the formation of Peat & Fossil coal," n.d. but probably 1830s, PRN.

22. Hildreth, "Observations," 1, 4, 6, 19, 123, 124; Benjamin Silliman to Samuel Hildreth, 2 November 1835, PSH.

23. Hildreth, "Observations," 105, 124.

24. Silliman, "Obituary," 312; Samuel George Morton to Samuel Hildreth, 25 January 1837, PSH; Samuel George Morton to Samuel Hildreth, 18 July 1838, PSH.

25. Lyell, *Principles*, 1: 2; Lyell, *Travels*, 2: 25.

26. E. Hitchcock, *Final Report*, 126; E. Hitchcock, *Elementary Geology* (1841), 63.

27. E. Hitchcock, "A Sketch," 55, 56; Benjamin Silliman to Edward Hitchcock, 25 May 1841, EOWHP; Benjamin Silliman to Edward Hitchcock, 27 June 1831, EOWHP.

28. E. Hitchcock, *Elementary Geology* (1841), 157; E. Hitchcock, "The Connection," 261–62; unidentified student's notes, "Lectures on Geology," September–November 1855, EOWHP; E. Hitchcock, *The Religion*, 68–69.

29. E. Hitchcock, "Report on Ichnolithology," 314; an engraving appears in his *Ichnology of New England*, plate IX, fig. 1.

30. Curwen, ed., *The Journal*, 167; Charles Darwin to Edward Hitchcock, 6 November 1845, EOWHP; Lyell, *Travels*, 1: 201.

31. E. Hitchcock, *Reminiscences of Amherst*, 60–64, 212; Pick, *Curious Footprints*, 11–13; Herbert and D'Arienzo, *Orra White Hitchcock*.

32. E. Hitchcock, *Elementary Geology* (1841), 164; E. Hitchcock, "The Connection," 263; E. Hitchcock, "The Law," 491.

33. E. Hitchcock, *Elementary Geology* (1841), 103–4, 163; on Bronn, see Nyhart, *Biology*, 110–21.

34. E. Hitchcock, "A Sketch," 56; E. Hitchcock, *Elementary Geology* (1841), 161, 163, 275; Rudwick, "Lyell," 8; Bartholomew, "The Non-Progress."

35. E. Hitchcock, *Elementary Geology* (1841), 117, 104; E. Hitchcock, "The Connection," 264–65.

36. E. Hitchcock, *Elementary Geology* (1841), 117; Rudwick, *Scenes from Deep Time*.

37. Rogers, *The Geology*, 1: 28; Taylor, "Notice," 88–89; Winchell, *Sketches of Creation*, 171. On coal models, which Lyell used in his American lectures, see S. Turner, "Thomas Sopwith"; Turner and Dearman, "Thomas Sopwith's Large Geological Models"; on models used in the Lowell Lectures, see Dott, "Lyell in America," 110. In a large literature on the unity of American art and science at this time, see Bedell, *The Anatomy*; Novak, *Nature and Culture*; DeLue, *George Inness*; Bell, *George Inness*; Elkins, "Art History."

38. Rogers, *The Geology*, 1: 24–25.

39. Wallace, *Palm Trees*, 6–8; D. Miller, *Dark Eden*.

40. H. Miller, *The Old Red Sandstone* (1851), 255. On Miller's popularity, see O'Connor, *The Earth on Show*.

41. E. Hitchcock, *Reminiscences of Amherst*, 348; Woods and Warren, *Glass Houses*; Downing, *Treatise*, 383; Loudon, *The Green-house Companion*, 123.

Chapter Four: The Oldest South

1. S. Beckert, *Empire of Cotton*, 100–101; Johnson, *River of Dark Dreams*, 10.

2. Nolan, "The Anatomy of the Myth"; Blight, *Race and Reunion*, 255–99; Foster, *Ghosts of the Confederacy*; Horsman, *Race and Manifest Destiny*.

3. "North America in the Cretaceous period," in Dana, *Manual of Geology*, 489; Tullos, "The Black Belt"; Washington, *Up from Slavery*, 108; Du Bois, *The Souls of Black Folk*, 113.

4. Featherstonhaugh, *Excursion through the Slave States*, 38.

5. The first entry in the *Oxford English Dictionary* for *palæontologist* is from one of the Philadelphia group: Conrad, *Fossils*, title page (this book will use the modern spelling, *paleontologist*, except in direct quotations); Jacob Elisa Doornik to Samuel George Morton, 19 June 1835, PSGM.

6. The paleontologists bought the pocket maps of New Jersey from the "Circulating Library" of A. T. Goodrich, 124 North Broadway, New York: Eaton, "Geological Journal. Commenced June 16th 1828," unpaginated 111, PAE; *New-York As It Is*, 159; Morton, "Synopsis," 276; Harlan, "Notice of the Plesiosaurus," 236.

7. Hodges, *Black New Jersey*, 59, 64; Rothman, *Slave Country*, 191.

8. Lea, *Contributions to Geology*, 13; Georges Cuvier and Alexandre Brongniart, *Essay on the Mineral Geography of the Environs of Paris* (1808), in Rudwick, *Georges Cuvier*, 127–56; on the interactions of South Carolina geologists with the Philadelphia group, see Stephens, *Science, Race, and Religion*, 127–45.

9. Barrande, *Système silurien*; "Joachim Barrande," 545.

10. Marcou, ed., *Life*, 2: 28. For Morton's past as a paleontologist and geologist, see Gerstner, "The 'Philadelphia School,'" ii; Gerstner, "The Influence of Samuel George Morton."

11. S. G. Morton, "Notice of the Fossil Teeth," 276. This date establishes the early international recognition of Morton as a paleontologist—not a craniologist. By now a set piece of historiography is that Louis Agassiz, who had just arrived in the United States in 1846, met Morton in Philadelphia, where he beheld Morton's collection of human crania. In combination with his shock at being served by black people, Agassiz was convinced by Morton's skull collection that racial hierarchies were real. A version of the account may have originated in Marcou, ed., *Life*, 2: 27–28. It is important to stress, however, that Agassiz had already long known of Morton as a paleontologist of fossil animals, since the English geologist Gideon Mantell had acted as intermediary for scientific exchanges between the two in the early 1830s: Gideon Mantell to Morton, 4 December 1834, PSGM.

12. Morton to Gideon Mantell, 14 November 1832, LSGM; Wetherill, "Observations," 13 (Plate 7).

13. Marcou, ed., *Life*, 2: 28; S. G. Morton, "Synopsis," 274; John Lawson to Morton, 30 December 1820, PSGM; S. G. Morton, "Geological Observations," 63; Kemmerly, "The Vanuxem Collection"; William Samuel Waithman Ruschenberger to Morton, 5 February 1836 (tiger skull), PSGM; Morton to Marmaduke Burrough [?], 16 August 1836, LSGM; Jordan, *Colonial Families of Philadelphia*, 2:1718.

14. Morton to José María Vargas, 1 January 1836, LSGM; Morton to Dr. John Emerson, 16 February 1836, LSGM. John Emerson was an army surgeon posted at Fort Armstrong at the time, and the owner of the slave Dred Scott.

15. S. G. Morton, "Account of a Craniological Collection," 217; Meigs, *Catalogue of Human Crania*, 3; Morton to Jacob Elisa Doornik, 13 July 1835, LSGM; Morton to William Maclure, 22 July 1836, LSGM.

16. The works listed here are in chronological order since the publication of Stanton's *The Leopard's Spots* (1960). The list is not comprehensive, but rather a guide to some of the more influential studies. Stanton's book appears to have been among the first books to critically assess Morton's human cranial studies. Its publication coincided with the civil rights movement and the new attention to the history of racial ideologies in the United States, exemplified by the rise of the term "scientific racism," which was not used in the nineteenth century. Daniels, *American Science in the Age of Jackson*; Stocking, *Race, Culture, and Evolution*; Fredrickson, *The Black Image in the White Mind*; Stocking, "Some Problems in the Understanding of Nineteenth Century Anthropology"; Horsman, *Race and Manifest Destiny*; Bieder, *Science Encounters the Indian*; Horsman, *Josiah Nott of Mobile*; Hinsley, *The Smithsonian and the American Indian*; Stephens, *Science, Race, and Religion*; Menand, *The Metaphysical Club*; Dain, *A Hideous Monster of the Mind*; Brace, *"Race" Is a Four-Letter Word*; Livingstone, *Adam's Ancestors*; Fabian, *The Skull Collectors*; Lewis, *A Democracy of Facts*; and Achim, "Skulls and Idols."

17. A small sampling of the scientific debate on the topic includes Gould, "Morton's Ranking of Races by Cranial Capacity"; Gould, *The Mismeasure of Man*; Michael, "A New Look at Morton's Craniological Research"; Cook, "The Old Physical Anthropology and the New World"; Lewis et al., "The Mismeasure of Science"; and "Editorial: Mismeasure for Mismeasure."

18. Darwin, *On the Origin of Species* (1859), 1. On the history of the idea of species, see Wilkins, *Species*; Bowler, *Evolution*; Lovejoy, "Buffon and the Problem of Species"; Riskin, *The Restless Clock*, 214–49; Beer, *Darwin's Plots*, 71–135.

19. Bowler, *Evolution*, 83–89; Corsi, *The Age of Lamarck*, 131; Appel, *The Cuvier-Geoffroy Debate*; Burkhardt, *The Spirit of System*; S. G. Morton, "Some Remarks," 81–82.

20. S. G. Morton, "Some Remarks," 82; Conrad, *Fossils Shells*, vii, viii; Lovejoy, "The Argument for Organic Evolution," 365.

21. Riskin, "The Naturalist and the Emperor."

22. S. G. Morton, "Geological Observations," 59.

23. S. G. Morton, "Synopsis," 290, 291; Morton to Georges Cuvier, 3 March 1832; Morton to Alexandre Brongniart, 3 March 1832; Morton to Henri Marie Ducrotay de Blainville, 4 March 1832; Morton to Marcel de Serres, 4 March 1832; Morton to Gideon Mantell, 19 May 1832, all in LSGM; Benjamin Silliman to Morton, 11 October 1832; Benjamin Silliman to Morton, 12 October 1828, PSGM.

24. S. G. Morton, "Synopsis," 290–91.

25. A. Brongniart, "Rapport," 460–61; Morton to Benjamin Silliman, 6 February 1832, PBS.

26. De la Beche, *Notes*, 3; Rudwick, *Worlds before Adam*, 26–27; Emling, *The Fossil Hunter*; William Daniel Conybeare to Henry De la Beche, 3 April 1824, in Sharpe and McCartney, *The Papers of H. T. De la Beche*, 33; "good conduct" medallion: Graphic Arts Collection, Special Collections, Firestone Library, Princeton University.

27. Chubb, "Sir Henry Thomas De la Beche," 17; De la Beche to William Conybeare, 13 May 1824, in Sharpe and McCartney, *The Papers of H. T. De la Beche*, 36; De la Beche, "Remarks on the Geology of Jamaica," 171, 174.

28. Rudwick, *Scenes from Deep Time*, 47.

29. Morton to James DeKay, 22 April 1832, LSGM; S. G. Morton, "Synopsis," 276; Morton to S. Wright, 8 July 1832, LSGM.

30. Moffat, "Charles Tait"; Conrad, *New Fresh Water Shells*, 23; Timothy Abbott Conrad to Morton, 28 December 1833, PSGM; Lea, *Contributions*, 21–22; Conrad, "Claiborne," 26.

31. Carey, *Map of Alabama*; Rothman, *Slave Country*, 40; Krauthamer, *Black Slaves*, 1–3; Nutt, "Depopulation of Countries," nu363 (5), PRN.

32. Featherstonhaugh, "Journals," vol. 7, entry for 6 September 1835, PGWF.

33. Lincecum, "Choctaw Traditions," 1–2; Lewis, "Choctaw."

34. Lea, *Contributions*, 21; Moffat, "Charles Tait," 227.

35. Morton to Isaac Lea, 21 September 1833, LSGM; Ford, "Timothy Abbott Conrad"; Marcou, ed., *Life*, 2: 29; Moffat, "Charles Tait," 228; Morton to Gideon Mantell, 16 September 1832, LSGM.

36. Morton to Charles Tait, 5 December 1832, LSGM; Morton to Timothy Abbott Conrad, 17 December 1832, LSGM; Timothy Abbott Conrad to Morton, 30 January 1832, PSGM.

37. Timothy Abbott Conrad to Morton, 20 April 1833, PSGM; Moffat, "Charles Tait," 229–230.

38. Conrad, *New Fresh Water Shells*, 23; Lyell, *A Second Visit*, 2: 47; Timothy Abbott Conrad to Morton, September 11, 1833, PSGM; Featherstonhaugh, *Excursion*, 119; Timothy Abbott Conrad to Morton, 6 December 1833, PSGM; Conrad to Morton, November 1833, PSGM; Lea, *Contributions to Geology*, 18; Strang, *Frontiers of Science*, 245–86.

39. Timothy Abbott Conrad to Morton, 3 March 1833, PSGM.

40. Timothy Abbott Conrad to Morton, 3 March 1833, PSGM.

41. Morton to Isaac Lea, 21 September 1833, LSGM; Morton to William Maclure, 20 May 1836, LSGM; Morton to William Maclure, 22 July 1836, LSGM.

42. Timothy Abbott Conrad to Morton, 26 October 1833, PSGM; Conrad, "Claiborne, Alabama," 29; Featherstonhaugh, "Journals," vol. 6, entry for 3 November 1834, PGWF; Lesueur, "Walnut Hills Fossil Shells, 1829," Charles Alexandre Lesueur Illustrations, ANSP; Dockery, "Lesueur's Walnut Hills Fossil Shells."

43. Nutt, "Geological remarks upon the appearance, position, and decomposition of granite and quartz, with their many varieties, as observed on a tour through a part of Georgia, Alabama, Arkansas, Louisiana, and Mississippi," nu 363 (8), PRN; Nutt, "Criticisms and explanations of many passages of the Bible," nu 363 (3), PRN; Dunbar, "Abstract," 58; Dunbar, "Description," 169.

44. Luarca-Shoaf, "The Mississippi River," 190–91; Lyons, "Panorama."

45. Morton to Gideon Mantell, 14 November 1832, LSGM.

46. S. G. Morton, *Synopsis*, 83–84.

47. Gideon Mantell to Morton, 19 December 1832; Gideon Mantell to Morton, 18 January 1834; Gideon Mantell to Morton, 16 January 1836, all in PSGM.

48. Samuel Hildreth to Morton, 29 May 1834, PSGM; Charles Lyell to Gideon Mantell, 29 October 1841, in K. Lyell, ed., *Life*, 2: 59; Lyell, *Travels*, 1: 78.

49. Mallet, *Cotton*, xii; Schulten, *Mapping the Nation*, 149–52; the map is available at http://www.mappingthenation.com/index.php/viewer/index/4/13.

50. Nott and Gliddon, *Types of Mankind*, 87.

Chapter Five: Mammals, the First Americans

1. Cope, "The Monster," 529; Osborn began offering a long-running course on mammal evolution at Columbia in 1892: "Museum Notes," 334; Gregory, "The Orders of Mammals," 3.

2. Price, *The Oglala People*, 31–32; Hämäläinen, *Lakota America*, chaps. 5–7; Dussias, "Science"; Bradley, *Dinosaurs*; Allen, "The American Bisons," vi.

3. Owen et al., *Report*, xxxiii, 197.

4. Leconte, *Elements*, 475.

5. Hayden, "On the Geology," 9; Marsh, "Introduction and Succession," 353; Richard Owen to Edward Drinker Cope, 31 June 1873, EDCP; Osborn, *The Age of Mammals*, 43. The extensive scholarship on the "bone wars" (a term coined after 1960) includes Shor, *The Fossil Feud*; Davison, *The Bone Sharp*; Warren, *Joseph Leidy*; D. R. Wallace, *The Bonehunters' Revenge*; Jaffe, *The Gilded Dinosaur*; Rea, *Bone Wars*; Thompson, *The Legacy of the Mastodon*; Brinkman, *The Second Jurassic Dinosaur Rush*; Lanham, *The Bone Hunters*; Dawson, *Show Me the Bone*; Michael Everhart, *Oceans of Kansas*; Dingus, *King of the Dinosaur Hunters*; and Rieppel, *Assembling the Dinosaur*. On women in American paleontology, see Turner, Burek, and Moody, "Forgotten Women."

6. Osborn, "Prehistoric Quadrupeds," 715; Mayor, *Fossil Legends*.

7. Webb, *Buffalo Land*, 344; Dana, *Manual of Geology*, 490; Cope, *The Vertebrata*, 42.

8. Taft, *Artists*, 117–28; Everhart, "William E. Webb," 184–86.

9. "March 8th," 9–10.

10. LeConte, *Elements*, 476.

11. Alvarez et al., "Extraterrestrial Cause"; Keller et al., "Main Deccan Volcanism"; Bosker, "The Nastiest Feud in Science"; Sepkoski, *Catastrophic Thinking*; Jaher, *Doubters and Dissenters*.

12. Dana, *Manual of Geology*, 502–503; Osborn, "The Causes," 770; Owen, "Termination," 366; Cope, "The Monster," 523; Darwin, *On the Origin of Species* (1859), 321–22.

13. Marsh, *Dinocerata*, 6; Osborn, "Prehistoric Quadrupeds," 715; Hatcher, "Origin"; Meek and Hayden, "Descriptions," 432; Marsh, *Dinocerata*, 6; Leidy, *The Ancient Fauna*, 79; Roosevelt and Heller, *Life-Histories*; Osborn, *The Titanotheres*, 2: 853.

14. Gunning, *Life-History*, 67, 69, 239–41, 248–49; Larrabee, "The Scientific Work," 529; Morris, "The Extinction of Species," 260; Leconte, *Elements*, 476; Cope, "The Monster," 534; Cope, "On the Extinct Cats." American paleontologists adopted Charles Lyell's subdivisions for this period. The ancient Greek-inspired names signaled each epoch's distance from the present and the proportion of currently living species found in each stratum. The earliest was the Eocene (dawn + recent); next came the Miocene (less + recent); the Pliocene (most + recent); and finally the Pleistocene (most + new).

15. Cope, *Theology*, 23; Osborn, "New"; Marsh, "Introduction and Succession," 338, 337; Bowler, *The Non-Darwinian Revolution*.

16. Marsh, "Introduction and Succession," 377; Morris, "The Extinction of Species," 259; Osborn, "Prehistoric Quadrupeds," 710–11; Marsh, *Dinocerata*, 58. On brains in nineteenth-century evolutionary thought, see Bowler, *Theories*, 152–85.

17. Leidy, "On the Fossil Horse"; Marsh, "Introduction and Succession," 358, 362; LeConte, *Elements*, 506; Osborn, "Prehistoric Quadrupeds," 711–12; Marsh, "Fossil Horses."

18. Bowler, *Theories*, 79.

19. W. B. Scott, *A History*, 302; Osborn, "Restorations," 85; Osborn, "Prehistoric Quadrupeds," 712.

20. Leidy, *The Ancient Fauna*, 7; Marsh, "Introduction and Succession," 375; Leidy, "On the Fossil Horse," 263.

21. Hatcher, "The Titanotherium Beds," 214; Osborn, "Prehistoric Quadrupeds," 714; Osborn et al., "New or Little Known Titanotheres"; Marsh, "On the Structure," 83; Dana, *Manual of Geology*, 515. On gender in animal specimen display, see Kohlstedt, "Nature by Design."

22. Marsh, "Restoration," 163; Allen, "The American Bisons," 67; Osborn, "The Cranial Evolution," 163; Marsh, *Dinocerata*, 58–63; Marsh, "Introduction and Succession," 361–62; Hornaday, "The Extermination," 433. On the extinction of the bison, see Isenberg, *The Destruction*; Smits, "The Frontier Army."

23. Dana, *Manual of Geology*, 573.

24. Spencer, *The Principles*, 246; Dussias, "Science"; Howard, "Dakota Winter Counts"; Nabokov, *A Forest of Time*; Schuller, "The Fossil"; Wise, "Time Discovered"; Luciano, "Tracking Prehistory."

25. Green and Thornton, *The Year the Stars Fell*, 1–26; Ewers, "Early White Influence."

26. Fletcher, *Brief Memoirs*, 6; Pearlstein, Brostoff, and Trentelman, "A Technical Study"; Mallery, *Picture-Writing*, 26, 28; Hinsley, "Zunis and Brahmins," 170; Conn, *Museums*; Conn, *History's Shadow*; O'Brien, *Firsting and Lasting*.

27. Mallery, "Pictographs," 91–92; Mallery, "A Calendar," 6, 23. Today this calendar is known as the Lone Dog Winter Count.

28. Dutton, *Tertiary History*; Spence, *Dispossessing*; Jacoby, *Crimes*; Pyne, *How the Canyon*.

Chapter Six: Glacial Progress

1. G. F. Wright, *The Ice Age*, 208–9.

2. Badè, *The Life and Letters*, 1: 358.

3. E. Hitchcock, *Elementary Geology* (1841), 163.

4. I have benefited here from conversation with Julia Adeney Thomas, 8 September 2023.

5. On perceptions of the New England winter before the nineteenth century, see Wickman, *Snowshoe Country*, 233–66.

6. Brown, *Benjamin Silliman*, 311; Zeilinga de Boer, *New Haven's Sentinels*, 5, 13.

7. Lyell, *Principles of Geology*, 1: 299–300; Benjamin Silliman, "Lecture X," 2 April 1835, PBS; Mills, "Darwin and the Iceberg Theory."

8. L. Agassiz, "Upon Glaciers"; Conrad, "Notes."

9. Lurie, *Louis Agassiz*, 60–63; Bowler, *Evolution*, 112–18; Debus and Debus, *Paleoimagery*, chap. 24.

10. L. Agassiz, *Études* [first], ii, v, 16, 22; Birkhold, "Measuring Ice."

11. Rudwick, *Worlds before Adam*, 517–39; Carozzi, "Agassiz's Amazing Geological Speculation."

12. Conrad, "Notes," 239, 240.

13. Lurie, *Louis Agassiz*, 94–96.

14. Gideon Mantell to Samuel George Morton, 4 December 1834, PSGM.

15. L. Agassiz, *Recherches sur les Poissons Fossiles*, vi; Lurie, *Louis Agassiz*, 79–87.

16. There is much controversy today over the meaning of *roches moutonnées*; my description is taken from how Americans in the nineteenth century understood the term. For flocks of sheep, see Borns and Maasch, eds., *Foot Steps*, 40–41; for sacks of wool, see E. Hitchcock, *Elementary Geology* (1841), 207.

17. Louis Agassiz to Benjamin Silliman, [illegible month and day], 1837, PBS; Benjamin Silliman, "Lecture 27," 1 July 1841, PBS; Benjamin Silliman, "Lecture 24," 29 June 1841, PBS; Benjamin Silliman to Edward Hitchcock, 25 May 1841, EOWHP; Benjamin Silliman to Edward Hitchcock, 15 June 1841, EOWHP.

18. E. C. Agassiz, *Louis Agassiz*, 1: 347; Churchill, "The Reception of Agassiz's Glacial Theory"; Carozzi, "Agassiz's Influence"; R. Silliman, "Agassiz vs. Lyell"; Aldrich, *New York State Natural History Survey*, 110ff.; Marché, "Edward Hitchcock."

19. Benjamin Silliman to Edward Hitchcock, 25 May 1841, EOWHP; E. Hitchcock, *Elementary Geology* (1841), 216.

20. E. Hitchcock, "First Anniversary Address," 252, 255.

21. E. Hitchcock, *Elementary Geology* (1841), 218.

22. E. Hitchcock, *Final Report*, a3, 228; unidentified student's notes, Edward Hitchcock's "Lectures on Geology," September to November 1855, EOWHP; E. Hitchcock, *Elementary Geology* (1841), 198.

23. L. Agassiz, *Geological Sketches*, 2nd series, 77.

24. L. Agassiz, *Geological Sketches*, 2nd series, 77.

25. E. C. Agassiz, *Louis Agassiz*, 2: 411, 426–27, 446.

26. Marcou, ed., *Life*, 2: 97, 22–23; James D. Dana to Louis Agassiz, 1 June 1847, LAC; James D. Dana to Louis Agassiz, 23 July 1853, LAC.

27. Marcou, ed., *Life*, 2: 7–8; Lurie, *Louis Agassiz*, 240.

28. Lurie, *Louis Agassiz*, 148.

29. Blum, *Picturing Nature*, 211. It was possible to draw the same static four categories of Cuvier and Agassiz in the form of a tree, as Anna Maria Redfield did in her "General View of the Animal Kingdom." For more on biological metaphors, see Archibald, *Aristotle's Ladder*; Lima, *The Book of Circles*; Lima, *The Book of Trees*.

30. Agassiz and Gould, *Principles of Zoology*, 211, 214, 220.

31. Wallis, "Black Bodies."

32. E. C. Agassiz, *Louis Agassiz*, 2: 463; L. Agassiz, *Lake Superior*, 36–37, 52, 56–57; Lurie, *Louis Agassiz*, 100.

33. E. C. Agassiz, *Louis Agassiz*, 2: 464; L. Agassiz, *Lake Superior*, 376–77.

34. L. Agassiz, *Lake Superior*, 398–99, 404, 410, 417.

35. Herber, ed., *Baird-Agassiz Letters*, 37, 158.

36. E. C. Agassiz, *Louis Agassiz*, 2: 469; G. W. White, "The First Appearance"; L. Agassiz, "On the Fishes of Lake Superior," 31.

37. *National Political Map of the United States*.

38. Lurie, *Louis Agassiz*, 142; Guyot, *The Earth and Man*, 31, 331.

39. https://avalon.law.yale.edu/19th_century/csa_missec.asp; Harvey, *The Voyage of the Icebergs*, 66.

40. Noble, *After Icebergs*, 177; Harvey, *The Civil War*, 54.

41. Carr, *Frederic Edwin Church*, 80–99; Harvey, *The Voyage of the Icebergs*, 66.

42. L. Agassiz, "America the Old World," 373; 381; L. Agassiz, "The Silurian Beach," 460; L. Agassiz "Ice-Period in America," 93. Much of the content of these articles appeared as *Geological Sketches* (1866) and *Geological Sketches*, 2nd series (1876).

43. E. C. Agassiz, *Louis Agassiz*, 2: 593, 608, 610; Louis Agassiz to Edward Drinker Cope, 5 February 1869, EDCP.

44. E. C. Agassiz, *Louis Agassiz*, 2: 548; L. Agassiz, "Ice-Period in America," 89; Bedell, *The Anatomy of Nature*, 121.

45. Borns and Maasch, eds., *Foot Steps*, 4–5, 76.

46. DeLaski, "Glacial Action," 334, 336, 340; Borns and Maasch, eds., *Foot Steps*, 24.

47. Borns and Maasch, eds., *Foot Steps*, 3, 7, 191, 193, 195, 196.

48. Winchell, *Sketches of Creation*, 229; Geikie, *The Great Ice Age*, 381; Croll, *Climate and Time*, 241; S.-M. Grant, *North over South*; Lawson, *Patriot Fires*.

49. Cruikshank, *Do Glaciers Listen?*, 3–9.

50. LeConte, "On Some of the Ancient Glaciers," 326; Badè, *The Life and Letters*, 1: 358.

51. Badè, *The Life and Letters*, 1: 348, 351; Muir, "Living Glaciers," 549; LeConte, "On Some of the Ancient Glaciers," 326; Stephens, *Joseph LeConte*, 122–23.

52. King, *Mountaineering*, 248.

53. Croll, *Climate and Time*, 3.

54. Among the earliest uses of the "Anthropocene" is Crutzen and Stoermer, "The 'Anthropocene.'"

Chapter Seven: The Dinosaurs Go to College

1. There may have been others that are undocumented or that have been destroyed, such as the colored ceiling fresco at the University of Georgia from the 1870s or 1880s that depicted, in concentric circles, "the evolution of life through all the geological or zoologic ages." Stephens, "Darwin's Disciple," 85.

2. Marsden, *The Soul*.

3. Woodrow Wilson, "Notes for Four Lectures on the Study of History, I: 'The Preliminary Age,'" c. September 24, 1885, in Link, ed., *The Papers of Woodrow Wilson Digital Edition*. Some of the first uses of the word *prehistoric* are D. Wilson, *The Archaeology* and *Prehistoric Man*; the first usage of *pre-history* may be Tylor, *Primitive Culture*, 2: 401. In a larger literature, see especially Trautmann, "The Revolution."

4. W. Barksdale Maynard, "Princeton and the Civil War," https://slavery.princeton.edu/stories/princeton-and-the-civil-war; Kimberly Klein, "The Civil War Comes to Princeton in 1861," https://slavery.princeton.edu/stories/the-civil-war-comes-to-princeton-in-1861.

5. W. B. Scott, *Some Memories*, 35, 44.

6. McCosh and Dickie, *Typical Forms*, 332–33, 345; on McCosh's thought, see Hoeveler, *James McCosh*, 188–92; Livingstone, "Science"; Gundlach, *Process*, 88–89.

7. Libbey, Jr., "The Life," 207; Guyot, *The Earth and Man*, 233; Guyot, *Physical Geography*, 114–15; Dana, "Memoir," 334–35; Means, "The Narrative," 337; on Guyot's departure from Alexander von Humboldt's more integrative vision, see Walls, *The Passage to Cosmos*, 193–94.

8. P. Wilson, "Arnold Guyot"; Dana, "Memoir," 335–36; Boutwell, "Pestalozzian System," 55–56; Winterer, "Avoiding," 292.

9. "Olla-Podrida" (1 April 1874), 343; S. E. Turner, "The E. M. Museum."

10. Guyot, "The Museum," 264. Among required studies for the junior and senior classes was "Physical Geography (or Geology)." See *Catalogue* (1868–69), 20–21; for library books, see *Subject-Catalogue*, 220–21, 273, 295.

11. Woodrow Wilson, "Editorial," *Princetonian* 3 (March 13, 1879), 197, in Link, ed., *The Papers of Woodrow Wilson Digital Edition*; Osborn, *Creative Education*, 25–26; *Catalogue* (1870–71), 77.

12. "Reporter," *Princetonian* 3, no. 12 (16 January 1879), 144; Guyot, "The Museum," 265; "Olla-Podrida," (1 June 1870), 58–59; Gundlach, *Process*, 135.

13. Williams, *The Handbook*, 41–47; "Olla-Podrida," (1 July 1874), 41–42; Arnold Guyot to Edward Drinker Cope, 12 December 1878, EDCP. On Ward, see Kohlstedt, "Henry A. Ward"; Barrow, "The Specimen Dealer"; Andrei, *Nature's Mirror*, 10–11.

14. Guyot, "The Museum," 267; Libbey, Jr., "The Life," 214; Guthrie, ed., *The Princeton University Art Museum*, xi–xii; "Olla-Podrida," (1 June 1870), 58. The prehistoric lake dwellings were discovered in 1854 by Ferdinand Keller, whose monograph was first translated into English in 1866 (Keller, *The Lake Dwellings*). Later, the Swiss architect Le Corbusier drew on the Swiss lake

dwellings for some of his "floating" designs, an example of the connection between human prehistory and Modernism that is explored further in chapter 8. See Vogt, "The Discovery of Lake Dwellings."

15. C. Hodge, *What Is Darwinism?*, 177. On Hodge's views, see Noll, ed., *The Princeton Theology*; Roberts, *Darwinism and the Divine*, 17.

16. C. Hodge, *What Is Darwinism?*, 4, 5, 30, 46, 144, 157.

17. S. C. Adams, *Adams' Synchronological Chart*; Rosenberg and Grafton, *Cartographies of Time*, 172–78.

18. Allan Marquand's student lecture notes for James McCosh's psychology course (1872), in Rosaco, "The Teaching of Art," 15.

19. Rudwick, *Scenes from Deep Time*, 144; Doyle, "A Vision of 'Deep Time.'"

20. R. Owen, *On the Nature of Limbs*, 2; Desmond, "Designing the Dinosaur"; Hawkins, *A Comparative View*, intro.; Hawkins, "On Visual Education," 444; Bramwell and Peck, *All in the Bones*, 24, 36; McCarthy, *The Crystal Palace Dinosaurs*, unpaginated 1 of chap. 4.

21. Bramwell and Peck, *All in the Bones*, 37; Nichols, *After 1851*, 259; Ryder, "Hawkins' Hadrosaurs," 169–70; Warren, *Joseph Leidy*, 32.

22. Leidy, *Cretaceous Reptiles*, 76; "Dec. 14th," 217; Hawkins and Wallis, *Comparative Anatomy*, 25.

23. James McCosh to Edward Drinker Cope, 17 December 1878, EDCP.

24. Guyot, "The Museum," 266.

25. Dates of paintings are from Smithsonian finding aid: https://siarchives.si.edu/collections/siris_arc_404381.

26. Wonders, *Habitat Dioramas*.

27. Guyot, *Creation*, xi.

28. Guyot, *Creation*, 102, 103; Lurie, *Louis Agassiz*, 245.

29. Bramwell and Peck, *All in the Bones*, 42; McKinsey, *Niagara Falls*.

30. Copied from H. Miller, *The Old Red Sandstone* (7th ed., 1857), according to Benjamin Waterhouse Hawkins Album Images, ANSP; Gossen, "The Victorians' Dinosaurs"; E. White, "On *Cephalaspis lyelli* Agassiz."

31. Guyot, *Creation*, 106.

32. Dana, *Manual*, 386.

33. Winchell, *Sketches of Creation*, 171.

34. R. Owen, *Geology*, 35–38; Winchell, *Sketches of Creation*, 173.

35. Doyle and Robinson, "The Victorian 'Geological Illustrations,'" 183; Doyle, "A Vision of 'Deep Time,'" 200, 204; *Routledge's Guide*, 197–98.

36. Parley, *Wonders*, 17.

37. *Routledge's Guide*, 197.

38. Dickens, *Bleak House*, 9.

39. Guyot, *Creation*, 115 and plate VII.

40. The creatures are labeled in Guyot, *Creation*, plate VII; "Dec. 14th," 214; "March 8th," 9.

41. Winchell, *Sketches of Creation*, 190; Cope, "August 21st," 279; Cope, "Synopsis," 114; "March 8th," 9–10.

42. Cope, "The Fossil Reptiles."

43. Falconer and Cautley, *Fauna Antiqua Sivalensis*, 2; Hugh Falconer to Samuel George Morton, 19 January 1837, PSGM; Falconer and Cautley, "Sivatherium Giganteum"; L. Agassiz, "The Silurian Beach," 53; Chakrabarti, *Inscriptions of Nature*.

44. Guyot, *Creation*, 116.

45. Huxley, *Evidence*, 155; K. Lyell, ed., *Life*, 2: 330. A small sampling of the extensive literature on this revolutionary moment includes Grayson, *The Establishment*; Van Riper, *Men*; Cohen, "Charles Lyell"; Trautmann, "The Revolution"; Bowler, *Theories*; Stocking, *Victorian Anthropology*; Kelley, "The Rise of Prehistory"; Trigger, *A History*.

46. W.B.R. King, "The Reputed Fossil Man"; Huxley, *Evidence*, 155.

47. Kohlstedt, "Nature by Design"; Moser, *Ancestral Images*, 107–56.

48. Guyot, "The Museum," 265–66; Guthrie, ed., *The Princeton University Art Museum*, xi.

49. Guyot, *Creation*, 124; Hawkins and Wallis, *Comparative Anatomy*, 9.

50. Guyot, *Creation*, 124, 126.

51. Wilson, "Princeton in the Nation's Service," 460, 464, 465; Wilson, "Address by the President," 142.

Chapter Eight: The Caveman within Us

1. Fielding, *The Caveman within Us*; Schmalzer, *The People's Peking Man*; Clark, *God—or Gorilla*, 1–16.

2. J. Turner, *The Liberal Education*; Lears, *No Place of Grace*; Winterer, *The Culture of Classicism*; Kern, *The Culture of Time and Space*.

3. Morgan, *Ancient Society*, iii, vi.

4. Ward, *Dynamic Sociology*, 1: 15; Ward, *Pure Sociology*, 458; Rafferty, *Apostle*, 71–140.

5. Ward, "Sketch of Paleobotany," 432.

6. Ward, "Sketch of Paleobotany," 444.

7. C. Hodge, *What Is Darwinism?*, 144.

8. Darwin, *The Descent of Man*, 203.

9. Cohen, "'How Nationality Influences Opinion'"; Gayon, "Darwin and Darwinism."

10. Saporta, *Aperçu géologique du terroir d'Aix-en-Provence*; Matheron, "Notice sur les reptiles fossiles," 356, 350–52; Athanassoglou-Kallmyer, *Cézanne and Provence*, chap. 4.

11. *Personnages et visages caricaturés, et annotations*: https://collections.louvre.fr/en/ark:/53355/cl020032794.

12. Athanassoglou-Kallmyer, *Cézanne and Provence*, 180, 279, n. 25.

13. Conry, ed., *Correspondance*, 88; Ward, "Saporta," 144; Ward to Saporta, 6 August 1888, PLFW; Ward to Saporta, 19 July 1886, PLFW; Saporta, *Le Monde des plantes*, 4, 50. Ward also corresponded with other fossil plant experts in Europe such as Alphonse de Candolle in France and Constantin Freiherr von Ettingshausen in Germany.

14. Ward, *Dynamic Sociology*, 1: 446–47.

15. Ward, *Dynamic Sociology*, 1: vi, vii, 9, 20–21.

16. Ward, *Pure Sociology*, 332; Ward, *The Psychic Factors*, 180; Ward, "Our Better Halves," 275.

17. Gilman, *The Living*, 187; Scharnhorst and Knight, "Charlotte Perkins Gilman's Library."

18. Gilman, "Similar Cases," 244–45; Scharnhorst, "Making Her Fame."

19. Gilman, *The Living*, 187; Scharnhorst, "Making Her Fame," 192–201; Gough and Rudd, eds., *A Very Different Story*.

20. Osborn, "Birth Selection," 175; Kennedy, "Philanthropy," 193.

21. Osborn, "Men of the Old Stone Age," 282–83; Rainger, *An Agenda for Antiquity*, 100, 144.

22. Marion, *Premières Observations*, 9; Breuil, *Beyond the Bounds of History*, 18.

23. L. Lartet, "Une sépulture des troglodytes"; L. Lartet, "Mémoire sur une sépulture"; E. Lartet and Christy, *Reliquiæ Aquitanicæ*, 22, 41.

24. Osborn, "Men of the Old Stone Age," 281; Wissler, "The Art of the Cave Man," 289, 292.

25. Rainger, *An Agenda*, 93, 98; Moser, *Ancestral Images*, chap. 6; Porter, "The Rise of Parnassus," 26; Clark, *God—or Gorilla*, 195–223.

26. Osborn, *Men of the Old Stone Age*, 293, 316, 358, 450.

27. Grant, *The Passing*, ix, 100–103; Kennedy, "Philanthropy," 207; Riskin and Winterer, *The Apes and Us*, 55–60.

28. Rainger, *An Agenda*, 169–70; Osborn, "The Hall of the Age of Man," 229, 236, 238.

29. Rainger, *An Agenda*, 89; Milner, *Charles R. Knight*, 10–11.

30. Czerkas and Glut, *Dinosaurs*, 8–9; Berman, "A Note"; Weinberg, *The Lure of Paris*; Boime, *The Academy*.

31. Osborn, "The Hall of the Age of Man," 236; Czerkas and Glut, *Dinosaurs*, 25, 26, 28, 66; Sommer, "Seriality in the Making," 466.

32. Knight, "Was This the First Man?" 40; Sommer, "Seriality in the Making," 474.

33. L. P. Osborn, *The Chain of Life*, 169–70; Ross, *The Old World in the New*, 285–86; Hammond, "The Expulsion of the Neanderthals"; Bowler, *Theories*, 75–77; Clark, "Evolution for John Doe"; Leonard, *Illiberal Reformers*, chap. 7.

34. Osborn, "The Hall of the Age of Man," 236; Wissler, *Man and Culture*, 224

35. Knight, *Prehistoric Man*, 4, 253–54; Knight, "Was This the First Man?" 40; Berman, "Bad Hair Days in the Paleolithic."

36. Osborn, "The Hall of the Age of Man," 236. Osborn called it the "bâton de commandement," a phrase he seems to have taken from the French paleontologist Gabriel de Mortillet; see Sollas, *Ancient Hunters*, 322.

37. Porter, "The Rise of Parnassus," 29; Moser, "Gender Stereotyping."

38. Osborn, "The Discovery of Tertiary Man," 3.

39. Osborn, *The Hall of the Age of Man*, 41; Osborn, *Huxley and Education*, 37; L. P. Osborn, *The Chain of Life*, 182.

40. Osborn, "Hesperopithecus, the First Anthropoid Ape"; Osborn, "Hesperopithecus, the Anthropoid Primate"; 464; Darwin, *The Descent of Man*, 207; Bowler, *Theories*, 125; G. Smith, "Hesperopithecus."

41. L. P. Osborn, *The Chain of Life*, 186; Gregory, "Hesperopithecus."

42. Wissler, "Existing and Extinct Races of Men," 98.

43. Auden, "In Memory of Sigmund Freud."

44. Freud, *The Interpretation of Dreams*, 486.

45. Freud, "The Aetiology of Hysteria," 191; Freud, *The Interpretation of Dreams*, 405, 447.

46. Romanes, *Mental Evolution*, 2; Jackson, "Remarks," 29, 31, 32; W. James, *The Principles of Psychology*, 1: 6, 81, 30.

47. Gamwell and Solms, *From Neurology to Psychoanalysis*, 8–10, 13; on mind and Christianity in the early twentieth century, see J. Roberts, "Psychoanalysis and American Christianity."

48. Hinkle, "Woman's Subjective Dependence," 196; Sherry, "Beatrice Hinkle," 493–95; Wittenstein, "The Feminist Uses"; Karier, "Art," 59; J. Roberts, "Psychoanalysis and American Christianity," 232.

49. Hinkle, *Psychology of the Unconscious*, ix.

50. Hinkle, *The Re-Creating*, 10–11; Hinkle, "Jung's Libido Theory," 1080; Hinkle, *Psychology of the Unconscious*, xxvi, xxxiii; Hinkle, "Woman's Subjective Dependence," 202, 199.

51. Hinkle, "Why Feminism?," 9; Hinkle, "Woman's Subjective Dependence," 194, 196.

52. Sherry, *The Jungian Strand*, 8; Jung, *The Red Book*, 97.

53. Hinkle, *The Re-Creating*, 127–8.

54. London, *The Book of Jack London*, 2: 323.

55. Lay, *The Child's Unconscious Mind*, 262; Fielding, *The Caveman within Us*, xii; Eichler, *The Customs of Mankind*, 60.

56. Auden, "Psychology and Art," 18; Meyer, *What Was Contemporary Art?*, 16–21; Nagel and Wood, *Anachronic Renaissance*, 9.

57. Frobenius and Fox, *Prehistoric Rock Pictures*, 9–10, 28; Meyer, *What Was Contemporary Art?*, 139–60.

58. Adès, *André Masson*, 16–20; MOMA press release, https://assets.moma.org/documents/moma_press-release_325089.pdf?_ga=2.176484838.297800010.1650646231-1550081573.1650646231; A. Barr, ed., *Fantastic Art*, #226, #234; Seibert, "'First Surrealists Were Cavemen,'" 18, 26–28, n. 3; Masson quotation from MOMA website: https://www.moma.org/collection/works/79309.

59. Milam, *Creatures of Cain*.

Chapter Nine: Pterodactyls in Eden

1. Roger Patterson, "Dinosaurs in Eden," *Answers Magazine*, July 1, 2015, https://answersingenesis.org/dinosaurs/when-did-dinosaurs-live/dinosaurs-eden/. On the Creation Museum, Trollinger and Trollinger, *Righting America*, 1.

2. Frank Newport, "In U.S., 46% Hold Creationist View of Human Origins" (June 1, 2012): https://news.gallup.com/poll/155003/Hold-Creationist-View-Human-Origins.aspx.

3. See also Worthen, *Apostles of Reason*.

4. W. H. Green, "Primeval Chronology," 303; Cramer, "The Theological," 511; C. Hodge, *Systematic Theology*, 1: 18; A. A. Hodge and Warfield, "Tractate on Inspiration," 37–38; Scofield, ed., *The Scofield Reference Bible*, 3; Numbers, *The Creationists*, 59–60; Numbers, "The Most Important," 270; Sandeen, *The Roots*, 222–24; Mangum and Sweetnam, *The Scofield Bible*.

5. Lyell, *Travels*, 1: 52; Darwin, *On the Origin of Species* (1860), 424; Lyell, *Geological Evidences*, 406; Owen, "Scientific Pursuits," 44.

6. White, *Spiritual Gifts*, 91; G.E.M. Price, *Outlines*, 13, 14, 17.

7. Baumgardner, "Catastrophic Plate Tectonics," 113.

8. "An Example of Circular Reasoning," June 4, 2007: https://answersingenesis.org/fossils/circular-reasoning/.

9. G.E.M. Price, *The New Geology*, 608.

10. Morris and Whitcomb, *The Genesis Flood*, 472. The book is discussed in Numbers, *The Creationists*, 213–38; Trollinger and Trollinger, *Righting America*, 8–9; Gordin, *The Pseudoscience Wars*, 135–37.

11. Sutton, *American Apocalypse*; Kruse, *One Nation*; Grem, *The Blessings*.

12. Creation Research Society, https://www.creationresearch.org/; Reed, "Time Warp I," 116; Meister, "Discovery of Trilobite Fossils," 98–100; Oard, "The Uinta Mountains and the Flood," 116. The Creation Research Society denied permission to reproduce the relevant images from Meister and Oard.

13. Permission to reproduce this image was denied, but it can be found here: https://answersingenesis.org/dinosaurs/humans/dinosaurs-ark-how-possible/.

Epilogue

1. Jorge Luis Borges, "The Book of Sand," trans. Norman Thomas de Giovanni, *New Yorker*, 25 October 1976, 38–39.

2. McPhee, *Basin and Range*, 21.

3. Serres and Latour, *Conversations*, 60.

BIBLIOGRAPHY

Archival Sources

Academy of Natural Sciences of Philadelphia (after 2011, of Drexel University)
    Charles Alexandre Lesueur Illustrations
    Papers of Timothy Abbott Conrad
American Philosophical Society
    Papers of Benjamin Smith Barton
    Papers of George William Featherstonhaugh
    Papers of Samuel George Morton
Amherst College, Archives and Special Collections
    Edward and Orra White Hitchcock Papers
Harvard University, Houghton Library
    Louis Agassiz Correspondence and Other Papers
Haverford College
    Edward Drinker Cope Papers
Huntington Library
    Papers of Rush Nutt
Marietta College, Special Collections
    Papers of Samuel Hildreth
New York Botanical Garden Archives
    Papers of John Torrey
New York State Library, Albany
    Papers of Amos Eaton
Princeton University, Seeley Mudd Manuscript Library
    Department of Geosciences Records
Princeton University, Special Collections Library
    Letterbook of Samuel George Morton
Smithsonian Institution Archives
    Papers of Lester Frank Ward
Yale University
    Papers of Benjamin Silliman

Printed Sources

Achim, Miruna. "Skulls and Idols: Anthropometrics, Antiquity Collections, and the Origin of American Man, 1810–1850." In *Nature and Antiquities: The Making of Archaeology in the Americas*, edited by Philip L. Kohl, Irina Podgorny, and Stefanie Gänger. Tucson: University of Arizona Press, 2014. 23–46.

Adams, Henry. *The Education of Henry Adams*. Boston: Houghton Mifflin, 1918.

Adams, John Quincy. *Memoirs of John Quincy Adams, Comprising Portions of His Diary from 1795 to 1848*, vol. 8. Edited by Charles Francis Adams. Philadelphia: J. B. Lippincott, 1876.

Adams, Sean Patrick. "Promotion, Competition, Captivity: The Political Economy of Coal." *Journal of Policy History* 18, no. 1 (Jan. 2006): 74–95.

Adams, Sean Patrick, ed. *The American Coal Industry, 1790–1902*. Vol. 3: *King Coal's Uneasy Throne in America, 1870–1902*. London: Pickering and Chatto, 2014.

Adams, Sebastian C. *Adams' Synchronological Chart of Universal History*. New York: Colby, 1871.

Adès, Dawn. *André Masson*. New York: Rizzoli, 1994.

Agassiz, Elizabeth Cary, ed. *Louis Agassiz: His Life and Correspondence*. 2 vols. Boston: Houghton Mifflin, 1885.

Agassiz, Louis. "America the Old World." *Atlantic Monthly*, March 1863, 373–82.

———. *Études sur les Glaciers*. Neuchâtel: Jent et Gassmann, 1840.

———. *Études sur les Glaciers: Dessinés d'après Nature et Lithographiés par J^{ph} Bettanier*. Neuchâtel: H. Nicolet, 1840.

———. "External Appearance of Glaciers." *Atlantic Monthly*, Jan. 1864, 56–65.

———. "The Fern Forests of the Carboniferous Period." *Atlantic Monthly*, May 1863, 615–25.

———. *Geological Sketches*. Boston: Ticknor and Fields, 1866.

———. *Geological Sketches*, 2nd ser. Boston: J. R. Osgood, 1876.

———. "Glacial Period." *Atlantic Monthly*, Feb. 1864, 224–32.

———. "Glacial Phenomena in Maine." *Atlantic Monthly*, Feb. 1867, 211–20; Mar. 1867, 281–87.

———. "The Growth of Continents." *Atlantic Monthly*, July 1863, 72–81.

———. "Ice-Period in America." *Atlantic Monthly*, July 1864, 86–93.

———. *Lake Superior: Its Physical Character, Vegetation, and Animals, Compared with Those of Other and Similar Regions; with a Narrative of the Tour, by J. Elliot Cabot and Contributions by Other Scientific Gentlemen*. Boston: Gould, Kendall, and Lincoln, 1850.

———. *Monographie des Poissons Fossiles du Vieux Grès Rouge, ou Système Dévonien (Old Red Sandstone) des Îles Britanniques et de Russie*. Neuchâtel: Self-published; Soleure: Jent et Gassmann, 1844.

———. "On the Fishes of Lake Superior." *Proceedings of the American Association for the Advancement of Science. First Meeting, Held at Philadelphia, September, 1848*. Philadelphia: John C. Clark, 1849. 30–32.

———. *Recherches sur les Poissons Fossiles*. Vol. 2: *Contentant l'Histoire de L'Ordre des Ganoïdes*. Neuchâtel: Imprimerie de Petitpierre, 1833–45.

———. "The Silurian Beach." *Atlantic Monthly*, April 1863, 460–71.

———. "The Tertiary Age and Its Characteristic Animals." *Atlantic Monthly*, Sept. 1863, 333–42.

———. "Upon Glaciers, Moraines, and Erratic Blocks; Being the Address Delivered at the Opening of the Helvetic Natural History Society, at Neuchatel, on the 24th of July 1837, by Its President, M. L. Agassiz." *Edinburgh New Philosophical Journal* 24, no. 48 (April 1838): 364–83.

Agassiz, Louis, and Augustus Gould. *Outlines of Comparative Physiology: Touching the Structure and Development of the Races of Animals, Living and Extinct; for the Use of Schools and Colleges; Edited from the Revised Edition, and Greatly Enlarged, by Thomas Wright, MD.* London: H. G. Bohn, 1855.

———. *Principles of Zoölogy.* Boston: Gould, Kendall, and Lincoln, 1848.

Aldrich, Michele. *New York State Natural History Survey, 1836–1845: A Chapter in the History of American Science.* Ithaca, NY: Paleontological Research Institution, 2000.

Allen, James A. "The American Bisons, Living and Extinct." *Memoirs of the Museum of Comparative Zoology at Harvard College* 4, no. 10 (1876): 1–221.

Alvarez, Luis W., Walter Alvarez, Frank Asaro, and Helen V. Michel. "Extraterrestrial Cause for the Cretaceous-Tertiary Extinction: Experimental Results and Theoretical Interpretation." *Science* 208, no. 4448 (June 6, 1980): 1095–1108.

American Paradise: The World of the Hudson River School. Introduction by John K. Howat. New York: Metropolitan Museum of Art, 1987.

Andrei, Mary Anne. *Nature's Mirror: How Taxidermists Shaped America's Natural History Museums and Saved Endangered Species.* Chicago: University of Chicago Press, 2020.

Appel, Toby. *The Cuvier-Geoffroy Debate: French Biology in the Decades before Darwin.* New York: Oxford University Press, 1987.

Archibald, J. David. *Aristotle's Ladder, Darwin's Tree: The Evolution of Visual Metaphors for Biological Order.* New York: Columbia University Press, 2014.

Athanassoglou-Kallmyer, Nina Maria. *Cézanne and Provence: The Painter in His Culture.* Chicago: University of Chicago Press, 2003.

Auden, W. H. "In Memory of Sigmund Freud." In *Another Time.* London: Faber and Faber, 1940.

———. "Psychology and Art." In *The Arts To-Day.* Edited with an Introduction by Geoffrey Grigson. London: John Lane, 1935. 1–21.

Auerbach, Eric. *Mimesis: The Representation of Reality in Western Literature.* 1946. Translated by Willard Trask with a new introduction by Edward Said. Princeton, NJ: Princeton University Press, 2003.

Badè, William. *The Life and Letters of John Muir.* 2 vols. Boston: Houghton Mifflin, 1924.

Barr, Alfred H., Jr., ed. *Fantastic Art, Dada, Surrealism.* New York: Museum of Modern Art, 1936.

Barr, James. "Why the World Was Created in 4004 BC: Archbishop Ussher and Biblical Chronology." *Bulletin of the John Rylands Library* 67, no. 2 (Jan. 1985): 575–608.

Barrande, Joachim. *Système silurien du centre de la Bohême.* 8 vols. Prague: Chez l'auteur et éditeur, 1852–1911.

Barrow, Mark, Jr. *Nature's Ghosts: Confronting Extinction from the Age of Jefferson to the Age of Ecology.* Chicago: University of Chicago Press, 2009.

———. "The Specimen Dealer: Entrepreneurial Natural History in America's Gilded Age." *Journal of the History of Biology* 33, no. 3 (Dec. 2000): 493–534.

Bartholomew, Michael. "The Non-Progress of Non-Progression: Two Responses to Lyell's Doctrine." *British Journal for the History of Science* 9, no. 2 (July 1976): 166–74.

Barton, Benjamin Smith. *New Views of the Origin of the Tribes and Nations of America.* Philadelphia: John Bioren, 1797.

Baumgardner, John R. "Catastrophic Plate Tectonics: The Physics behind the Genesis Flood." *Proceedings of the International Conference on Creationism* 5, no. 13 (2003): 113–26.

Beckert, Jens. *Imagined Futures: Fictional Expectations and Capitalist Dynamics.* Cambridge, MA: Harvard University Press, 2016.

Beckert, Sven. *Empire of Cotton: A Global History.* New York: Knopf, 2014.

Bedell, Rebecca. *The Anatomy of Nature: Geology and American Landscape Painting, 1825–1875.* Princeton, NJ: Princeton University Press, 2001.

Beer, Gillian. *Darwin's Plots: Evolutionary Narrative in Darwin, George Eliot and Nineteenth-Century Fiction.* 1983. 3rd ed. Cambridge: Cambridge University Press, 2009.

Bell, Adrienne Baxter. *George Inness and the Visionary Landscape.* New York: National Academy of Design, 2003.

Berman, Judith C. "Bad Hair Days in the Paleolithic: Modern (Re)Constructions of the Cave Man." *American Anthropologist* 101, no. 2 (June 1999): 288–304.

———. "A Note on the Paintings of Prehistoric Ancestors by Charles R. Knight." *American Anthropologist* 105, no. 1 (March 2003): 143–46.

Bieder, Robert E. *Science Encounters the Indian, 1820–1880: The Early Years of American Ethnology.* Norman: University of Oklahoma Press, 1986.

Bigsby, J. J. "Description of a New Species of Trilobite." *Journal of the Academy of Natural Sciences of Philadelphia* 4, pt. 2 (1825): 365–68.

Birkhold, Matthew. "Measuring Ice: How Swiss Peasants Discovered the Ice Age." *Germanic Review: Literature, Culture, Theory* 94, 3 (2019): 194–208.

Bjornerud, Marcia. *Timefulness: How Thinking Like a Geologist Can Help Save the World.* Princeton, NJ: Princeton University Press, 2018.

Blight, David. *Race and Reunion: The Civil War in American Memory.* Cambridge, MA: Belknap Press of Harvard University Press, 2001.

Blum, Ann Shelby. "'A Better Style of Art': The Illustrations of the *Paleontology of New York*." *Earth Sciences History* 6, no. 1 (1987): 72–85.

———. *Picturing Nature: American Nineteenth-Century Zoological Illustration.* Princeton, NJ: Princeton University Press, 1993.

Boime, Albert. *The Academy and French Painting in the Nineteenth Century.* New Haven, CT: Yale University Press, 1986.

Boltwood, Bertram B. "On the Ultimate Disintegration Products of the Radio-active Elements. Part II: The Disintegration Products of Uranium." *American Journal of Science* 4th ser., 23, no. 173 (Feb. 1907): 77–88.

Borns, Harold W., Jr., and Kirk Allen Maasch, eds. *Foot Steps of the Ancient Great Glacier of North America: A Long Lost Document of a Revolution in 19th Century Geological Theory.* New York: Springer, 2015.

Bosker, Bianca. "The Nastiest Feud in Science." *Atlantic*, September 2018. https://www.theatlantic.com/magazine/archive/2018/09/dinosaur-extinction-debate/565769/.

Boutwell, George S. "Pestalozzian System." *Popular Science Monthly* 44 (Nov. 1893): 55–56.

Bowler, Peter J. *Evolution: The History of an Idea*. 1984. Rev. ed. Berkeley: University of California Press, 1989.
———. *Fossils and Progress: Paleontology and the Idea of Progressive Evolution in the Nineteenth Century*. New York: Science History Publications, 1976.
———. *The Non-Darwinian Revolution: Reinterpreting a Historical Myth*. Baltimore: Johns Hopkins University Press, 1988.
———. *Theories of Human Evolution: A Century of Debate, 1844–1944*. Baltimore: Johns Hopkins University Press, 1986.
Bozeman, Theodore Dwight. *Protestants in an Age of Science: The Baconian Ideal in Antebellum Religious Thought*. Chapel Hill: University of North Carolina Press, 1977.
Brace, C. Loring. *"Race" Is a Four-Letter Word: The Genesis of the Concept*. New York: Oxford University Press, 2005.
Bradley, Lawrence. *Dinosaurs and Indians: Paleontology Resource Dispossession from Sioux Lands*. Denver: Outskirts Press, 2014.
Bramwell, Valerie, and Robert M. Peck. *All in the Bones: A Biography of Benjamin Waterhouse Hawkins*. Philadelphia: Academy of Natural Sciences, 2008.
Breuil, Henri. *Beyond the Bounds of History: Scenes from the Old Stone Age*. Translated by Mary E. Boyle. London: P. R. Gawthorn, 1949.
Brinkman, Paul. *The Second Jurassic Dinosaur Rush: Museums and Paleontology in America at the Turn of the Twentieth Century*. Chicago: University of Chicago Press, 2010.
Brongniart, Adolphe-Théodore. *Histoire des Végétaux Fossiles, ou Recherches Botaniques et Géologiques sur les Végétaux Renfermés dans les Diverses Couches du Globe*. Paris: G. Dufour et Ed. d'Ocagne, 1828.
Brongniart, Alexandre. "Rapport sur un Mémoire de M. Dufrénoy, Ingénieur des Mines, ayant pour titre: Des Caractères particuliers que présente le terrain de Craie dans le Sud de la France et sur les pentes des Pyrénées." *Annales des sciences naturelles* 22 (1831): 436–63.
———. *Traité des Arts Céramiques ou des Poteries, Considérées dans Leur Histoire, Leur Pratique et Leur Théorie*. 2 vols. Paris: Béchet Jeune and A. Mathias, 1844.
Brongniart, Alexandre, and Anselme-Gaëtan Desmarest, *Histoire Naturelle des Crustacés Fossiles, sous les Rapports Zoologiques et Géologiques*. Paris: F.-G. Levrault, 1822.
Brown, Chandos Michael. *Benjamin Silliman: A Life in the Young Republic*. Princeton, NJ: Princeton University Press, 1990.
Brush, Stephen G. "The Nebular Hypothesis and the Evolutionary Worldview." *History of Science* 25, no. 3 (Sept. 1987): 245–78.
Buckland, William. "Eyes of Trilobites." In *Geology and Mineralogy Considered with Reference to Natural Theology: A New Edition, with Supplementary Notes*, vol. 1. Rev. ed. Philadelphia: Lea and Blanchard, 1841. 299–304.
———. *The Bridgewater Treatises: On the Power, Wisdom, and Goodness of God as Manifested in the Creation. Treatise VI: Geology and Mineralogy Considered with Reference to Natural Theology*. 2 vols. London: William Pickering, 1836.
———. *Reliquiæ Diluvianæ; or, Observations on the Organic Remains Contained in Caves, Fissures, and Diluvial Gravel, and on Other Geological Phenomena, Attesting the Action of an Universal Deluge*. London: J. Murray, 1823.

Buffon, George Louis LeClerc, Comte de. *Natural History, General and Particular*, vol. 5. Translated by William Smellie. London: W. Strahan and T. Cadell, 1781.

Burkhardt, Richard W. *The Spirit of System: Lamarck and Evolutionary Biology; Now with "Lamarck in 1995."* Cambridge, MA: Harvard University Press, 1995.

Burnet, Thomas. *The Theory of the Earth: Containing an Account of the Original of the Earth, and of all the General Changes which it Hath Already Undergone, or is to Undergo Till the Consummation of All Things.* 3d ed. London: R. N. for Walter Kettilby, 1697.

Carey, Henry C. *Geographical, Statistical, and Historical Map of Alabama*. Philadelphia: H. C. Carey and I. Lea, 1822.

Carozzi, Albert V. "Agassiz's Amazing Geological Speculation: The Ice-Age." *Studies in Romanticism* 5, no. 2 (Jan. 1966): 57–83.

———. "Agassiz's Influence on Geological Thinking in the Americas." *Archives de Sciences* 27, no. 1 (Jan. 1974): 5–38.

Carr, Gerald L. *Frederic Edwin Church: The Icebergs*. Dallas: Dallas Museum of Fine Arts, 1980.

Castelnau, François de. *Essai sur le système silurien de l'Amérique septentrionale*. Paris: P. Bertrand, 1843.

Catalogue of the Officers and Students of the College of New Jersey, for the Academical Year, 1868–69. Princeton, NJ: Standard Office, 1869.

Catalogue of the Officers and Students of the College of New Jersey, for the Academical Year, 1870–71. Princeton, NJ: Standard Office, 1871.

Chadarevian, Soraya de, and Nick Hopwood, eds. *Models: The Third Dimension of Science*. Stanford, CA: Stanford University Press, 2004.

Chakrabarti, Pratik. *Inscriptions of Nature: Geology and the Naturalization of Antiquity*. Baltimore: Johns Hopkins University Press, 2020.

Chamberlin, Thomas C. "The Extent and Significance of the Wisconsin Kettle Moraine." *Transactions of the Wisconsin Academy of Sciences, Arts and Letters* 4 (1876–77): 201–34.

———. "The Method of Multiple Working Hypotheses." *Science* 15, no. 366 (Feb. 1890): 92–96.

———. "Preliminary Paper on the Terminal Moraine of the Second Glacial Epoch." In *Third Annual Report of the United States Geological Survey to the Secretary of the Interior, 1881–'82*, edited by John Wesley Powell. Washington, DC: United States Geological Survey, 1883. 291–402.

———. "The Rock-Scorings of the Great Ice Invasions." In *Seventh Annual Report of the United States Geological Survey to the Secretary of the Interior, 1885–'86*, edited by John Wesley Powell. Washington, DC: Government Printing Office, 1888. 155–248.

Charwat, Pat. "Treasures in Our Collections—Jacob Green's Trilobite Casts." *American Paleontologist* 17, no. 4. (Winter 2009): 5: online supplement. http://www.museumoftheearth.org/files/pubtext/supplements/suppl_557b.pdf (link no longer functional at time of publication).

Chubb, Lawrence J. "Sir Henry Thomas De la Beche." In *Jamaican Rock Stars, 1823–1971: The Geologists Who Explored Jamaica*, edited by Stephen K. Donovan. Boulder, CO: Geological Society of America, 2010. 9–28.

Churchill, Frederick B. "The Reception of Agassiz's Glacial Theory by American Geologists: A Study in the Maturity of Mid-Nineteenth Century Science." MA thesis, Columbia University, 1961.

Clark, Constance Areson. "Evolution for John Doe: Pictures, the Public, and the Scopes Trial Debate." *Journal of American History* 87, no. 4 (March 2001): 1275–1303.

———. *God—or Gorilla: Images of Evolution in the Jazz Age*. Baltimore: Johns Hopkins University Press, 2008.

Cohen, Claudine. "Charles Lyell and the Evidences of the Antiquity of Man." In *Lyell: The Past Is the Key to the Present*, edited by Derek J. Blundell and Andrew C. Scott. Geological Society Special Publication 143. London: Geological Society, 1998. 83–93.

———. *The Fate of the Mammoth: Fossils, Myth, and History*. Chicago: University of Chicago Press, 2002.

———. "'How Nationality Influences Opinion': Darwinism and Palaeontology in France (1859–1914)." *Studies in History and Philosophy of Biological and Biomedical Sciences* 66 (Dec. 2017): 8–17.

Cohn, Norman. *Noah's Flood: The Genesis Story in Western Thought*. New Haven, CT: Yale University Press, 1996.

Collie, Michael. *Science on Four Wheels: The European Travels of Roderick Murchison, 1840–45*. Bethesda, MD: Academica Press, 2010.

Collie, Michael, and John Dimer, eds. *Murchison's Wanderings in Russia: His Geological Exploration of Russia in Europe and the Ural Mountains, 1840 and 1841*. Keyworth, UK: British Geological Survey, 2004.

Comstock, John Lee. *Outlines of Geology: Intended as a Popular Treatise on the Most Interesting Parts of the Science; Together with an Examination of the Question, Whether the Days of Creation Were Indefinite Periods*. Hartford, CT: D. F. Robinson, 1834.

Conn, Steven. *History's Shadow: Native Americans and Historical Consciousness in the Nineteenth Century*. Chicago: University of Chicago Press, 2004.

———. *Museums and American Intellectual Life, 1876–1926*. Chicago: University of Chicago Press, 1998.

Conrad, Timothy Abbott. "Claiborne, Alabama: From the Note Book of a Traveller." *The Advocate of Science, and Annals of Natural History* 1, no. 1 (August 1834): 26–31.

———. *Fossil Shells of the Tertiary Formations of North America, Illustrated by Figures Drawn on Stone, from Nature*. 3 vols. Philadelphia: Judah Dobson, 1832–1835.

———. *Fossils of the Tertiary Formations of North America, Illustrated by Figures Drawn on Stone, from Nature*. Philadelphia: Judah Dobson, 1838.

———. *New Fresh Water Shells of the United States, with Coloured Illustrations, and a Monograph of the Genus Anculotus of Say; Also a Synopsis of the American Naiades*. Philadelphia: Judah Dobson, 1834.

———. "Notes on American Geology: Observations on Characteristic Fossils, and Upon a Fall of Temperature in Different Geological Epochs." *American Journal of Science and Arts* 35, no. 2 (Jan. 1839): 237–51.

———. "Observations on the Silurian and Devonian Systems of the United States, with Descriptions of New Organic Remains." *Journal of the Academy of Natural Sciences of Philadelphia* 8, no. 2 (1842): 228–35.

———. "Third Annual Report of T. A. Conrad, on the Paleontological Department of the Survey." In *Communication from the Governor, Transmitting Several Reports Relative to the*

Geology Survey of the State [1836–1840], vol. 4. Albany, NY: Natural History Survey, 1840. 199–207.

Conry, Yvette, ed. *Correspondance entre Charles Darwin et Gaston de Saporta. Précédée de "Histoire de la paléobotanique en France au XIXe siècle."* Paris: Presses Universitaires de France, 1972.

Conybeare, William Daniel, and William Phillips. *Outlines of the Geology of England and Wales.* London: William Phillips, 1822.

Cook, Della C. "The Old Physical Anthropology and the New World: A Look at the Accomplishments of an Antiquated Paradigm." In *Bioarchaeology: The Contextual Analysis of Human Remains*, edited by Jane E. Buikstra and Lane A. Beck. Amsterdam: Elsevier, 2006. 27–72.

Cope, Edward Drinker. "August 21st. The President, Dr. Hays, in the Chair." *Proceedings of the Academy of Natural Sciences of Philadelphia* 18 (1866): 275–79.

———. "The Fossil Reptiles of New Jersey (Continued)." *American Naturalist* 3, no. 2 (1869): 84–91.

———. "The Monster of Mammoth Buttes." *Penn Monthly* 4 (Aug. 1873): 521–34.

———. "On the Extinct Cats of North America." *American Naturalist* 14, no. 12 (Dec. 1880): 833–58.

———. "Synopsis of the Extinct Batrachia, Reptilia and Aves of North America." *Transactions of the American Philosophical Society* 14, no. 1 (1870): 1–252.

———. *Theology of Evolution: A Lecture.* Philadelphia: Arnold, 1887.

———. *The Vertebrata of the Cretaceous Formations of the West.* Washington, DC: Government Printing Office, 1875.

Corgan, James X., ed. *The Geological Sciences in the Antebellum South.* Tuscaloosa: University of Alabama Press, 1982.

Corsi, Pietro. *The Age of Lamarck: Evolutionary Theories in France 1790–1830.* Translated by Jonathan Mandelbaum. Rev. ed. Berkeley: University of California Press, 1988.

Cramer, Frank. "The Theological and Scientific Theories of the Origin of Man." *Bibliotheca Sacra* 48 (1891): 510–16.

Crèvecoeur, J. Hector St. Jean de. *Letters from an American Farmer.* Dublin: John Exshaw, 1782.

Croll, James. *Climate and Time in Their Geological Relations: A Theory of Secular Changes of the Earth's Climate.* London: Daldy, Isbister, 1875.

Cruikshank, Julie. *Do Glaciers Listen? Local Knowledge, Colonial Encounters, and Social Imagination.* Chicago: University of Chicago Press, 2005.

Crutzen, Paul J., and Eugene F. Stoermer. "The 'Anthropocene.'" *Global Change Newsletter* 41 (May 2000): 17–18.

Curwen, Eliot Cecil, ed. *The Journal of Gideon Mantell, Surgeon and Geologist: Covering the Years 1818–1852.* New York: Oxford University Press, 1940.

Cuvier, Georges, and Alexandre Brongniart. *Essai sur la Géographie Minéralogique des Environs de Paris.* Paris: Baudoin, 1811.

Czerkas, Sylvia Massey, and Donald F. Glut. *Dinosaurs, Mammoths, and Cavemen: The Art of Charles R. Knight.* Foreword by Edwin H. Colbert. New York: E. P. Dutton, 1982.

Dain, Bruce. *A Hideous Monster of the Mind: American Race Theory in the Early Republic.* Cambridge, MA: Harvard University Press, 2002.

Dal Prete, Ivano. *On the Edge of Eternity: The Antiquity of the Earth in Medieval and Early Modern Europe*. New York: Oxford University Press, 2022.

Dalrymple, G. Brent. *Ancient Earth, Ancient Skies: The Age of the Earth and Its Cosmic Surroundings*. Stanford, CA: Stanford University Press, 2004.

Dana, James Dwight. *Manual of Geology: Treating of the Principles of the Science with Special Reference to American Geological History, for the Use of Colleges, Academies, and Schools of Science; Illustrated by a Chart of the World, and Over One Thousand Figures, Mostly from American Sources*. Philadelphia: Theodore Bliss, 1863.

———. "Memoir of Arnold Guyot, 1807–1884." *Biographical Memoirs of the National Academy of Sciences* 2 (1886): 309–47.

Daniels, George H. *American Science in the Age of Jackson*. New York: Columbia University Press, 1968.

Darwin, Charles. *The Descent of Man, and Selection in Relation to Sex*. 2 vols. London: J. Murray, 1871.

———. *On the Origin of Species by Means of Natural Selection, or The Preservation of Favoured Races in the Struggle for Life*. London: J. Murray, 1859.

———. *On the Origin of Species by Means of Natural Selection; or The Preservation of Favoured Races in the Struggle for Life*. Rev. ed. New York: D. Appleton, 1860.

Davison, Jane Pierce. *The Bone Sharp: The Life of Edward Drinker Cope*. Philadelphia: Academy of Natural Sciences of Philadelphia, 1997.

Dawson, Gowan. *Show Me the Bone: Reconstructing Prehistoric Monsters in Nineteenth-Century Britain and America*. Chicago: University of Chicago Press, 2016.

Dawson, John Williams. *The Story of the Earth and Man*. New York: Harper and Brothers, 1873.

De la Beche, Henry. *Notes on the Present Condition of the Negroes in Jamaica*. London: T. Cadell, 1825.

———. "Remarks on the Geology of Jamaica." *Transactions of the Geological Society* 2, no. 2 (1827): 143–94.

De Waal Malefijt, Annemarie. *Images of Man: A History of Anthropological Thought*. New York: Alfred A. Knopf, 1974.

Debus, Allen A., and Diane E. Debus. *Paleoimagery: The Evolution of Dinosaurs in Art*. Jefferson, NC: McFarland, 2002.

"Dec. 14th. Vice-President Bridges in the Chair." *Proceedings of the Academy of Natural Sciences of Philadelphia* 10 (1858): 213–22.

Dekay, James E. "Observations on the Structure of Trilobites, and Description of an Apparently New Genus; with Notes on the Geology of Trenton Falls by James Renwick." *Annals of the Lyceum of Natural History of New York* 1 (1824): 174–89.

DeLaski, John K. "Glacial Action about Penobscot Bay." *American Journal of Science* 2nd ser., vol. 37, no. 111 (May 1864): 335–44.

DeLue, Rachael Ziady. *George Inness and the Science of Landscape*. Chicago: University of Chicago Press, 2005.

Desmond, Adrian J. "Designing the Dinosaur: Richard Owen's Response to Robert Edmond Grant." *Isis* 70, no. 2 (June 1979): 224–34.

Dewey, Chester. "Sketch of the Mineralogy and Geology of the Vicinity of Williams' College, Williamstown, Mass." *American Journal of Science and Arts* 1, no. 3 (Nov. 1819): 337–46.

Dickens, Charles. *Bleak House*. London: Bradbury and Evans, 1853.

Dickinson, Emily, and R. W. Franklin. *The Poems of Emily Dickinson: Reading Edition.* Cambridge, MA: Belknap Press of Harvard University Press, 2005.

Dingus, Lowell. *King of the Dinosaur Hunters: The Life of John Bell Hatcher and the Discoveries that Shaped Paleontology.* New York: Pegasus, 2018.

Dockery, David T. "Lesueur's Walnut Hills Fossil Shells." *Mississippi Geology* 2 (March 1982): 7–13.

Dott, Robert, Jr. "Charles Lyell's Debt to North America: His Lectures and Travels from 1841 to 1853." In *Lyell: The Past Is the Key to the Present*, edited by Derek J. Blundell and Andrew C. Scott. London: Geological Society, 1998. 53–69.

———. "Lyell in America—His Lectures, Field Work, and Mutual Influences, 1841–1853." *Earth Sciences History* 15, no. 2 (1996): 101–40.

Downing, Andrew Jackson. *A Treatise on the Theory and Practice of Landscape Gardening, Adapted to North America.* New York: Wiley and Putnam, 1841.

Doyle, Peter. "A Vision of 'Deep Time': The 'Geological Illustrations' of Crystal Palace Park, London." In *The History of Geoconservation*, edited by C. V. Burek and C. D. Prosser. Geological Society Special Publication 300. London: Geological Society, 2008. 197–206.

Doyle, Peter, and Eric Robinson. "The Victorian 'Geological Illustrations' of Crystal Palace Park." *Proceedings of the Geologists' Association* 104, no. 3 (1993): 181–94.

Draper, John William. *History of the American Civil War.* 3 vols. New York: Harper and Brothers, 1867–1870.

Du Bois, W.E.B. *The Souls of Black Folk: Essays and Sketches.* Chicago: A. C. McClurg, 1903.

Dugatkin, Lee Alan. *Mr. Jefferson and the Giant Moose: Natural History in Early America.* Chicago: University of Chicago Press, 2009.

Dunbar, William. "Abstract of a Communication from Mr. Martin Duralde, Relative to Fossil Bones, &c. of the Country of Apelousas West of the Mississippi to Mr. William Dunbar of the Natchez, and by Him Transmitted to the Society." *Transactions of the American Philosophical Society* 6 (1809): 55–58.

———. "Description of the River Mississippi and Its Delta, with That of the Adjacent Parts of Louisiana." *Transactions of the American Philosophical Society* 6 (1809): 165–87.

Dupree, A. Hunter. *Asa Gray, 1810–1888.* Cambridge, MA: Belknap Press of Harvard University Press, 1959.

Dussias, Allison. "Science, Sovereignty, and the Sacred Text: Paleontological Resources and Native American Rights." *Maryland Law Review* 55, no. 1 (1996): 84–159.

Dutton, Clarence E. *Tertiary History of the Grand Cañon District: [with] Atlas to Accompany the Monograph on the Tertiary History of the Grand Cañon District.* Text, Washington DC: Government Printing Office; atlas, New York: Julius Bien, 1882.

Eaton, Amos. "The Globe Had a Beginning." *American Journal of Science and Arts* 3 (1821): 238.

"Editorial." *Princetonian*, March 13, 1879, 197.

"Editorial: Mismeasure for Mismeasure." *Nature* 474 (June 23, 2011): 419.

Eichler, Lillian. *The Customs of Mankind.* New York: Doubleday, Page, 1924.

Elias, Norbert. *Time: An Essay.* Translated by Edmund Jephcott. 1992. Repr. Oxford: Blackwell, 1993.

Elkins, James. "Art History and Images that Are Not Art." *Art Bulletin* 77, no. 4 (December 1995): 553–71.

Emling, Shelley. *The Fossil Hunter: Dinosaurs, Evolution, and the Woman Whose Discoveries Changed the World*. New York: St. Martin's Griffin, 2009.

Evans, Lewis. *A Map of Pensilvania, New-Jersey, New-York, and the Three Delaware Counties; by Lewis Evans* (n.p., 1749).

Everhart, Michael. *Oceans of Kansas: A Natural History of the Western Interior Sea*. 2nd ed. Bloomington: Indiana University Press, 2017.

———. "William E. Webb: Civil War Correspondent, Railroad Land Baron, Town Founder, Kansas Legislator, Adventurer, Fossil Collector, Author." *Transactions of the Kansas Academy of Science* 119, no. 2 (Spring 2016): 179–92.

Ewers, John Canfield. "Early White Influence upon Plains Indian Painting: George Catlin and Carl Bodmer among the Mandan, 1832–34." *Smithsonian Miscellaneous Collections* 134, no. 7 (April 1957): 1–11.

Fabian, Ann. *The Skull Collectors: Race, Science, and America's Unburied Dead*. Chicago: University of Chicago Press, 2010.

Falconer, Hugh, and Proby T. Cautley. *Fauna Antiqua Sivalensis, Being the Fossil Zoology of the Sewalik Hills, in the North of India; Part I: Proboscidea*. London: Smith, Elder, 1846.

———. "Sivatherium Giganteum, a New Fossil Ruminant Genus, from the Valley of the Markanda, in the Siválik Branch of the Sub-Himálayan Mountains." *Journal of the Asiatic Society of Bengal* 5 (Jan. 1836): 38–50.

Featherstonhaugh, George. *Excursion through the Slave States*. New York: Harper and Brothers, 1844.

Fielding, William J. *The Caveman within Us: His Peculiarities and Powers; How We Can Enlist His Aid for Health and Efficiency*. New York: E. P. Dutton, 1922.

Fletcher, Robert. *Brief Memoirs of Colonel Garrick Mallery, U.S.A., Who Died October 24, 1894*. Washington, DC: Judd and Detweiler, 1895.

Ford, J. "Timothy Abbott Conrad." *Nautilus* 10, no. 10 (Feb. 1897): 110–12.

Foster, Gaines. *Ghosts of the Confederacy: Defeat, the Lost Cause, and the Emergence of the New South, 1865–1913*. New York: Oxford University Press, 1985.

Foulke, William Parker. "A Statement regarding the Fossil Bones, Shells and Wood Presented by Him to the Academy This Evening." *Proceedings of the Academy of Natural Sciences of Philadelphia* 10 (1858): 213–22.

Fredrickson, George M. *The Black Image in the White Mind: The Debate on Afro-American Character and Destiny, 1817–1914*. New York: Harper and Row, 1971.

Freud, Sigmund. "The Aetiology of Hysteria." In *The Standard Edition of the Complete Psychological Works of Sigmund Freud; Volume III (1893–1899): Early Psycho-Analytic Publications*, edited by James Strachey, Anna Freud, Alix Strachey, and Alan Tyson. 1893. Reprint, London: Hogarth Press, Institute of Psycho-analysis, 1962. 187–221.

———. *The Interpretation of Dreams*. Authorized translation of third edition. Translated and with introduction by A. A. Brill. New York: Macmillan, 1913.

Frobenius, Leo, and Douglas A. Fox. *Prehistoric Rock Pictures in Europe and Africa: From Material in the Archives of the Research Institute for the Morphology of Civilization, Frankfort-on-Main*. New York: Museum of Modern Art, 1937.

Gamwell, Lynn, and Mark Solms. *From Neurology to Psychoanalysis: Sigmund Freud's Neurological Drawings and Diagrams of the Mind*. Binghamton, NY: Binghamton University Art Museum, 2006.

Gayon, Jean. "Darwin and Darwinism in France before 1900." In *The Cambridge Encyclopedia of Darwin and Evolutionary Thought*, edited by Michael Ruse. Cambridge: Cambridge University Press, 2013. 243–49.

Geikie, James. *The Great Ice Age, and Its Relation to the Antiquity of Man*. New York: D. Appleton, 1874.

Gerbi, Antonello. *The Dispute of the New World: The History of a Polemic, 1750–1900*. Translated by Jeremy Moyle. Rev. ed. Pittsburgh: University of Pittsburgh Press, 1973.

Gerstner, Patsy A. "The Influence of Samuel George Morton on American Geology." In *Beyond History of Science: Essays in Honor of Robert E. Schofield*, edited by Elizabeth Garber. Bethlehem, PA: Lehigh University Press, 1990. 126–36.

———. "The 'Philadelphia School' of Paleontology, 1820–1845." PhD diss., Case Western Reserve University, 1967.

Gibbes, Robert W. *The Present Earth, the Remains of a Former World: A Lecture Delivered before the South Carolina Institute, September 6, 1849*. Columbia: A. S. Johnston, 1849.

Gillispie, Charles Coulston. *Genesis and Geology: A Study of the Relations of Scientific Thought, Natural Theology, and Social Opinion in Great Britain, 1790–1850*. Foreword by Nicolaas A. Rupke. Rev. ed. Cambridge, MA: Harvard University Press, 1996.

Gilman, Charlotte Perkins [Stetson]. *The Living of Charlotte Perkins Gilman: An Autobiography*. New York: Arno Press, 1972.

———. "Similar Cases." *New Californian: A Theosophic Journal Devoted to the Practical Realization of Universal Brotherhood* 1, no. 8 (Jan. 1892): 244–45.

———. *Women and Economics: A Study of the Economic Relation between Men and Women as a Factor in Social Evolution*. Boston: Small, Maynard, 1898.

Glen, William. "On the Mass-Extinction Debates: An Interview with Stephen J. Gould." In *The Mass-Extinction Debates: How Science Works in a Crisis*, edited by William Glen. Stanford, CA: Stanford University Press, 1994. 253–68.

Goodman, Ruth. *The Domestic Revolution: How the Introduction of Coal into Victorian Homes Changed Everything*. New York: Liveright, 2020.

Gordin, Michael. *The Pseudoscience Wars: Immanuel Velikovsky and the Birth of the Modern Fringe*. Chicago: University of Chicago Press, 2012.

Gossen, Anne. "The Victorians' Dinosaurs." *Garden State Legacy*, December 2010, unpaginated.

Gough, Val, and Jill Rudd, eds. *A Very Different Story: Studies on the Fiction of Charlotte Perkins Gilman*. Liverpool: Liverpool University Press, 1998.

Gould, Stephen J. *The Mismeasure of Man*. New York: W. W. Norton, 1981.

———. "Morton's Ranking of Races by Cranial Capacity: Unconscious Manipulation of Data May Be a Scientific Norm." *Science* 200, no. 4341 (May 5, 1978): 503–9.

Grafton, Anthony. *Joseph Scaliger: A Study in the History of Classical Scholarship*. Vol. 1: *Textual Criticism and Exegesis*. Oxford: Clarendon, 1983.

Graham, Billy. *World Aflame*. New York: Doubleday, 1965.

Grant, Madison. *The Passing of the Great Race; or the Racial Basis of European History*. New York: Charles Scribner's Sons, 1916.

Grant, Susan-Mary. *North over South: Northern Nationalism and American Identity in the Antebellum Era*. Lawrence: University Press of Kansas, 2000.

Grayson, Donald. *The Establishment of Human Antiquity*. London: Academic, 1983.
Green, Ashbel. "Biographical Sketch of Jacob Green, M.D." In *A History of the Jefferson Medical College of Philadelphia, with Biographical Sketches of the Early Professors*, by James F. Gayley. Philadelphia: Joseph M. Wilson, 1858. 31–34.
Green, Candace S., and Russell Thornton, eds. *The Year the Stars Fell: Lakota Winter Counts at the Smithsonian*. Lincoln: University of Nebraska Press, 2007.
Green, Jacob. *The Inferior Surface of the Trilobite Discovered: Illustrated with Coloured Models*. Philadelphia: Judah Dobson, 1839.
———. *A Monograph of the Trilobites of North America: With Coloured Models of the Species*. Philadelphia: Joseph Brano, 1832.
———. *Notes of a Traveller, during a Tour through England, France, and Switzerland, in 1828*. 3 vols. New York: G. and C. and H. Carvill, 1831.
Green, William Henry. "Primeval Chronology." *Bibliotheca Sacra* 47 (1890): 285–303.
Greene, Mott T. *Geology in the Nineteenth Century: Changing Views of a Changing World*. Ithaca, NY: Cornell University Press, 1982.
Gregory, William K. "Hesperopithecus Apparently Not an Ape Nor a Man." *Science* 66, 1720 (16 Dec. 1927): 579–81.
———. "The Orders of Mammals." *Bulletin of the American Museum of Natural History* 27 (Feb. 1910): 1–469.
Grem, David. *The Blessings of Business: How Corporations Shaped Conservative Christianity*. New York: Oxford University Press, 2016.
Guettard, Jean-Étienne. "Carte minéralogique, où l'on voit la nature des terrains du Canada et de la Louisiane." In "Mémoire dans lequel on compare le Canada à la Suisse, par rapport à ses minéraux," June 7, 1752. *Histoire de l'Académie Royale des Sciences* (1752): 189–220, 323–60.
Gundlach, Bradley. *Process and Providence: The Evolution Question at Princeton, 1845–1929*. Grand Rapids, MI: Eerdmans, 2013.
Gunning, William D. *Life-History of Our Planet*. W. B. Keen, Cooke, 1876.
Guralnick, Stanley M. "Geology and Religion before Darwin: The Case of Edward Hitchcock, Theologian and Geologist (1793–1864)." *Isis* 63, no. 4 (1972): 529–43.
Guthrie, Jill, ed. *The Princeton University Art Museum: Handbook of the Collections*. Princeton, NJ: Princeton University Art Museum, 2007.
Guyot, Arnold. *Creation; or, The Biblical Cosmogony in the Light of Modern Science*. New York: Charles Scribner's Sons, 1884.
———. *The Earth and Man: Lectures on Comparative Physical Geography, in Its Relation to the History of Mankind*. Translated by Cornelius Conway Felton. Rev. ed. Boston: Gould, Kendall, and Lincoln, 1850.
———. "The Museum of Geology and Archaeology." In *The Princeton Book*. Boston: Houghton, Osgood, 1879. 264–67.
———. *Physical Geography*. New York: Charles Scribner's Sons, 1873.
Haber, Francis C. *The Age of the World: Moses to Darwin*. Baltimore: Johns Hopkins University Press, 1959.
Hämäläinen, Pekka. *Lakota America: A New History of Indigenous Power*. New Haven, CT: Yale University Press, 2019.

Hamlin, Kimberly. *From Eve to Evolution: Darwin, Science, and Women's Rights in Gilded Age America*. Chicago: University of Chicago Press, 2014.

Hammond, Michael. "The Expulsion of the Neanderthals from Human Ancestry: Marcellin Boule and the Social Context of Scientific Research." *Social Studies of Science* 12, no. 1 (1982): 1–36.

Hamy, Ernest-Théodore. *Les voyages du naturaliste Ch. Alex. Lesueur dans l'Amérique du Nord*. Paris: Au Siège de la Société des Américanistes de Paris, 1904.

Harlan, Richard. "Notice of the Plesiosaurus, and Other Fossil *Reliqiuæ*, from the State of New Jersey." *Journal of the Academy of Natural Science of Philadelphia* 4, no. 2 (1825): 232–36.

Harvey, Eleanor Jones. *The Civil War and American Art*. New Haven, CT: Yale University Press in Association with the Smithsonian American Art Museum, 2012.

———. *The Voyage of the Icebergs: Frederic Church's Arctic Masterpiece*. Dallas: Dallas Museum of Art; New Haven, CT: Yale University Press, 2002.

Hatcher, John Bell. "Origin of the Oligocene and Miocene Deposits of the Great Plains." *Proceedings of the American Philosophical Society* 41, no. 169 (April 1902): 113–31.

———. "The Titanotherium Beds." *American Naturalist* 27, no. 315 (March 1893): 204–21.

Hattem, Michael. *Past and Prologue: Politics and Memory in the American Revolution*. New Haven, CT: Yale University Press, 2020.

Hawkins, Benjamin Waterhouse. *A Comparative View of the Human and Animal Frame*. London: Chapman and Hall, 1860.

———. "On Visual Education as Applied to Geology." *Journal of the Society of Arts* 2, no. 78 (19 May 1854): 444–49.

Hawkins, Benjamin Waterhouse, and George Wallis. *Comparative Anatomy as Applied to the Purposes of the Artist*. London: Winsor and Newton, 1883.

Hayden, Ferdinand Vandeveer. "On the Geology of the Tertiary Formations of Dakota and Nebraska." In *The Extinct Mammalian Fauna of Dakota and Nebraska: Including an Account of Some Allied Forms from Other Localities, Together with a Synopsis of the Mammalian Remains of North America*, edited by Joseph Leidy. Philadelphia: J. B. Lippincott, 1869. 9–21.

Hendel, Ronald. *The Book of Genesis: A Biography*. Princeton, NJ: Princeton University Press, 2012.

Herber, Elmer Charles, ed. *Baird-Agassiz Letters: Correspondence between Spencer Fullerton Baird and Louis Agassiz, Two Pioneer American Naturalists*. Washington, DC: Smithsonian Institution, 1963.

Herbert, Robert L., and Daria D'Arienzo, *Orra White Hitchcock: An Amherst Woman of Art and Science*. Amherst, MA: Mead Art Museum, 2011.

Hildreth, Samuel P. "Observations on the Bituminous Coal Deposits of the Valley of the Ohio, and the Accompanying Rock Strata; with Notices of the Fossil Organic Remains and the Relics of Vegetable and Animal Bodies, Illustrated by a Geological Map, by Numerous Drawings of Plants and Shells, and by Views of Interesting Scenery." *American Journal of Science and Arts* 29, no. 1 (January 1836): 1–154.

Hinkle, Beatrice. "Jung's Libido Theory and the Bergsonian Philosophy." *New York Medical Journal* 99, no. 22 (May 30, 1914): 1080–86.

———, trans. *Psychology of the Unconscious: A Study of the Transformations and Symbolism of the Libido. A Contribution to the History of the Evolution of Thought. By Dr. C. G. Jung*. New York: Moffat, Yard, 1916.

———. *The Re-Creating of the Individual: A Study of Psychological Types and Their Relation to Psychoanalysis*. New York: Harcourt, Brace, 1923.

———. "Why Feminism?" *Nation* 125, no. 3235 (July 6, 1927): 8–9.

———. "Woman's Subjective Dependence upon Man." *Harper's Monthly Magazine* 164 (January 1932): 196–205.

Hinsley, Curtis M. *The Smithsonian and the American Indian: Making a Moral Anthropology in Victorian America*. Washington, DC: Smithsonian Institution Press, 1994.

———. "Zunis and Brahmins: Cultural Ambivalence in the Gilded Age." In *Romantic Motives: Essays on Anthropological Sensibility*, edited by George W. Stocking. Madison: University of Wisconsin Press, 1989. 169–207.

Hitchcock, Charles. "The Earlier Forms of Life." *Popular Science Monthly* 10 (January 1877): 257–72.

Hitchcock, Edward. "The Connection between Geology and the Mosaic History of the Creation." *Biblical Repository and Quarterly Observer* 5, no. 18 (April 1835): 439–51; 6, no. 20 (October 1835): 261–332.

———. *Elementary Geology*. Amherst, MA: J. S. and C. Adams, 1840.

———. *Elementary Geology*. 2nd ed. Amherst, MA: J. S. and C. Adams, 1841.

———. *Final Report on the Geology of Massachusetts*. Vol. 1. Northampton, MA: J. H. Butler, 1841.

———. "First Anniversary Address before the Association of American Geologists, at Their Second Annual Meeting in Philadelphia, April 5, 1841." *American Journal of Science and Arts* 41, no. 2 (1841): 232–76.

———. *Ichnology of New England: A Report on the Sandstone of the Connecticut Valley, Especially Its Fossil Footmarks, Made to the Government of the Commonwealth of Massachusetts*. Boston: William White, 1858.

———. "The Law of Nature's Constancy Subordinate to the Higher Law of Change." *Bibliotheca Sacra* 20, no. 79 (July 1863): 489–561.

———. "Ornithichnology. Description of the Foot Marks of Birds, (Ornithichnites) on New Red Sandstone in Massachusetts." *American Journal of Science and Arts* 29, no. 2 (Jan. 1836): 307–40.

———. *The Religion of Geology and Its Connected Sciences*. Boston: Phillips, Sampson, 1851.

———. *Reminiscences of Amherst College, Historical, Scientific, Biographical and Autobiographical: Also, of Other and Wider Life Experiences*. Northampton, MA: Bridgman and Childs, 1863.

———. "Report on Ichnolithology, or Fossil Footmarks, with a Description of Several New Species, and the Coprolites of Birds, from the Valley of Connecticut River, and of a Supposed Footmark from the Valley of Hudson River." *American Journal of Science and Arts* 47, no. 2 (1844): 292–322.

———. "A Sketch of the Geology, Mineralogy, and Scenery of the Regions Contiguous to the River Connecticut; with a Geological Map and Drawings of Organic Remains; and Occasional Botanical Notices." *American Journal of Science and Arts* 6 (1823): 1–86.

Hodge, Archibald A., and Benjamin Warfield. "Tractate on Inspiration" (1881). Reprinted in *Westminster Doctrine anent Holy Scripture: Tractates by Professors A. A. Hodge and Warfield, with Notes on Recent Discussions*, edited by Robert Howie. Glasgow: David Bryce and Son, 1891. 27–60.

Hodge, Charles. *Systematic Theology*. 3 vols. New York: Charles Scribner, 1871–1873.

———. *What Is Darwinism?* New York: Scribner, Armstrong, 1874.

Hodges, Graham Russell Gao. *Black New Jersey: 1664 to the Present Day*. New Brunswick, NJ: Rutgers University Press, 2018.

Hoeveler, J. David. *James McCosh and the Scottish Intellectual Tradition: From Glasgow to Princeton*. Princeton, NJ: Princeton University Press, 1981.

Holland, John. *The History and Description of Fossil Fuel, the Collieries, and Coal Trade of Great Britain*. London: Whittaker, 1835.

Hornaday, William Temple. "The Extermination of the American Bison, with a Sketch of Its Discovery and Life History." In *Report of the United States Museum for the Year Ending June 30, 1887*. Washington, DC: Government Printing Office, 1889. 367–548.

Horsman, Reginald. *Josiah Nott of Mobile: Southerner, Physician, and Racial Theorist*. Baton Rouge: Louisiana State University Press, 1987.

———. *Race and Manifest Destiny: The Origins of American Racial Anglo-Saxonism*. Cambridge, MA: Harvard University Press, 1981.

Howard, James. "Dakota Winter Counts as a Source of Plains History." *Bureau of American Ethnology Bulletin* 173, Anthropological Papers no. 61 (1960): 335–416.

Hutton, James. "Theory of the Earth; or, An Investigation of the Laws Observable in the Composition, Dissolution, and Restoration of Land upon the Globe." *Transactions of the Royal Society of Edinburgh* 1 (1788): 209–304.

Huxley, Thomas Henry. *Evidence as to Man's Place in Nature*. London: Williams and Norgate, 1863.

Isenberg, Andrew C. *The Destruction of the Bison: An Environmental History, 1750–1920*. Cambridge: Cambridge University Press, 2000.

Jackson, J. Hughlings. "Remarks on Evolution and Dissolution of the Nervous System." *Journal of Mental Science* 33, no. 141 (April 1887): 25–48.

Jacoby, Karl. *Crimes against Nature: Squatters, Poachers, Thieves, and the Hidden History of American Conservation*. Berkeley: University of California Press, 2014.

Jaffe, Mark. *The Gilded Dinosaur: The Fossil War between E. D. Cope and O. C. Marsh and the Rise of American Science*. New York: Crown, 2000.

Jaher, Frederic Cople. *Doubters and Dissenters: Cataclysmic Thought in America, 1883–1918*. New York: Free Press of Glencoe, 1964.

James, Edwin. *Account of an Expedition from Pittsburgh to the Rocky Mountains, Performed in the Years 1819 and '20 by Order of the Hon. J. C. Calhoun, Sec'y of War, under the Command of Major Stephen H. Long; From the Notes of Major Long, Mr. T. Say, and Other Gentlemen of the Exploring Party; Compiled by Edwin James, Botanist and Geologist for the Expedition*. 2 vols. Philadelphia: H. C. Carey and I. Lea, 1822–1823.

James, William. *Pragmatism: A New Name for Some Old Ways of Thinking*. New York: Longmans, Green, 1907.

———. *The Principles of Psychology*. 2 vols. New York: Henry Holt, 1890.

Jevons, William Stanley. *The Coal Question: An Inquiry concerning the Progress of the Nation, and the Probable Exhaustion of Our Coal-Mines*. 1865. 2nd ed. London: Macmillan, 1866.

"Joachim Barrande." *Proceedings of the American Academy of Arts and Sciences* 19 (May 1883–84): 539–45.

Johnson, Walter. *River of Dark Dreams: Slavery and Empire in the Cotton Kingdom.* Cambridge, MA: Harvard University Press, 2013.

Jordan, John W., ed. *Colonial Families of Philadelphia.* 2 vols. New York: Lewis, 1911.

Jung, Carl. *The Red Book =Liber novus.* Edited by Sonu Shamdasani. Preface by Ulrich Hoerni. Translated by Mark Kyburz, John Peck, and Sonu Shamdasani. New York: W. W. Norton, 2009.

Kaalund, Nanna Katrine Luders. "Of Rocks and 'Men': The Cosmogony of John William Dawson." In *Historicizing Humans: Deep Time, Evolution, and Race in Nineteenth-Century British Sciences,* edited by Efram Sera-Shriar. Pittsburgh: University of Pittsburgh Press, 2018. 44–67.

Karier, Clarence. "Art in a Therapeutic Age, Part I." *Journal of Aesthetic Education* 13, no. 3 (July 1979): 51–66.

Keith, Arthur. *The Antiquity of Man.* London: Williams and Norgate, 1915.

Keller, Ferdinand. *The Lake Dwellings of Switzerland and Other Parts of Europe.* Translated by John Edward Lee. London: Longmans, Green, 1866.

Keller, Gerta, Thierry Adatte, Silvia Gardin, Anna Bartolini, and Sunil Bajpai. "Main Deccan Volcanism Phase Ends Near the K–T Boundary: Evidence from the Krishna–Godavari Basin, SE India." *Earth and Planetary Science Letters* 268, nos. 3–4 (2008): 293–311.

Kelley, Donald R. "The Rise of Prehistory." *Journal of World History* 14, no. 1 (March 2003): 17–36.

Kemmerly, Phillip R. "The Vanuxem Collection." *Earth Sciences History* 1 (1982): 32–35.

Kennedy, John Michael. "Philanthropy and Science in New York City: The American Museum of Natural History, 1868–1968." PhD diss., Yale University, 1968.

Kern, Stephen. *The Culture of Time and Space, 1880–1918.* 1983. Reprint, Cambridge, MA: Harvard University Press, 2003.

King, Clarence. *Mountaineering in the Sierra Nevada.* Boston: J. R. Osgood, 1872.

King, W.B.R. "The Reputed Fossil Man of the Neanderthal." *Quarterly Journal of Science* 1 (1864): 88–97.

Kingsley, William L. *Yale College, a Sketch of Its History.* 2 vols. New York: H. Holt, 1879.

Knight, Charles R. *Prehistoric Man: The Great Adventurer.* New York: Appleton-Century-Crofts, 1949.

———. "Was This the First Man?" *Popular Science Monthly* 98 (June 1921): 40–41.

Kohlstedt, Sally Gregory. "Henry A. Ward: The Merchant Naturalist and American Museum Development." *Journal of the Society for the Bibliography of Natural History* 9 (1980): 647–61.

———. "Nature by Design: Masculinity and Animal Display in Nineteenth-Century America." In *Figuring It Out: Science, Gender, and Visual Culture,* edited by Ann Shteir and Bernard Lightman. Hanover, NH: Dartmouth College Press, 2006. 110–39.

Krauthamer, Barbara. *Black Slaves, Indian Masters: Slavery, Emancipation, and Citizenship in the Native American South.* Chapel Hill: University of North Carolina Press, 2013.

Kruse, Kevin. *One Nation, Under God: How Corporate America Invented Christian America.* New York: Basic Books, 2015.

Lajoix, Anne. "Alexandre Brongniart: Scholar and Member of the Institut de France." In *The Sèvres Porcelain Manufactory: Alexandre Brongniart and the Triumph of Art and Industry,*

1800–1847, edited by Derek E. Ostergard. New Haven, CT: Yale University Press for the Bard Graduate Center for Studies in the Decorative Arts, 1997. 25–41.

Lanham, Url. *The Bone Hunters: The Heroic Age of Paleontology in the American West*. 1973. Reprint, New York: Dover, 1991.

Larrabee, William H. "The Scientific Work of W. D. Gunning." *Popular Science Monthly* 50 (Feb. 1897): 526–30.

Lartet, Edouard, and Henry Christy. *Reliquiæ Aquitanicæ; Being Contributions to the Archæology and Palæontology of Périgord and the Adjoining Provinces of Southern France*. Edited by Thomas Rupert Jones. London: Williams and Norgate, 1875.

Lartet, Louis. "Mémoire sur une sépulture des anciens troglodytes du Périgord." *Annales des sciences naturelles: Zoologie et paléontologie* ser. 5, 10 (1868): 133–45.

———. "Une sépulture des troglodytes du Périgord." *Bulletins de la Société d'Anthropologie de Paris* 3 (1868): 335–49.

Lawson, Melinda. *Patriot Fires: Forging a New American Nationalism in the Civil War North*. Lawrence: University Press of Kansas, 2002.

Lay, Wilfrid. *The Child's Unconscious Mind: The Relations of Psychoanalysis to Education; a Book for Teachers and Parents*. New York: Dodd, Mead, 1919.

Lea, Isaac. *Contributions to Geology*. Philadelphia: Carey, Lea and Blanchard, 1833.

Lears, T. J. Jackson. *No Place of Grace: Antimodernism and the Transformation of American Culture, 1880–1920*. 1981. Reprint, Chicago: University of Chicago Press, 2021.

LeConte, Joseph. *Elements of Geology: A Text-Book for Colleges and for the General Reader*. New York: D. Appleton, 1878.

———. "On Some of the Ancient Glaciers of the Sierras." *American Journal of Science and Arts* 3rd ser., 5, no. 29 (May 1873): 325–42.

Leidy, Joseph. *The Ancient Fauna of Nebraska: Or, A Description of Remains of Extinct Mammalia and Chelonia, from the Mauvaises Terres of Nebraska*. Smithsonian Contributions to Knowledge 6. Washington, DC: Smithsonian Institution, 1853.

———. *Cretaceous Reptiles of the United States*. Smithsonian Contributions to Knowledge, vol. 14, no. 192. Washington, DC: Smithsonian Institution, 1865.

———. "On the Fossil Horse of America." *Proceedings of the Academy of Natural Sciences of Philadelphia* 3, no. 11 (Sept.–Oct. 1847): 262–66.

Leonard, Thomas C. *Illiberal Reformers: Race, Eugenics, and American Economics in the Progressive Era*. Princeton, NJ: Princeton University Press, 2016.

Lesley, Joseph Peter. *Manual of Coal and Its Topography. Illustrated by Original Drawings, Chiefly of Facts in the Geology of the Appalachian Region of the United States of North America*. Philadelphia: J. B. Lippincott, 1856.

Lewis, Andrew J. *A Democracy of Facts: Natural History in the Early Republic*. Philadelphia: University of Pennsylvania Press, 2011.

Lewis, Jackson. "Choctaw: Choctaw Migration." In *Creation Myths and Legends of the Creek Indians*, edited by Bill Grantham. Gainesville: University Press of Florida, 2002. 157–58.

Lewis, Jason, David DeGusta, Marc Meyer, Janet Monge, Alan Mann, and Ralph Holloway. "The Mismeasure of Science: Stephen Jay Gould versus Samuel George Morton on Skulls and Bias." *PLoS Biology* 9, no. 6 (June 7, 2011).

Libbey, William, Jr. "The Life and Scientific Work of Arnold Guyot." *Journal of the American Geographical Society of New York* 16 (1884): 194–221.

Lima, Manuel. *The Book of Circles: Visualizing Spheres of Knowledge*. New York: Princeton Architectural Press, 2017.

———. *The Book of Trees: Visualizing Branches of Knowledge*. New York: Princeton Architectural Press, 2014.

Lincecum, Gideon. "Choctaw Traditions about Their Settlement in Mississippi and the Origin of Their Mounds." In *Pushmataha: A Choctaw Leader and His People*. Introduction by Greg O'Brien. Tuscaloosa: University of Alabama Press, 2004. 1–25.

Lindley, John, and William Hutton. *The Fossil Flora of Great Britain; or, Figures and Descriptions of the Vegetable Remains Found in a Fossil State in this Country*. 3 vols. London: J. Ridgway and Sons, 1831–1837.

Link, Arthur S., ed. *The Papers of Woodrow Wilson Digital Edition*. Charlottesville: University of Virginia Press, Rotunda, 2017.

Lippitt, Professor. "A Trip to Siluria." *Ladies' Repository* 13, no. 1 (Jan. 1853): 34–36.

Livingstone, David. *Adam's Ancestors: Race, Religion, and the Politics of Human Origins*. Baltimore: Johns Hopkins University Press, 2008.

———. "Science, Region, and Religion: The Reception of Darwinism in Princeton, Belfast, and Edinburgh." In *Disseminating Darwinism: The Role of Place, Race, Religion, and Gender*, edited by Ronald Numbers and John Stenhouse. Cambridge: Cambridge University Press, 1999. 7–38.

London, Charmian. *The Book of Jack London*. 2 vols. New York: Century, 1921.

Loudon, John Claudius. *The Green-house Companion*. 3rd ed. London: Whittaker, Treacher, 1832.

Lovejoy, Arthur. "The Argument for Organic Evolution before the *Origin of Species*, 1830–1858." In *Forerunners of Darwin: 1745–1859*, edited by Bentley Glass, Owsei Temkin, and William L. Straus, Jr. Baltimore: Johns Hopkins University Press, 1959. 356–414.

———. "Buffon and the Problem of Species." In *Forerunners of Darwin: 1745–1859*, edited by Bentley Glass, Owsei Temkin, and William L. Straus, Jr. Baltimore: Johns Hopkins University Press, 1959. 84–113.

Luarca-Shoaf, Nenette. "The Mississippi River in Antebellum Visual Culture." PhD diss., University of Delaware, 2012.

Lubbock, John. *Pre-historic Times as Illustrated by Ancient Remains, and the Manners and Customs of Modern Savages*. Edinburgh: Williams and Norgate, 1865.

Luciano, Dana. "Tracking Prehistory." *J19: Journal of Nineteenth-Century Americanists* 3, no. 1 (Spring 2015): 173–81.

Lunbeck, Elizabeth. *The Psychiatric Persuasion: Knowledge, Gender, and Power in Modern America*. Princeton, NJ: Princeton University Press, 1994.

Lurie, Edward. *Louis Agassiz: A Life in Science*. Chicago: University of Chicago Press, 1960.

Lyell, Charles. *Geological Evidences of the Antiquity of Man, with Remarks on Theories of the Origin of Species by Variation*. Philadelphia: George W. Childs, 1863.

———. *Principles of Geology: Being an Attempt to Explain the Former Changes of the Earth's Surface, by Reference to Causes Now in Operation*. 3 vols. London: J. Murray, 1830–1833.

———. *A Second Visit to the United States of North America*. 2 vols. New York: Harper and Brothers, 1849.

———. *Travels in North America, in the Years 1841–2; with Geological Observations on the United States, Canada, and Nova Scotia.* 2 vols. New York: Wiley and Putnam, 1845.

Lyell, Katharine Murray, ed. *Life, Letters and Journals of Sir Charles Lyell, Bart.* 2 vols. London: J. Murray, 1881.

Lyons, Lisa. "Panorama of the Monumental Grandeur of the Mississippi Valley." *Design Quarterly* 101/102 (1976): 32–34.

MacDuffie, Allen. *Victorian Literature, Energy, and the Ecological Imagination.* Cambridge: Cambridge University Press, 2014.

Maclure, William. "Observations on the Geology of the United States, Explanatory of a Geological Map." *Transactions of the American Philosophical Society* 6 (1809): 411–28.

Mallery, Garrick. "A Calendar of the Dakota Nation." Edited by Ferdinand V. Hayden. *Bulletin of the United States Geological and Geographical Survey of the Territories* 3, no. 1 (5 April 1877): 3–26.

———. "Pictographs of the North American Indians: A Preliminary Paper." In *Fourth Annual Report of the Bureau of Ethnology to the Secretary of the Smithsonian Institution, 1882–'83.* Washington, DC: Government Printing Office, 1886. 3–256.

———. *Picture-Writing of the American Indians: Extract from the Tenth Annual Report of the Bureau of Ethnology.* Washington, DC: Government Printing Office, 1894.

Mallet, John William. *Cotton: The Chemical, Geological, and Meteorological Conditions Involved in Its Successful Cultivation; With an Account of the Actual Conditions and Practice of Culture in the Southern or Cotton States of North America.* London: Chapman and Hall, 1862.

Malm, Andreas. *Fossil Capital: The Rise of Steam Power and the Roots of Global Warming.* New York: Verso, 2016.

Mangum, R. Todd, and Mark S. Sweetnam. *The Scofield Bible: Its History and Impact on the Evangelical Church.* Colorado Springs: Paternoster, 2009.

"March 8th, Dr. Carson, Vice-President, in the Chair." *Proceedings of the Academy of Natural Sciences of Philadelphia* 22, no. 1 (Jan.–Apr. 1870): 9–11.

Marchand, Suzanne L. *Porcelain: A History from the Heart of Europe.* Princeton, NJ: Princeton University Press, 2020.

Marché, Jordan D., II. "Edward Hitchcock, Roderick Murchison, and Rejection of the Alpine Glacial Theory (1840–1845)." *Earth Sciences History* 37, no. 2 (2018): 380–402.

Marcou, Jules, ed. *Life, Letters, and Words of Louis Agassiz.* 2 vols. New York: Macmillan, 1896.

Marion, Antoine-Fortuné. *Premières observations sur l'ancienneté de l'homme dans les Bouches-du-Rhône.* Aix-en-Provence: Remondet-Aubin, 1867.

Marsden, George. *The Soul of the American University: From Protestant Establishment to Established Nonbelief.* New York: Oxford University Press, 1994.

Marsh, Othniel. "American Jurassic Mammals." *American Journal of Science and Arts* 3rd ser., 33, no. 196 (April 1887): 327–48.

———. *Dinocerata: A Monograph of an Extinct Order of Gigantic Mammals.* Monographs of the United States Geological Survey, vol. 10. Washington, DC: Government Printing Office, 1886.

———. "Fossil Horses in America." *American Naturalist* 8, no. 5 (May 1874): 288–94.

———. "Introduction and Succession of Vertebrate Life in America." *American Journal of Science and Arts,* 3rd ser., 14, no. 83 (Nov. 1877): 337–78.

———. "On the Structure and Affinities of the Brontotheridae." *American Naturalist* 8, no. 2 (Feb. 1874): 79–85.

———. "Restoration of *Brontops robustus*, from the Miocene of America." *American Journal of Science*, 3rd ser., 37, no. 218 (Feb. 1889): 163–65.

Matheron, Philippe. "Notice sur les reptiles fossiles des depots fluvio-lacustres Crétacés du bassin a lignite de Fuveau." In *Mémoires de l'Académie des Sciences, Belles-lettres et Arts de Marseille: Années 1868–1869*. Marseille: Barlatier-Feissat Père et Fils, 1869. 345–80.

Matthews, Kara J., R. Dietmar Müller, and David T. Sandwell. "Oceanic Microplate Formation Records the Onset of India-Eurasia Collision." *Earth and Planetary Science Letters*, 433 (1 January 2016): 204–14.

Mayor, Adrienne. *Fossil Legends of the First Americans*. Princeton, NJ: Princeton University Press, 2005.

McCarthy, Steve, ed. *The Crystal Palace Dinosaurs: The Story of the World's First Prehistoric Sculptures*. Designed and produced by Mike Gilbert. London: Crystal Palace Foundation, 1994.

McCosh, James and George Dickie. *Typical Forms and Special Ends in Creation*. New York: Robert Carter and Brothers, 1856.

McKinsey, Elizabeth. *Niagara Falls: Icon of the American Sublime*. Cambridge: Cambridge University Press, 1985.

McPhee, John. *Basin and Range*. New York: Farrar, Strauss, and Giroux, 1981.

Means, John O. "The Narrative of the Creation in Genesis." *Bibliotheca Sacra* 12, no. 45 (January 1855): 83–130; 12, no. 46 (April 1855): 323–38.

Meek, Fielding B., and Ferdinand V. Hayden. "Descriptions of New Lower Silurian, (Primordial), Jurassic, Cretaceous, and Tertiary Fossils, Collected in Nebraska, by the Exploring Expedition under the Command of Capt. Wm F. Reynolds, U.S. Top. Engineers, with Some Remarks on the Rocks from Which They Were Obtained." *Proceedings of the Academy of Natural Sciences of Philadelphia* 13 (1861): 415–47.

Meigs, J. Aitken. *Catalogue of Human Crania, in the Collection of the Academy of Natural Sciences of Philadelphia: Based upon the Third Edition of Dr. Morton's "Catalogue of Skulls," &c*. Philadelphia: J. B. Lippincott, 1857.

Meister, William J., Sr. "Discovery of Trilobite Fossils in Shod Footprint of Human in 'Trilobite Beds'—A Cambrian Formation, Antelope Springs, Utah." *Creation Research Society Quarterly* 5, no. 3 (December 1968): 97–101.

Meltzer, David J. *The Great Paleolithic War: How Science Forged an Understanding of America's Ice Age Past*. Chicago: University of Chicago Press, 2015.

Menand, Louis. *The Metaphysical Club: A Story of Ideas in America*. New York: Farrar, Straus, and Giroux, 2001.

Meyer, Richard. *What Was Contemporary Art?* Cambridge, MA: MIT Press, 2013.

Michael, J. S. "A New Look at Morton's Craniological Research." *Current Anthropology* 29, no. 2 (April 1988): 349–54.

Milam, Erika Lorraine. *Creatures of Cain: The Hunt for Human Nature in Cold War America*. Princeton, NJ: Princeton University Press, 2019.

Miller, Angela L. *The Empire of the Eye: Landscape Representation and American Cultural Politics, 1825–1875*. Ithaca, NY: Cornell University Press, 1993.

Miller, David. *Dark Eden: The Swamp in Nineteenth-Century American Culture*. New York: Cambridge University Press, 1989.

Miller, Hugh. *The Old Red Sandstone; or, New Walks in an Old Field*. 4th ed. Boston: Gould and Lincoln, 1851.

Mills, William. "Darwin and the Iceberg Theory." *Notes and Records of the Royal Society Journal of the History of Science* 38, no. 1 (Aug. 1983): 109–27.

Milner, Richard. *Charles R. Knight: The Artist Who Saw through Time*. New York: Abrams, 2012.

Moffat, Charles H. "Charles Tait, Planter, Politician, and Scientist of the Old South." *Journal of Southern History* 14, no. 2 (May 1948): 206–33.

Moore, James R. "Geologists and Interpreters of Genesis." In *God and Nature: Historical Essays on the Encounter between Christianity and Science*, edited by David Lindberg and Ronald Numbers. Berkeley: University of California Press, 1986. 322–51.

Morgan, Lewis Henry. *Ancient Society, or Researches in the Lines of Human Progress from Savagery, through Barbarism to Civilization*. New York: Henry Holt, 1877.

Morris, Charles. "The Extinction of Species." *Proceedings of the Academy of Natural Sciences of Philadelphia* 47 (1895): 253–63.

Morris, Henry M., and John C. Whitcomb. *The Genesis Flood: The Biblical Flood and Its Scientific Implications*. Phillipsburg, NJ: Presbyterian and Reformed, 1961.

Morton, John L. *King of Siluria: How Roderick Murchison Changed the Face of Geology*. Horsham, UK: Brocken Spectre, 2004.

Morton, Samuel George. "Account of a Craniological Collection; with Remarks on the Classification of Some Families of the Human Race." *Transactions of the American Ethnological Society* 2 (1848): 216–22.

———. *Crania Ægyptiaca; or, Observations on Egyptian Ethnography, Derived from Anatomy, History and the Monuments*. Philadelphia: John Penington, 1844.

———. *Crania Americana; or, A Comparative View of the Skulls of Various Aboriginal Nations of North and South America: To Which Is Prefixed an Essay on the Varieties of the Human Species*. Philadelphia: J. Dobson, 1839.

———. "Geological Observations on the Secondary, Tertiary, and Alluvial Formations of the Atlantic Coast of the United States of America; Arranged from the Notes of Lardner Vanuxem." *Journal of the Academy of Natural Sciences of Philadelphia* 6, no. 1 (1829): 59–71.

———. *Letter to the Rev. John Bachman, D.D., on the Question of Hybridity in Animals, Considered in Reference to the Unity of the Human Species*. Charleston, SC: Walker and James, 1850.

———. "Notice of the Fossil Teeth of Fishes of the United States, the Discovery of the Galt in Alabama, and a Proposed Division of the American Cretaceous Group." *American Journal of Science and Arts* 28, no. 2 (1835): 276–78.

———. "Some Remarks on the Value of the Word *Species* in Zoology." *Proceedings of the Academy of Natural Sciences of Philadelphia* 5, no. 5 (1850–51): 81–82.

———. *Synopsis of the Organic Remains of the Cretaceous Group of the United States. Illustrated by 19 Plates, to Which Is Added an Appendix, Containing a Tabular View of the Tertiary Fossils Hitherto Discovered in North America*. Philadelphia: Key and Biddle, 1834.

———. "Synopsis of the Organic Remains of the Ferruginous Sand Formation of the United States; with Geological Remarks." *American Journal of Science* 17, no. 2 (Jan. 1830): 274–95.

Moser, Stephanie. *Ancestral Images: The Iconography of Human Origins.* Ithaca, NY: Cornell University Press, 1998.

———. "Gender Stereotyping in Pictorial Reconstructions of Human Origins." In *Women in Archaeology: A Feminist Critique,* edited by Hilary du Cros and Laurajane Smith. Canberra: Department of Prehistory, Research School of Pacific Studies, 1993. 75–92.

Muir, John. "Living Glaciers of California." *Overland Monthly,* December 1872, 547–49.

Mundus novus. Paris: Felix Baligault and Jehan Lambert, 1503.

Murchison, Roderick Impey. *Siluria: The History of the Oldest Known Rocks Containing Organic Remains, with a Brief Sketch of the Distribution of Gold over the Earth.* London: J. Murray, 1854.

———. *The Silurian System, Founded on Geological Researches in the Counties of Salop, Hereford, Radnor, Montgomery, Caermarthen, Brecon, Pembroke, Monmouth, Gloucester, Worcester, and Stafford; with Descriptions of the Coalfields and Overlying Formations.* 3 vols. London: J. Murray, 1839.

Murchison, Roderick Impey, Edouard de Verneuil, and Alexandre de Keyserling. *Geology of Russia in Europe and the Ural Mountains.* 2 vols. London: J. Murray, 1845.

———. *On the Geological Structure of the Central and Southern Regions of Russia in Europe, and of the Ural Mountains.* London: R. and J. E. Taylor, 1842.

Murphy, William. "The Ancient History of Plants." *Ladies' Repository,* October 1858, 593–97.

"Museum Notes." *American Museum Journal* 16, no. 5 (May 1916): 333–36.

Nabokov, Peter. *A Forest of Time: American Indian Ways of History.* Cambridge: Cambridge University Press, 2002.

Nagel, Alexander, and Christopher S. Wood. *Anachronic Renaissance.* New York: Zone, 2010.

National Political Map of the United States. New York: Adolphus Ranney, 1856.

New-York as It Is, in 1835. New York: J. Disturnell, 1835.

Nichols, Kate, and Sarah Victoria Turner, eds. *After 1851: The Material and Visual Cultures of the Crystal Palace at Sydenham.* Manchester: Manchester University Press, 2017.

Nieuwland, Ilja. *American Dinosaur Abroad: A Cultural History of Carnegie's Plaster Diplodocus.* Pittsburgh: University of Pittsburgh Press, 2019.

Noble, Louis. *After Icebergs with a Painter: A Summer Voyage to Labrador and around Newfoundland.* New York: D. Appleton, 1861.

Nolan, Alan T. "The Anatomy of the Myth." In *The Myth of the Lost Cause and Civil War History,* edited by Gary Gallagher and Alan Nolan. Bloomington: Indiana University Press, 2000. 11–35.

Noll, Mark. *Princeton and the Republic, 1768–1822: The Search for Christian Enlightenment in the Era of Samuel Stanhope Smith.* Princeton, NJ: Princeton University Press, 1989.

———, ed. *The Princeton Theology, 1812–1921: Scripture, Science, and Theological Method from Archibald Alexander to Benjamin Breckinridge Warfield.* Grand Rapids, MI: Baker Academic, 2001.

Nott, Josiah, and George R. Gliddon. *Types of Mankind.* Philadelphia: Lippincott, Grambo, 1854.

Novak, Barbara. *Nature and Culture: American Landscape and Painting, 1825–1875.* Rev. ed. New York: Oxford University Press, 1995.

Numbers, Ronald L. *Creation by Natural Law: Laplace's Nebular Hypothesis in American Thought.* Seattle: University of Washington Press, 1977.

———. *The Creationists: From Scientific Creationism to Intelligent Design*. Expanded Edition. Cambridge, MA: Harvard University Press, 2006.

———. "'The Most Important Biblical Discovery of Our Time': William Henry Green and the Demise of Ussher's Chronology." *Church History* 69, no. 2 (June 2000): 257-76.

Nyhart, Lynn K. *Biology Takes Form: Animal Morphology and the German Universities, 1800–1900*. Chicago: University of Chicago Press, 1995.

Oard, Michael J. "The Uinta Mountains and the Flood." *Creation Research Society Quarterly* 49, no. 2 (2012): 109–21.

O'Brien, Jean. *Firsting and Lasting: Writing Indians Out of Existence in New England*. Minneapolis: University of Minnesota Press, 2010.

O'Connor, Ralph. *The Earth on Show: Fossils and the Poetics of Popular Science, 1802–1856*. Chicago: University of Chicago Press, 2007..

Ogle, Vanessa. *The Global Transformation of Time: 1870–1950*. Cambridge, MA: Harvard University Press, 2015.

"Olla-Podrida." *Nassau Literary Magazine*, June 1, 1870, 58–59.

"Olla-Podrida." *Nassau Literary Magazine*, April 1, 1874, 341–73.

"Olla-Podrida." *Nassau Literary Magazine*, July 1, 1874, 39–58.

Osborn, Henry Fairfield. *The Age of Mammals in Europe, Asia and North America*. New York: Macmillan, 1910.

———. "Birth Selection *versus* Birth Control." *Science* 76, no. 1965 (August 26, 1932): 173–79.

———. "The Causes of Extinction in Mammalia." *American Naturalist* 40, no. 479 (Nov. 1906): 769–95.

———. "The Cranial Evolution of Titanotherium." *Bulletin of the American Museum of Natural History* 8 (1896): 157–97.

———. *Creative Education in School, College, University, and Museum*. New York: Charles Scribner's Sons, 1927.

———. "The Discovery of Tertiary Man." *Science* 71, no. 1827 (Jan. 3, 1930): 1–7.

———. *The Hall of the Age of Man*. Guide Leaflet Series, no. 52. 3d. ed. New York: American Museum of Natural History, 1925.

———. "The Hall of the Age of Man in the American Museum." *Natural History* 20, 3 (May–June 1920): 229–46.

———. "Hesperopithecus, the Anthropoid Primate of Western Nebraska." *Nature* 110 (26 August 1922): 281–83.

———. "Hesperopithecus, the First Anthropoid Primate Found in North America." *Science* 55, no. 1427 (May 5, 1922): 463–65.

———. *Huxley and Education: Address at the Opening of the College Year, Columbia University, September 28, 1910*. New York: Charles Scribner's Sons, 1910.

———. "Men of the Old Stone Age: With an Account of a Motor Tour." *American Museum Journal* 12, no. 8 (1912): 279–95.

———. "Prehistoric Quadrupeds of the Rockies." *Century Illustrated Monthly Magazine* 52, n.s. vol. 30 (September 1896), 705–15.

———. "Restorations and Models of the Extinct North American Mammals." *American Museum Journal* 1, no. 6 (Jan. 1901): 85–87.

———. *The Titanotheres of Ancient Wyoming, Dakota, and Nebraska*. 2 vols. United States Geological Survey Monograph 55. Washington, DC: Government Printing Office, 1929.

Osborn, Henry Fairfield, William K. Gregory, Jacob L. Wortman, Olof A. Peterson, William D. Matthew, and Walter Granger. "New or Little Known Titanotheres from the Eocene and Oligocene." *Bulletin of the American Museum of Natural History* 24 (1908): 599–617.

Osborn, Lucretia Perry. *The Chain of Life*. New York: Charles Scribner's Sons, 1925.

Owen, David Dale. "Scientific Pursuits: Introduction to Geology." *Quarterly Journal and Review* (Cincinnati, OH) 1 (Jan. 1846): 43–46.

———. "Termination of the Palaeozoic Period, and Commencement of the Mesozoic." *American Journal of Science and Arts* 2nd ser., 3, no. 9 (May 1847): 365–68.

Owen, David Dale, Joseph Leidy, J. G. Norwood, C. C. Parry, Henry Pratten, B. F. Shumard, and Charles Whittlesey. *Report of a Geological Survey of Wisconsin, Iowa, and Minnesota; and Incidentally of a Portion of Nebraska Territory*. Philadelphia: Lippincott, Grambo, 1852.

Owen, Richard. *Geology and Inhabitants of the Ancient World*. London: Crystal Palace Library and Bradbury and Evans, 1855.

———. *On the Nature of Limbs*. London: John Van Voorst, 1849.

Oxford University Museum of Natural History, ed. *Strata: William Smith's Geological Maps*. Foreword by Robert Macfarlane. Introduction by Douglas Palmer. Chicago: University of Chicago Press, 2020.

Paredes, Liana. *Sèvres Then and Now: Tradition and Innovation in Porcelain, 1750–2000*. London: D. Giles, 2009.

Parley, Peter [Samuel Griswold Goodrich]. *Peter Parley's Wonders of the Earth, Sea, and Sky*. New York: S. Colman, 1840.

Peale, Charles Willson. *A Scientific and Descriptive Catalogue of Peale's Museum*. Philadelphia: Samuel Smith, 1796.

Peale, Rembrandt. *An Historical Disquisition on the Mammoth; or, Great American Incognitum, an Extinct, Immense, Carnivorous Animal, Whose Fossil Remains Have Been Found in North America*. London: E. Lawrence, 1803.

Pearlstein, Ellen, Lynn Brostoff, and Karen Trentelman. "A Technical Study of the Rosebud Winter Count." *Plains Anthropologist* 54, no. 209 (Feb. 2009): 3–17.

Perrin, Carleton. "The Chemical Revolution." In *Companion to the History of Modern Science*, edited by R. C. Olby, G. N. Cantor, J.R.R. Christie, and M.J.S. Hodge. London: Routledge, 1990. 264–77.

Phillips, John S. "Donations to the Museum." *Proceedings of the Academy of Natural Science Philadelphia* 1, nos. 11 and 12 (Feb.–March 1842): 147–48.

Pick, Nancy. *Curious Footprints: Professor Hitchcock's Dinosaur Tracks and Other Natural History Treasures at Amherst College*. Photographs by Frank Ward and afterword by Ben Lifson. Amherst, MA: Amherst College Press, 2006.

Porter, Charlotte. "The Rise of Parnassus: Henry Field Osborn and the Hall of the Age of Man." *Museum Studies Journal* 1 (Spring 1983): 26–34.

Price, Catherine. *The Oglala People, 1841–1879: A Political History*. Lincoln: University of Nebraska Press, 1996.

Price, George E. McCready. *The New Geology: A Textbook for Colleges, Normal Schools, and Training Schools; and for the General Reader*. Mountain View, CA: Pacific Press, 1923.

———. *Outlines of Modern Christianity and Modern Science*. Oakland, CA: Pacific Press, 1902.

Pyne, Stephen. *How the Canyon Became Grand: A Short History*. New York: Penguin, 1999.

Rafferty, Edward C. *Apostle of Human Progress: Lester Frank Ward and American Political Thought, 1841–1913*. Lanham, MD: Rowman and Littlefield, 2003.

Rainger, Ronald. *An Agenda for Antiquity: Henry Fairfield Osborn and Vertebrate Paleontology at the American Museum of Natural History, 1890–1935*. Tuscaloosa: University of Alabama Press, 1991.

Ratner-Rosenhagen, Jennifer. *American Nietzsche: A History of an Icon and His Ideas*. Chicago: University of Chicago Press, 2012.

Rea, Tom. *Bone Wars: The Excavation of Andrew Carnegie's Dinosaur*. Pittsburgh: University of Pittsburgh Press, 2001.

Redfield, Anna Maria. "General View of the Animal Kingdom." In *Zoölogical Science; or, Nature in Living Forms, Illustrated by Numerous Plates; Adapted to Elucidate the Chart of the Animal Kingdom, by A. M. Redfield; and Designed for the Higher Seminaries, Common Schools, Libraries, and the Family Circle*. New York: E. B. and E. C. Kellogg, 1858.

Reed, John K. "Time Warp I: The Permian-Triassic Boundary in the Texas Panhandle." *Creation Research Society Quarterly* 39, no. 2 (2002): 116–19.

Regal, Brian. *Henry Fairfield Osborn: Race and the Search for the Origins of Man*. Surrey, UK: Ashgate, 2002.

"Reporter." *Princetonian*, January 16, 1879, 142–44.

Rieppel, Lukas. *Assembling the Dinosaur: Fossil Hunters, Tycoons, and the Making of a Spectacle*. Cambridge, MA: Harvard University Press, 2019.

Riskin, Jessica. "The Naturalist and the Emperor, a Tragedy in Three Acts; or, How History Fell Out of Favor as a Way of Knowing Nature." *Know* 2, no. 1 (Spring 2018): 85–110.

———. *The Restless Clock: A History of the Centuries-Long Argument over What Makes Living Things Tick*. Chicago: University of Chicago Press, 2016.

Riskin, Jessica, and Caroline Winterer. *The Apes and Us: A Century of Representations of Our Closest Relatives*. With a foreword by Henry Lowood. Stanford, CA: Stanford University Libraries, Silicon Valley Archives, 2024.

Roberts, Jon. *Darwinism and the Divine in America: Protestant Intellectuals and Organic Evolution, 1859–1900*. Madison: University of Wisconsin Press, 1988.

———. "Psychoanalysis and American Christianity, 1900–1945." In *When Science and Christianity Meet*, edited by David C. Lindberg and Ronald L. Numbers. Chicago: University of Chicago Press, 2003. 225–44.

Roberts, Michael B. "Genesis Chapter 1 and Geological Time from Hugo Grotius and Marin Mersenne to William Conybeare and Thomas Chalmers (1620–1825)." In *Myth and Geology*, edited by Luigi Piccardi and W. Bruce Masse. Geological Society Special Publication 273. London: Geological Society, 2007. 39–49.

Robson, John M. "The Fiat and Finger of God: The Bridgewater Treatises." In *Victorian Faith in Crisis: Essays on Continuity and Change in Nineteenth-Century Religious Belief*, edited by Richard J. Helmstadter and Bernard V. Lightman. Houndmills, UK: Macmillan, 1990. 71–125.

Rogers, Henry Darwin. *The Geology of Pennsylvania: A Government Survey.* 2 vols. Philadelphia: J. B. Lippincott, 1858.

Romanes, George John. *Mental Evolution in Man.* London: Kegan Paul, Trench, 1888.

Roosevelt, Theodore, and Edmund Heller. *Life-Histories of African Game Animals.* 2 vols. New York: Charles Scribner's Sons, 1914.

Rosasco, Betsy. "The Teaching of Art and the Museum Tradition: Joseph Henry to Allan Marquand." *Record of the Art Museum, Princeton University* 55, no. 1/2 (1996): 7–52.

Rosenberg, Daniel, and Anthony Grafton. *Cartographies of Time: A History of the Timeline.* Princeton, NJ: Princeton Architectural Press, 2010.

Ross, Edward Alsworth. *The Old World in the New: The Significance of Past and Present Immigration to the American People.* New York: Century, 1914.

Rossi, Paolo. *The Dark Abyss of Time: The History of the Earth and the History of Nations from Hooke to Vico.* Translated by Lydia Cochrane. Chicago: University of Chicago Press, 1984.

Rothman, Adam. *Slave Country: American Expansion and the Origins of the Deep South.* Cambridge, MA: Harvard University Press, 2005.

Routledge's Guide to the Crystal Palace and Park at Sydenham. London: George Routledge, 1854.

Rudwick, Martin S. J. *Bursting the Limits of Time: The Reconstruction of Geohistory in the Age of Revolution.* Chicago: University of Chicago Press, 2005.

———. "Charles Lyell Speaks in the Lecture Theatre." *British Journal for the History of Science* 9, no. 2 (July 1976): 147–55.

———. *Georges Cuvier, Fossil Bones, and Geological Catastrophes: New Translations and Interpretations of the Primary Texts.* Chicago: University of Chicago Press, 1997.

———. *The Great Devonian Controversy: The Shaping of Scientific Knowledge among Gentlemanly Specialists.* Chicago: University of Chicago Press, 1985.

———. "Lyell and the *Principles of Geology*." In *Lyell: The Past Is the Key to the Present*, edited by Derek J. Blundell and Andrew C. Scott. Geological Society Special Publication 143. London: Geological Society, 1998. 3–15.

———. *Scenes from Deep Time: Early Pictorial Representations of the Prehistoric World.* Chicago: University of Chicago Press, 1992.

———. "The Strategy of Charles Lyell's *Principles of Geology*." *Isis* 61, no. 1 (Spring 1970): 5–33.

———. *Worlds before Adam: The Reconstruction of Geohistory in the Age of Reform.* Chicago: University of Chicago Press, 2008.

Rupke, Nicolaas A. *Richard Owen: Biology without Darwin.* Chicago: University of Chicago Press, 2009.

———. *Richard Owen: Victorian Naturalist.* New Haven, CT: Yale University Press, 1994.

Ryder, Richard C. "Hawkins' Hadrosaurs: The Stereographic Record." *Mosasaur: Journal of the Delaware Valley Paleontological Society* 3 (Nov. 1986): 169–80.

Sandeen, Ernest R. *The Roots of Fundamentalism: British and American Millenarianism, 1800–1930.* Chicago: University of Chicago Press, 1970.

Saporta, Gaston de. *Aperçu géologique du terroir d'Aix-en-Provence.* Aix-en-Provence: A. Makaire, 1881.

———. *Le Monde des plantes avant l'apparition de l'homme.* Paris: G. Masson, 1879.

Scharnhorst, Gary. "Making Her Fame: Charlotte Perkins Gilman in California." *California History* 64, no. 3 (Summer 1985): 192–201.

Scharnhorst, Gary, and Denise D. Knight. "Charlotte Perkins Gilman's Library: A Reconstruction." *Resources for American Literary Studies* 23, no. 2 (1997): 181–219.

Schmalzer, Sigrid. *The People's Peking Man: Popular Science and Human Identity in Twentieth-Century China*. Chicago: University of Chicago Press, 2008.

Schneer, Cecil J. "Ebenezer Emmons and the Foundation of American Geology." *Isis* 60, no. 4 (Winter 1969): 439–50.

———. "The Great Taconic Controversy." *Isis* 69, no. 2 (June 1978): 173–91.

Schuller, Kyla. "The Fossil and the Photograph: Red Cloud, Prehistoric Media, and Dispossession in Perpetuity." *Configurations* 24, no. 2 (Spring 2016): 229–61.

Schulten, Susan. *Mapping the Nation: History and Cartography in Nineteenth-Century America*. Chicago: University of Chicago Press, 2012.

Scofield, Cyrus, ed. *The Scofield Reference Bible*. New York: Oxford University Press, 1917.

Scott, Andrew C. "The Legacy of Charles Lyell: Advances in Our Knowledge of Coal and Coal-Bearing Strata." In *Lyell: The Past Is the Key to the Present*, edited by Derek J. Blundell and Andrew C. Scott. Geological Society Special Publication 143. London: Geological Society, 1998. 243–60.

Scott, William Berryman. *A History of Land Mammals in the Western Hemisphere*. New York: Macmillan, 1913.

———. *Some Memories of a Palaeontologist*. Princeton, NJ: Princeton University Press, 1939.

Scully, Vincent. "*The Age of Reptiles* as a Work of Art." In *The Age of Reptiles: The Art and Science of Rudolph Zallinger's Great Dinosaur Mural at Yale*, compiled and edited by Rosemary Volpe. 2d ed. New Haven, CT: Peabody Museum of Natural History, Yale University, 2007. 63–72.

Secord, James. *Controversy in Victorian Geology: The Cambrian-Silurian Dispute*. Princeton, NJ: Princeton University Press, 1986.

———. "King of Siluria: Roderick Murchison and the Imperial Theme in Nineteenth-Century British Geology." *Victorian Studies* 25, no. 4 (Summer 1982): 413–42.

Seibert, Elke. "'First Surrealists Were Cavemen': The American Abstract Artists and Their Appropriation of Prehistoric Rock Pictures in 1937." *Getty Research Journal* 11 (2019): 17–38.

Sepkoski, David. *Catastrophic Thinking: Extinction and the Value of Diversity from Darwin to the Anthropocene*. Chicago: University of Chicago Press, 2020.

Serres, Michel, with Bruno Latour. *Conversations on Science, Culture, and Time*. Translated by Roxanne Lapidus. Ann Arbor: University of Michigan Press, 1995.

Sharpe, Tom, and P. J. McCartney. *The Papers of H. T. De la Beche (1796–1855) in the National Museum of Wales*. Geological Series no. 17. Cardiff: National Museum of Wales, 1998.

Sheehan, Jonathan. "The Stamp of Time Elapsed: Anthropology and the Flood in the Seventeenth Century." In *Sintflut und Gedächtnis*, edited by Martin Mulsow and Jan Assmann. Munich: Wilhelm Fink Verlag, 2006. 321–36.

Sheets-Pyenson, Susan. *John William Dawson: Faith, Hope and Science*. Montreal: McGill-Queen's University Press, 1996.

Sheriff, Carol. *The Artificial River: The Erie Canal and the Paradox of Progress, 1817–1862.* New York: Hill and Wang, 1996.
Sherry, Jay. "Beatrice Hinkle and the Early History of Jungian Psychology in New York." *Behavioral Sciences* (Basel) 3, no. 3 (September 2013): 492–500.
———. *The Jungian Strand in Transatlantic Modernism.* New York: Palgrave Macmillan, 2018.
Shor, Elizabeth. *The Fossil Feud between E. D. Cope and O. C. Marsh.* Detroit: Exposition, 1974.
Shteir, Ann. *Cultivating Women, Cultivating Science: Flora's Daughters and Botany in England, 1760–1860.* Baltimore: Johns Hopkins University Press, 1996.
Siegfried, André. *America Comes of Age.* Trans. H. H. Hemming and Doris Hemming. New York: Harcourt, Brace, 1927.
Silliman, Benjamin. "Obituary [of Samuel Prescott Hildreth]." *American Journal of Arts and Sciences* 2nd ser., 36, no. 107 (Nov. 1863): 312–13.
———. "Professor Jacob Green's Monograph of the Trilobites of North America, with Colored Models of the Species." *American Journal of Science and Arts* 23, 2 (Jan. 1833): 395–98.
Silliman, Robert H. "Agassiz vs. Lyell: Authority in the Assessment of the Diluvium-Drift Problem by North American Geologists, with Particular Reference to Edward Hitchcock." *Earth Sciences History* 13, no. 2 (1994): 180–86.
Sime, John. "The Illustration of Nature Recast: Jacob Green's Models of American Trilobites." Paper presented at the Geological Society of America Annual Meeting, Baltimore, MD, November 2, 2015. Manuscript in possession of author.
Smith, G. Elliott. "Hesperopithecus: The Ape-Man of the Western World." *Illustrated London News* 160, 4340 (June 24, 1922): 942–44.
Smith, Pamela. *From Lived Experience to the Written Word: Reconstructing Practical Knowledge in the Early Modern World.* Chicago: University of Chicago Press, 2022.
Smith, William. *Strata Identified by Organized Fossils, Containing Prints on Colored Paper of the Most Characteristic Specimens in Each Stratum.* London: W. Arding, 1816.
Smits, David. "The Frontier Army and the Destruction of the Buffalo, 1865–1883." *Western Historical Quarterly* 25, no. 3 (Autumn 1994): 312–38.
Socolow, Arthur A., ed. *The State Geological Surveys: A History.* N.p.: Association of American State Geologists, 1988.
Sollas, William Johnson. *Ancient Hunters and Their Modern Representatives.* London: Macmillan, 1911.
Sommer, Marianne. "Seriality in the Making: The Osborn-Knight Restorations of Evolutionary History." *History of Science* 48, no. 3–4 (2010): 461–82.
Spanagel, David I. *DeWitt Clinton and Amos Eaton: Geology and Power in Early New York.* Baltimore: Johns Hopkins University Press, 2014.
Spence, Mark David. *Dispossessing the Wilderness: Indian Removal and the Making of National Parks.* New York: Oxford University Press, 1999.
Spencer, Herbert. *The Principles of Psychology.* London: Longman, Brown, Green, and Longmans, 1855.
Stafford, Robert A. *Scientist of Empire: Sir Roderick Murchison, Scientific Exploration, and Victorian Imperialism.* Cambridge: Cambridge University Press, 1989.

Stagg, J.C.A., ed. *The Papers of James Madison Digital Edition*. Charlottesville: University of Virginia Press, Rotunda, 2010.

Stanton, William Ragan. *The Leopard's Spots: Scientific Attitudes toward Race in America, 1815–59*. Chicago: University of Chicago Press, 1960.

"Stated Meeting, February 1, 1842. Mr. Phillips in the Chair. Donations to the Museum." *Proceedings of the Academy of Natural Sciences of Philadelphia* 1, 11 (February 1842): 147–49.

Stephens, Lester D. "Darwin's Disciple in Georgia: Henry Clay White, 1875–1927." *Georgia Historical Quarterly* 78, no. 1 (Spring 1994): 66–91.

———. *Joseph LeConte: Gentle Prophet of Evolution*. Baton Rouge: Louisiana State University Press, 1982.

———. *Science, Race, and Religion in the American South: John Bachman and the Charleston Circle of Naturalists, 1815–1895*. Chapel Hill: University of North Carolina Press, 2000.

Stiling, Rodney L. "Scriptural Geology in America." In *Evangelicals and Science in Historical Perspective*, edited by David Livingstone, D. G. Hart, and Mark A. Noll. New York: Oxford University Press, 1999. 177–92.

Stocking, George W. *Race, Culture, and Evolution: Essays in the History of Anthropology*. New York: Free Press, 1968.

———. "Some Problems in the Understanding of Nineteenth Century Anthropology." In *Readings in the History of Anthropology*, edited by Regna Darnell. New York: Harper and Row, 1974. 407–25.

———. *Victorian Anthropology*. New York: Free Press, 1987.

Strang, Cameron B. *Frontiers of Science: Imperialism and Natural Knowledge in the Gulf South Borderlands, 1500–1850*. Chapel Hill: University of North Carolina Press, 2018.

Subject-catalogue of the Library of the College of New Jersey, at Princeton. New York: Chas. M Green, 1884.

Sutton, Matthew Avery. *American Apocalypse: A History of Modern Evangelicalism*. Cambridge, MA: Belknap Press of Harvard University Press, 2017.

Taft, Robert. *Artists and Illustrators of the Old West, 1850–1900*. New York: Charles Scribner's Sons, 1953.

Taylor, Richard C. "Notice of a Model of the Western Portion of the Schuylkill or Southern Coalfield of Pennsylvania, in Illustration of an Address to the Association of American Geologists, on the Most Appropriate Modes for Representing Geological Phenomena." In *Reports of the First, Second, and Third Meetings of the Association of American Geologists and Naturalists, at Philadelphia, in 1840 and 1841, and at Boston in 1842*. Boston: Gould, Kendall, and Lincoln, 1843. 81–94.

———. *Statistics of Coal*. 1848. 2d ed. Philadelphia: J. W. Moore, 1855.

Thackray, J. C. "R. I. Murchison's *Siluria* (1854 and Later)." *Archives of Natural History* 10, no. 1 (1981): 37–43.

Thompson, Keith Stewart. *The Legacy of the Mastodon: The Golden Age of Fossils in America*. New Haven, CT: Yale University Press, 2008.

Thrailkill, Jane F. "Fables of Extinction: Geologist Edward Hitchcock and the Literary Response to Darwin." In *Amherst in the World*, edited by Martha Saxton. Amherst, MA: Amherst College Press, 2020. 217–33.

Trautmann, Thomas R. "The Revolution in Ethnological Time." *Man* n.s., 27, no. 2 (June 1992): 379–97.
Tresch, John. *The Romantic Machine: Utopian Science and Technology after Napoleon*. Chicago: University of Chicago Press, 2012.
Trigger, Bruce. *A History of Archaeological Thought*. 2nd ed. Cambridge: Cambridge University Press, 2006.
Trollinger, Susan L., and William Vance Trollinger. *Righting America at the Creation Museum*. Baltimore: Johns Hopkins University Press, 2016.
Truettner, William. "The Genesis of Frederic Edwin Church's Aurora Borealis." *Art Quarterly* 31, no. 3 (1968): 266–83.
Tullos, Allen. "The Black Belt." *Southern Spaces*, April 19, 2004, https://southernspaces.org/2004/black-belt/.
Turner, James. *The Liberal Education of Charles Eliot Norton*. Baltimore: Johns Hopkins University Press, 1999.
Turner, Sara E. "The E. M. Museum: Building and Breaking an Interdisciplinary Collection." *Princeton University Library Chronicle* 65, 2 (Winter 2004): 237–64.
Turner, Susan. "Thomas Sopwith, Miner's Friend: His Contributions to the Geological Model-Making Tradition." In *History of Research in Mineral Resources*, edited by J. E. Ortiz, O. Puche, I. Rábano, and L. F. Mazadiego. Cuadernos del Museo Geominero no. 13. Madrid: Instituto Geológico y Minero de España, 2011. 177–92.
Turner, Susan, Cynthia Burek, and Richard Moody. "Forgotten Women in an Extinct Saurian (Man's) World." In *Dinosaurs and Other Extinct Saurians: A Historical Perspective*, edited by Richard Moody, E. Buffetaut, D. Naish, and D. M. Martill. Special Publications 343. London: Geological Society of London, 2010. 111–53.
Turner, Susan, and W. R. Dearman. "Thomas Sopwith's Large Geological Models." *Proceedings of the Yorkshire Geological Society* 44 (July 1982): 1–28.
Tylor, Edward Burnett. *Primitive Culture: Researches into the Development of Mythology, Philosophy, Religion, Art, and Custom*. 2 vols. London: John Murray, 1871.
Tyndall, John. "The Glaciers and Their Investigators." *Popular Science Monthly*, October 1873, 746–56.
Unger, Franz. *Die Urwelt in ihren verschiedenen Bildungsperioden: 14 landschaftliche Darstellungen mit erläuterndem Texte*. 2d ed. Leipzig: T. O. Weigel, 1858.
Ussher, James. *The Annals of the World*. London: E. Tyler for J. Crook and G. Bedell, 1658.
Vail, R.W.G. *The American Sketchbooks of Charles Alexandre Lesueur, 1816–1837*. Worcester, MA: American Antiquarian Society, 1938.
Van Riper, A. Bowdoin. *Men among the Mammoths: Victorian Science and the Discovery of Human Prehistory*. Chicago: University of Chicago Press, 1993.
Vespucci, Amerigo. *Alberic[us] Vespucci[us] Laure[n]tio Petri Francisci de Medicis salutem plurima[m] dicit*. Paris: Felix Baligault and Jehan Lambert, 1503.
Vogt, Adolf Max. "The Discovery of Lake Dwellings on the Lake of Zurich (1854)." In *Le Corbusier, the Noble Savage: Toward an Archaeology of Modernism*. Translated by Radka Donnell. Cambridge, MA: MIT Press, 1998. 225–29.
Wallace, Alfred Russel. *Palm Trees of the Amazon and Their Uses*. London: John Van Voorst, 1853.

Wallace, David Rains. *The Bonehunters' Revenge: Dinosaurs, Greed, and the Greatest Scientific Feud of the Gilded Age*. New York: Houghton Mifflin, 1999.

Wallis, Brian. "Black Bodies, White Science: The Slave Daguerreotypes of Louis Agassiz." *Journal of Blacks in Higher Education* 12 (Summer 1996): 102–6.

Walls, Laura Dassow. *The Passage to Cosmos: Alexander von Humboldt and the Shaping of America*. Chicago: University of Chicago Press, 2009.

Ward, Lester Frank. *Dynamic Sociology, or Applied Social Science, as Based upon Statistical Sociology and the Less Complex Sciences*. 2 vols. New York: D. Appleton, 1883.

———. "Our Better Halves." *Forum*, November 1888, 266–75.

———. *The Psychic Factors of Civilization*. Boston: Ginn, 1893.

———. *Pure Sociology: A Treatise on the Origin and Spontaneous Development of Society*. 1903. Reprint, New York: Macmillan, 1914.

———. "Saporta and Williamson and Their Work in Paleobotany." *Science* n.s. 2, no. 32 (Aug. 9, 1895): 141–50.

———. "Sketch of Paleobotany." In *The Fifth Annual Report of United States Geological Survey to the Secretary of the Interior 1883–'84*, by J. W. Powell. Washington, DC: Government Printing Office, 1885. 363–452.

Warren, Leonard. *Joseph Leidy: The Last Man Who Knew Everything*. New Haven, CT: Yale University Press, 1998.

Washington, Booker T. *Up from Slavery: An Autobiography*. New York: A. L. Burt, 1901.

Webb, William Edward. *Buffalo Land: An Authentic Account of the Discoveries, Adventures, and Mishaps of a Scientific and Sporting Party in the Wild West*. Cincinnati and Chicago: E. Hannaford, 1872.

Webster, Daniel. *An Address Delivered at the Laying of the Corner Stone of the Bunker Hill Monument*. 5th ed. Boston: Cummings, Hilliard, 1825.

Weeks, Edward. *The Lowells and Their Institute*. Boston: Little, Brown, 1966.

Weinberg, H. Barbara. *The Lure of Paris: Nineteenth-Century American Painters and Their French Teachers*. New York: Abbeville, 1991.

Wetherill, John P. "Observations on the Geology, Mineralogy, &c. of the Perkiomen Lead Mine, in Pennsylvania." *Journal of the Academy of Natural Sciences of Philadelphia* 5, 12 (1826): 1–13.

White, Ellen G. *Spiritual Gifts: Important Facts of Faith, in Connection with the History of Holy Men of Old*. Vol. 3. Battle Creek, MI: Steam Press of the Seventh Day Adventist Church, 1864.

White, Errol. "On *Cephalaspis lyelli* Agassiz." *Palaeontology* 1, no. 2 (1958): 99–105.

White, George W. "The First Appearance in Ohio of the Theory of Continental Glaciation." *Ohio Journal of Science* 67, no. 4 (July 1967): 210–17.

Wickman, Thomas M. *Snowshoe Country: An Environmental and Cultural History of Winter in the Early American Northeast*. New York: Cambridge University Press, 2018.

Wilcox, Donald J. *The Measure of Times Past: Pre-Newtonian Chronologies and the Rhetoric of Relative Time*. Chicago: University of Chicago Press, 1987.

Wilkins, John S. *Species: A History of the Idea*. Berkeley: University of California Press, 2009.

Willard, Emma. *Universal History in Perspective*. 2nd. ed. Philadelphia: A. S. Barnes, 1845.

Williams, John Rogers. "Plan of Main Hall." In *The Handbook of Princeton*. New York: Grafton Press, 1905. Between 42 and 43.

Willis, Bailey, and George Willis Stose. "Geologic Map of North America." Insets: Aleutian Islands with Windward Islands. Compiled by the United States Geological Survey in cooperation with the Geological Survey of Canada and Instituto Geológico de México under the supervision of Bailey Willis and George W. Stose. Geologic drafting by Henry S. Selden. Professional paper 71. Washington, DC: US Geological Survey, 1911.

Wilson, Daniel. *The Archaeology and Prehistoric Annals of Scotland*. Edinburgh: Sutherland and Knox, 1851.

———. *Prehistoric Man: Researches into the Origin of Civilisation in the Old and the New World*. 2 vols. Cambridge: Macmillan, 1862.

Wilson, Leonard G. "John Jeremiah Bigsby, MD: British Army Physician and Pioneer North American Geologist." In *A History of Geology and Medicine*, edited by C. J. Duffin, R.T.J. Moody, and C. Gardner-Thorpe. Geological Society Special Publication 375. London: Geological Society, 2013. 375–94.

———. *Lyell in America: Transatlantic Geology, 1841–1853*. Baltimore: Johns Hopkins University Press, 1998.

Wilson, Phillip K. "Arnold Guyot (1807–1884) and the Pestalozzian Approach to Geology Education." *Eclogae Geologicae Helvetiae* 92 (1999): 321–25.

Wilson, Woodrow. "Address by the President of the United States." In *Centennial Celebration of the United States Coast and Geodetic Survey, April 5 and 6, 1916*. Washington, DC: Government Printing Office, 1916. 141–44.

———. "Princeton in the Nation's Service." *Forum* 22 (Dec. 1896): 447–66.

Winchell, Alexander. *Sketches of Creation: A Popular View of Some of the Grand Conclusions of the Sciences in Reference to the History of Matter and of Life*. New York: Harper and Brothers, 1870.

Winchell, Newton. "Nature and Origin of the Drift-Deposits of the Northwest." *Popular Science Monthly* 3 (July 1873): 202–10.

Winsor, Mary P. *Reading the Shape of Nature: Comparative Zoology at the Agassiz Museum*. Chicago: University of Chicago Press, 1991.

Winterer, Caroline. *American Enlightenments: Pursuing Happiness in the Age of Reason*. New Haven, CT: Yale University Press, 2016.

———. "Avoiding a 'Hothouse System of Education': Nineteenth-Century Early Childhood Education from the Infant Schools to the Kindergartens." *History of Education Quarterly* 32, no. 3 (Fall 1992): 289–314.

———. *The Culture of Classicism: Ancient Greece and Rome in American Intellectual Life, 1780–1910*. Baltimore: Johns Hopkins University Press, 2001.

———. "The First American Maps of Deep Time." In *Time in Maps: From the Age of Discovery to Our Digital Era*, edited by Kären Wigen and Caroline Winterer. Chicago: University of Chicago Press, 2020. 149–70.

Wise, M. Norton. *Aesthetics, Industry, and Science: Hermann Von Helmholtz and the Berlin Physical Society*. Chicago: University of Chicago Press, 2018.

———. "Time Discovered and Time Gendered in Victorian Science and Culture." In *From Energy to Information: Representation in Science and Technology, Art, and Literature*, edited by Bruce Clarke and Linda Dalrymple Henderson. Stanford, CA: Stanford University Press, 2002. 39–58.

Wissler, Clark. "The Art of the Cave Man." *American Museum Journal* 12, no. 12 (December 1912): 289–96.

———. "Existing and Extinct Races of Men." In *Fifty-First Annual Report of the Trustees of the American Museum of Natural History for the Year 1919*. New York: American Museum of Natural History, 1920. 98.

———. *Man and Culture*. New York: Thomas Y. Crowell, 1923.

Witham, Henry Thornton Maire. *The Internal Structure of Fossil Vegetables Found in the Carboniferous and Oolitic Deposits of Great Britain, Described and Illustrated*. Edinburgh: Adam and Charles Black, 1833.

———. *Observations on Fossil Vegetables, Accompanied by Representations of Their Internal Structure as Seen through the Microscope*. Edinburgh: T. Blackwood, 1831.

———. "On the Vegetation of the First Period of an Ancient World." *American Journal of Science and Arts* 18, no. 1 (July 1830): 110–17.

Wittenstein, Kate. "The Feminist Uses of Psychoanalysis: Beatrice M. Hinkle and the Foreshadowing of Modern Feminism in the United States." *Journal of Women's History* 10, 2 (1998): 38–62.

Wonders, Karen. *Habitat Dioramas: Illusions of Wilderness in Museums of Natural History*. Uppsala: Almqvist and Wiksell, 1993.

Woodbridge, William C. *Isothermal Chart, or View of Climates & Productions; Drawn from the Accounts of Humboldt & Others*. Hartford, CT: Belknap and Hamersley, 1837.

Woods, May, and Arete Swartz Warren. *Glass Houses: A History of Greenhouses, Orangeries, and Conservatories*. New York: Rizzoli, 1988.

Worthen, Molly. *Apostles of Reason: The Crisis of Authority in American Evangelicalism*. New York: Oxford University Press, 2016.

Wosk, Julie. *Breaking Frame: Technology, Art, and Design in the Nineteenth Century*. New Brunswick, NJ: Rutgers University Press, 1992.

Wright, Chauncey. "Sir Charles Lyell (1797–1875)." *Nation*, March 4, 1875, 146–47.

Wright, Conrad. "The Religion of Geology." *New England Quarterly* 14, no. 2 (June 1941): 335–58.

Wright, G. Frederick. *The Ice Age in North America, and Its Bearings upon the Antiquity of Man*. New York: D. Appleton, 1889.

Zeilinga de Boer, Jelle. *New Haven's Sentinels: The Art and Science of East Rock and West Rock*. Middletown, CT: Wesleyan University Press, 2013.

Zeller, Suzanne. *Inventing Canada: Early Victorian Science and the Idea of a Transcontinental Nation*. Montreal: McGill-Queen's University Press, 2009.

ILLUSTRATION CREDITS

I.1. *The Age of Reptiles*, a mural by Rudolph F. Zallinger, fresco secco, 16 × 110 ft., 1947. Copyright Peabody Museum of Natural History, Yale University.

1.1. University of Toronto, Wenceslaus Hollar Collection. Digital ID Hollar_k_0013. Thomas Fisher Rare Book Library, University of Toronto.

1.2. Thomas Burnet, *The Theory of the Earth*, 3d ed. (London: R. N. for Walter Kettilby, 1697), opposite page 101. Department of Special Collections, Stanford University Libraries.

1.3. "A Map of the United States of America" by Samuel G. Lewis. In William Maclure, "Observations on the Geology of the United States, Explanatory of a Geological Map," *Transactions of the American Philosophical Society 6* (1809): 411–28. David Rumsey Map Collection, David Rumsey Map Center, Stanford Libraries. PURL: http://purl.stanford.edu/dk539gf3381.

1.4. Charles Willson Peale (1741–1827), self-portrait with mastodon bone, 1824; oil on canvas, 26 1/4 × 22 in. Purchase, James B. Wilbur Fund. New-York Historical Society, 1940.202.

1.5. Papers of Amos Eaton, 1797–1846 (SC10685), Box 2, Folder 2, "An enlarged view of the geological strata as presented in Genesee River," August 7, 1823, from the collections of the New York State Library, Manuscripts and Special Collections, Albany, New York.

1.6. "Orra White Hitchcock drawing of strata, Gill, Massachusetts," Orra White Hitchcock (1796–1863), date created 1828–1840, 1 drawing: pen and ink on linen; 32 × 105 cm. Orra White Hitchcock Classroom Drawings, Edward and Orra White Hitchcock Family Papers, Archives and Special Collections, Amherst College.

1.7. De Witt Clinton Boutelle (American, 1820–1884), *Trenton Falls Near Utica, New York*, 1873, oil on canvas, 50 1/8 × 40 1/8 in. High Museum of Art, Atlanta, gift of the West Foundation in honor of Gudmund Vigtel, 2010.107.

1.8. Baily Willis and George Willis Stose, "Geologic Map of North America. Professional paper 71. plate 1A-D," 1911. Composite map, 200 × 152 cm. US Geological Survey, Washington DC. David Rumsey Map Collection, David

Rumsey Map Center, Stanford Libraries. PURL: https://purl.stanford.edu/pq244dt7635.

2.1. ANSP Library and Archives Collection 136, Charles Alexander Lesueur papers and drawings, 1817–1827. "Paradoxus boltoni," one of the earliest drawings of American trilobites, c. 1825. Collection 136B. Academy of Natural Sciences of Drexel University.

2.2. N. P. Willis, *American Scenery*, 2 vols. (London: George Virtue, 1840), 1: 118–19. Department of Special Collections, Stanford University Libraries.

2.3. Amos Eaton Papers, 1797–1846 (SC10685), Box 1, Folder 6, "Section of the Strata 1/4 Mile East of Schohari C. House," 22 September 1834, from the collections of the New York State Library, Manuscripts and Special Collections, Albany, New York.

2.4. Alexandre Brongniart and Anselme-Gaëtan Desmarest, *Histoire Naturelle des Crustacés Fossiles, Sous Les Rapports Zoologiques et Géologiques* (Paris: F.-G. Levrault, 1822). Plate IV. Branner Earth Sciences Library and Map Collections, Stanford Libraries.

2.5. Three of Jacob Green and Joseph Brano's painted trilobite casts. Photograph by Daouda Njie, 2023. Academy of Natural Sciences of Drexel University/IP-GreenCast57.

2.6. John Sartain after Hugh Bridport, Jacob Green, mid-19th century. Mezzotint and etching on off-white wove paper, 4 11/16 × 3 7/8 in. Bequest of Dr. Paul J. Sartain. Pennsylvania Academy of the Fine Arts, 1948.23.338.

2.7. Tray, from a breakfast service (Plateau; Déjeuner Mosaïque Florentine); possibly by Charles-Louis Constans; attributed to Alexandre Brongniart (French, 1770–1847); manufactured by Sèvres Porcelain Manufactory (France); painted by Pierre Huard (French, active 1811–1847); France; hard paste porcelain, vitreous enamel, gold; H × W × D: 2.2 × 42.8 × 34 cm (7/8 × 16 7/8 × 13 3/8 in.). Gift of Katrina H. Becker in memory of her parents, Mr. and Mrs. Charles V. Hickox; 1981-38-1. Cooper Hewitt, Smithsonian Design Museum.

2.8. "Orra White Hitchcock drawing of invertebrate fossils," Orra White Hitchcock (1796–1863), date created: 1828–1840; 1 drawing: pen and ink on linen; 60 × 89 cm. Orra White Hitchcock Classroom Drawings, Edward and Orra White Hitchcock Family Papers, Archives and Special Collections, Amherst College.

2.9. Emma Willard, "Picture of Nations, or, Perspective Sketch of the Course of Empire." In *Atlas, to Accompany a System of Universal History*. Hartford, CT: F. J. Huntington, 1836. Courtesy of the David Rumsey Map Collection, David Rumsey Map Center, Stanford Libraries. PURL: http://purl.stanford.edu/gk585tb8931.

2.10. Claude-Marie Dubufe. *Adam et Eve*, 1827. Domaine public—Musée d'arts de Nantes. © Art Digital Studio.

2.11. Portrait of Roderick Impey Murchison. Photograph [carte de visite] by C. Silvy. Image ID: SIL-SIL14-m006-13. Courtesy of the Dibner Library of the History of Science and Technology, Smithsonian Libraries and Archives, https://library.si.edu/image-gallery/73748.

2.12. Frontispiece map in François de Castelnau, *Essai sur le Système Silurien de l'Amérique Septentrionale* (Paris: P. Bertrand, 1843). Courtesy of Ernst Mayr Library of the Museum of Comparative Zoology, Harvard University.

2.13. Magic lantern slide of the "Transition Period," as taught in the College of Wooster's "Historical Geology" class, post 1866. T. H. McAllister, Optician, NY, image fixed on glass bolted into a thin slab of wood with metal rings, chromolithograph slides 4 × 8 in. Courtesy of The College of Wooster, Department of Earth Sciences.

2.14. *Eozoön canadense*, in John Dawson, *The Story of the Earth and Man* (1873), 24. Stanford University Libraries.

3.1. William Stanley Jevons, *The Coal Question* (1866), frontispiece. Digitized by Stanford University Libraries.

3.2. Orra White Hitchcock, "Plate 27: Fossil Vegetables from Wrentham," in Edward Hitchcock, *Final Report on the Geology of Massachusetts* (Northampton: J. H. Butler, 1841). Branner Earth Sciences Library & Map Collections, Stanford Libraries.

3.3. Richard Taylor, *Statistics of Coal* (1848; repr. Philadelphia: J. W. Moore, 1855). Stanford University Libraries.

3.4. "Structure of Coal," from Henry Thornton Maire Witham, *Internal Structure of Fossil Vegetables* (Edinburgh: A. & C. Black, 1833), Plate XI. Department of Special Collections, Stanford University Libraries.

3.5. Frontispiece to Samuel Hildreth, "Observations on the Bituminous Coal Deposits of the Valley of the Ohio," *American Journal of Science and Arts* 29, no. 1 (1836): 1–154. Stanford University Libraries.

3.6. Modern photograph of Edward Hitchcock's "Middletown specimen." Penny Leveritt, Pocumtuck Valley Memorial Association. Beneski Museum of Natural History, Amherst College.

3.7. Attributed to Robert Peckham, American (1785–1877), "Professor Edward Hitchcock Returning from a Journey," c. 1838, oil on canvas mounted on board, frame size: 28 7/8 × 30 7/16 × 1 1/2 in.; 73.3 × 77.3 × 3.8 cm; canvas/mount: 25 × 26 1/4 in.; 63.5 × 66.7 cm. Mead Art Museum, Amherst College, Gift of Lucy and Caroline Hitchcock Acc. No.: P.1940.1.

3.8. "Orra White Hitchcock drawing of coal strata," Orra White Hitchcock (1796–1863), date created 1828–1840; 1 drawing: pen and ink on linen; 39 × 83 cm. Orra White Hitchcock Classroom Drawings, Edward and Orra White Hitchcock Family Papers, Archives and Special Collections, Amherst College.

3.9. "Lepidodendron," in Edward Hitchcock, *Elementary Geology* (Amherst: J. S. & C. Adams, 1840), 107. Branner Earth Sciences Library & Map Collections, Stanford Libraries.

3.10. Paleontological Chart from Edward Hitchcock, *Elementary Geology* (Amherst: J. S. & C. Adams, 1840), fold-out frontispiece. Branner Earth Sciences Library & Map Collections, Stanford Libraries.

3.11. "Subterranean Forest: Isle of Portland," in Edward Hitchcock, *Elementary Geology* (Amherst: J. S. & C. Adams, 1840), 111. Branner Earth Sciences Library & Map Collections, Stanford Libraries.

3.12. George Inness (artist) American, 1825–1894, "The Lackawanna Valley, c. 1856," oil on canvas, overall: 86 × 127.5 cm (33 7/8 × 50 3/16 in.); framed: 120.3 × 161.6 × 15.2 cm (47 3/8 × 63 5/8 × 6 in.). Gift of Mrs. Huttleston Rogers, accession number 1945.4.1. Courtesy National Gallery of Art, Washington.

3.13. "View of the great Coal Seam on the Monongahela at Brownsville, Pennsylvania," in Charles Lyell, *Travels in North America*. 2 vols. (New York: Wiley and Putnam, 1845), 2: facing 22. Branner Earth Sciences Library & Map Collections, Stanford Libraries.

3.14. *Bayou Teche* (1874). Joseph Meeker (1827–1887), oil on canvas, 20 × 36 in. The Johnson Collection, Spartanburg, South Carolina.

3.15. James Dwight Dana, *Manual of Geology* (Philadelphia: Theodore Bliss & Co., 1863), frontispiece. Branner Earth Sciences Library & Map Collections, Stanford Libraries.

3.16. "Horticultural Hall Fountain," 1915. DOR Archives, collection ID: City Archives-706-0-, asset ID: 5207. Photo courtesy of PhillyHistory.org, a project of the Philadelphia Department of Records.

4.1. "North America in the Cretaceous period, MO, Upper Missouri Region," from James Dwight Dana, *Manual of Geology* (1863), 489. Branner Earth Sciences Library & Map Collections, Stanford Libraries.

4.2. Lakin, J. H, photographer. *Picking Cotton Near Montgomery, Alabama* [186-]. 1 photographic print on stereocard: stereograph, albumen. https://lccn.loc.gov/2012648057.

4.3. ANSP Archives Collection 2011–58, Samuel George Morton portrait, c. 1851, by Paul Weber. The Academy of Natural Sciences of Drexel University.

4.4. "Life in the Jurassic Sea 'Duria Antiquior' (An Earlier Dorset)," c. 1850, Robert B. Farren (1832–1912), oil on canvas. © 2023. Sedgwick Museum of Earth Sciences, University of Cambridge. Reproduced with permission.

4.5. Samuel George Morton, *Crania Americana* (Philadelphia: J. Dobson, 1839), Plate 21. Department of Special Collections, Stanford University Libraries.

4.6. T. A. [Timothy Abbott] Conrad, Letter to S[amuel] G[eorge] Morton, March 3, 1833. Samuel George Morton Papers: Series I. Correspondence, Mss.B.M843. Courtesy of the American Philosophical Society.

4.7. John J. Egan, American (born Ireland), active mid-19th century; *Ferguson Group and Landing of General Jackson*, Scene 18 from *Panorama of the Monumental Grandeur of the Mississippi Valley*, c. 1850; distemper on cotton muslin; 90 in. × 348 ft. Saint Louis Art Museum, Eliza McMillan Trust 34:1953.

4.8. Detail of John J. Egan, American (born Ireland), active mid-19th century; *Ferguson Group and Landing of General Jackson*, Scene 18 from *Panorama of the Monumental Grandeur of the Mississippi Valley*, c. 1850; distemper on cotton muslin; 90 in. × 348 ft. Saint Louis Art Museum, Eliza McMillan Trust 34:1953.

4.9. Samuel George Morton, *Crania Americana* (Philadelphia: J. Dobson, 1839), 260. Department of Special Collections, Stanford University Libraries.

5.1. J. A. Allen, "The American Bisons, Living and Extinct." *Memoirs of the Museum of Comparative Zoology at Harvard College* 4, no. 10 (1876): 1–221. Stanford University Libraries.

5.2. David Dale Owen et al., *Report of a Geological Survey of Wisconsin, Iowa, and Minnesota: And Incidentally of a Portion of Nebraska Territory* (Philadelphia: Lippincott, Grambo, 1852), 196. Branner Earth Sciences Library & Map Collections, Stanford Libraries.

5.3. William Edward Webb, *Buffalo Land* (Cincinnati and Chicago: E. Hannaford, 1872), 357. Stanford University Libraries.

5.4. "Wyoming in the Later Eocene," frontispiece to William D. Gunning, *Life-History of Our Planet* (Chicago: W. B. Keen, Cooke, 1876). University of Oregon Libraries.

5.5. "Loxolophodon Cornutus Cope," from Edward D. Cope, "The Monster of Mammoth Buttes," *Penn Monthly* (August 1873), frontispiece. Courtesy of the University of Minnesota Libraries.

5.6. William Berryman Scott, *A History of Land Mammals in the Western Hemisphere* (New York: Macmillan, 1913), 447. Stanford University Libraries.

5.7. William Lathrop Kingsley, *Yale College, a Sketch of Its History* (New York: H. Holt, 1879), 2: 180–81. Department of Special Collections, Stanford University Libraries.

5.8. Charles Robert Knight (1874–1953), "Eohippus," 1905, painting, 13.25 × 19.25 in., asset ID: ptc-656, AMNH Archives, Art Survey No. 791: Knight, Charles Robert, 1874–1953. Eohippus, 1905. Courtesy of American Museum of Natural History.

5.9. Charles R. Knight, "Titanothere Family—Bull, Cow, and Calf—of the South Dakota Lake Basin." In Henry Fairfield Osborn, "Prehistoric Quadrupeds of the Rockies." *Century Illustrated Monthly Magazine* 52, n.s. vol. 30 (September 1896), 709. Stanford University Libraries.

5.10. "Buffalo Cow, Calf (Four Months Old), and Yearling." From "The Extermination of the American Bison, with a Sketch of Its Discovery and

Life History." In *Report of the United States Museum for the Year Ending June 30, 1887* (Washington, DC: Government Printing Office, 1889), 367–548, Plate IV, between 398 and 399. Department of Special Collections, Stanford University Libraries.

5.11. James Dwight Dana, *Manual of Geology* (Philadelphia: Theodore Bliss, 1863), 128. Branner Earth Sciences Library & Map Collections, Stanford Libraries

5.12. Garrick Mallery, "The Calendar of the Dakota Nation," *U.S. Geological and Geographical Survey Bulletin* 3, no. 1 (5 April 1877): opposite page 3. Branner Earth Sciences Library & Map Collections, courtesy Stanford Libraries.

5.13. William Henry Holmes, "Panorama from Point Sublime. Part 1. Looking East." In Department of the Interior, United States Geological Survey, J. W. Powell, Director, *Atlas to Accompany the Monograph on the Tertiary History of the Grand Cañon District by Capt. Clarence E. Dutton U.S.A. Washington 1882* (New York: Julius Bien Lith., 1882). David Rumsey Map Collection, David Rumsey Map Center, Stanford Libraries. PURL: http://purl.stanford.edu/ch301mb9884.

6.1. Samuel Finley Breese Morse, "Benjamin Silliman," 1825, oil on canvas, 55 1/4 × 44 1/4 in. (140.3 × 112.4 cm). Yale University Art Gallery. Gift of Bartlett Arkell, B.A. 1886, M.A. 1898, to Silliman College, 1940.117.

6.2. Louis Agassiz, *Études sur les Glaciers, Dessinés d'après Nature et Lithographiés par Jph Bettanier* (Neuchâtel: H. Nicolet, 1840), Plate 6. Department of Special Collections, Stanford University Libraries.

6.3. Edward Hitchcock, *Elementary Geology* (Amherst: J. S. & C. Adams, 1841), 197. Stanford University Libraries.

6.4. Edward Hitchcock, *Elementary Geology* (Amherst: J. S. & C. Adams, 1841), 201. Stanford University Libraries.

6.5. "Orra White Hitchcock drawing of diluvial elevations and depressions, Amherst, Massachusetts," Orra White Hitchcock (1796–1863), date created 1828–1840, 1 drawing: pen and ink on linen; 58 × 122 cm. Orra White Hitchcock Classroom Drawings, Edward and Orra White Hitchcock Family Papers, Archives and Special Collections, Amherst College.

6.6. George H. Durrie (American, 1820–1863). *Summer Landscape Near New Haven*, c. 1849. Oil on canvas, 35 7/16 × 49 3/8 in. (90 × 125.4 cm). Brooklyn Museum, Dick S. Ramsay Fund, 46.162.

6.7. Shriramk, "Grave of Louis Agassiz, front view," 19 August 2007 (Wikimedia Commons, Creative Commons Attribution-ShareAlike 2.5).

6.8. Frontispiece of Louis Agassiz and A. A. Gould, *Principles of Zoology* (1848; repr. Boston: Gould and Lincoln, 1851). Stanford University Libraries.

6.9. "Louis Agassiz full-length portrait standing next to chalk board with drawings of six invertebrates," c. 1871. Photograph. https://www.loc.gov/item/2013651563/.

ILLUSTRATION CREDITS   337

6.10. Louis Agassiz, *Lake Superior* (Boston: Gould, Kendall, and Lincoln, 1850), 106. Branner Earth Sciences Library & Map Collections, Stanford Libraries.

6.11. Detail of William C. Woodbridge, "Isothermal chart, productions." In *Modern Atlas, on a New Plan, to Accompany the System of Universal Geography; A New Edition, Improved.* Hartford, CT: Belknap and Hamersley, 1837. Courtesy of the David Rumsey Map Collection, David Rumsey Map Center, Stanford Libraries. PURL: http://purl.stanford.edu/qj510pw7607.

6.12. *The Icebergs*, 1861 (originally *The North*), Frederic Edwin Church, Oil on canvas, 64 1/2 × 112 1/2 in. Dallas Museum of Art, gift of Norma and Lamar Hunt. 1979.28. Image courtesy Dallas Museum of Art.

6.13. William Stanley Haseltine, *Rocks at Nahant*, 1864, oil on canvas, 22 3/8 × 40 1/2 in. Terra Foundation for American Art, Daniel J. Terra Collection, 1999.65. Photography © Terra Foundation for American Art, Chicago.

6.14. Jervis McEntee, American, 1828–1891, *Mount Desert Island, Maine*, 1864, oil on canvas, overall: 27.1 × 40.4 cm (10 11/16 × 15 7/8 in.). Gift of John Wilmerding in honor of Jo Ann and Julian Ganz, accession no. 2016.141.1. Courtesy National Gallery of Art, Washington.

6.15. Albert Bierstadt, American (born in Germany), 1830–1902, *Valley of the Yosemite*, 1864, oil on paperboard, 30.16 × 48.89 cm (11 7/8 × 19 1/4 in.). Museum of Fine Arts, Boston. Gift of Martha C. Karolik for the M. and M. Karolik Collection of American Paintings, 1815–1865. 47.1236.

7.1. Pach Brothers, *E. M. Museum of Geology and Archaeology*. 1886. albumen print; 1 photograph (bet. 5 × 7 in. and 11 × 17 in.). Mudd Manuscript Library, Historical Photograph Collection, Grounds and Buildings Series, Nassau Hall Faculty Room, Mudd, Box MP42, Item 1256. Courtesy of Princeton University Library.

7.2. Guyot Wall Hangings. Unknown date. Photo. American Institute of Geonomy and Natural Resources Princeton University. Department of Geosciences. Mudd Manuscript Library, Department of Geosciences Records AC139. ark:/88435/dc1831cx538.

7.3. Sebastian C. Adams, *Adams' Synchronological Chart of Universal History* (New York: Colby, 1881). Detail. David Rumsey Map Collection, David Rumsey Map Center, Stanford Libraries. PURL: https://purl.stanford.edu/tm858mb7248.

7.4. "Diagram of the Geological Restorations at the Crystal Palace," from Benjamin Waterhouse Hawkins, "On Visual Education as Applied to Geology." *Journal of the Society of Arts* 2, 78 (19 May 1854), 446. Stanford University Libraries.

7.5. *Silurian Shore at Low Tide*, 1875. Benjamin Waterhouse Hawkins (1807–1894; born and died London, UK). Oil on canvas; 80 × 222.8 cm,

82 × 224.5 × 4.9 cm (frame). Princeton University, Department of Geosciences, Guyot Hall. (PP334)

7.6. Royal H. Rose, E. M. Museum of Geology and Archaeology: Interior. Carte-de-visite. 1 photograph (approx. 5 × 7 in. or smaller). E. M. Museum of Geology and Archaeology. Mudd Manuscript Library, Historical Photograph Collection, Grounds and Buildings Series, Mudd, Box SP03, Item 654.

7.7. *Devonian Life of the Old Red Sandstone*, 1876. Benjamin Waterhouse Hawkins (1807–1894; born and died London, United Kingdom). Oil on canvas; 78.4 × 220 cm, 91.4 × 233.7 × 6.3 cm (frame). Princeton University, Department of Geosciences, Guyot Hall. (PP337)

7.8. Hugh Miller, *The Old Red Sandstone* (7th ed., London: J. M. Dent, 1857), plate 13. Branner Earth Sciences Library & Map Collections, Stanford Libraries.

7.9. *Carboniferous Coal Swamp*, 1875. Benjamin Waterhouse Hawkins (1807–1894; born and died London, United Kingdom). Oil on canvas; 79.3 × 157.3 × 3.5 cm. Princeton University, Department of Geosciences, Guyot Hall. (PP335)

7.10. *Triassic Life of Germany*. Benjamin Waterhouse Hawkins (1807–1894; born and died London, United Kingdom). Oil on canvas; 78.4 × 154.6 cm, 91.4 × 168.3 × 6.3 cm (frame). Princeton University, Department of Geosciences, Guyot Hall. (PP326)

7.11. *Early Jurassic Marine Reptiles*, 1876. Benjamin Waterhouse Hawkins (1807–1894; born and died London, United Kingdom). Oil on canvas; 80.3 × 155.5 × 4.4 cm. Princeton University, Department of Geosciences, Guyot Hall. (PP329)

7.12. Samuel Griswold Goodrich, *Peter Parley's Wonders of the Earth, Sea and Sky* (London: Darton and Clark, Holborn Hill, [between 1837 and 1841]), frontispiece, 1, 7. Thomas Fisher Rare Book Library, University of Toronto.

7.13. *Jurassic Life of Europe*, 1877. Benjamin Waterhouse Hawkins (1807–1894; born and died London, United Kingdom). Oil on canvas; 80.3 × 219.4 cm, 94.3 × 233.7 × 7.6 cm (frame). Princeton University, Department of Geosciences, Guyot Hall. (PP340)

7.14. *Cretaceous Life of New Jersey*, 1877. Benjamin Waterhouse Hawkins (1807–1894; born and died London, United Kingdom). Oil on canvas; 81 × 221.6 cm, 92.7 × 236.2 × 6.3 cm (frame). Princeton University, Department of Geosciences, Guyot Hall. (PP336)

7.15. Edward Drinker Cope, "Cope on Fossil Reptiles of New Jersey," *American Naturalist* 3 (1869): 84–91. Stanford University Libraries.

7.16. *Pleistocene Fauna of Asia*. Benjamin Waterhouse Hawkins (1807–1894; born and died London, United Kingdom). Oil on canvas; 79.7 × 156.2 cm,

80.6 × 158.1 × 3.8 cm (frame, partial sides missing). Princeton University, Department of Geosciences, Guyot Hall. (PP339)

7.17. *Irish Elk and Palaeolithic Hunter*. Benjamin Waterhouse Hawkins (1807–1894; born and died London, United Kingdom). Oil on canvas; 80 × 48.2 cm, 82 × 49.3 × 4.4 cm (frame). Princeton University, Department of Geosciences, Guyot Hall. (PP332)

7.18. E. M. Museum of Geology and Archaeology: Interior. Photographic prints. 1 photograph (approx. 5 × 7 in. or smaller). E. M. Museum of Geology and Archaeology. Mudd Manuscript Library, Historical Photograph Collection, Grounds and Buildings Series, Mudd, Box SP03, Item 655.

7.19. *George Washington at the Battle of Princeton*, 1783–84. Charles Willson Peale, 1741–1827; born Chester, MD; died Philadelphia, PA. Oil on canvas, 237 × 145 cm (93 5/16 × 57 1/16 in.), frame: 275 × 179 × 10 cm (108 1/4 × 70 1/2 × 3 15/16 in.), Princeton University, commissioned by the Trustees. (PP222) Courtesy of Princeton University Art Museum.

8.1. Specimen Ridge, overlooking Lamar River, Yellowstone National Park. Fossil tree trunks in volcanic tuffs and breccias. Lester F. Ward in photo, almost surely made in 1887. 736.C.2 The Jesse Earl Hyde Collection, Case Western Reserve University (CWRU) Department of Earth, Environmental and Planetary Science. https://artscimedia.case.edu/wp-content/uploads/sites/190/2016/07/14224242/736.C.2.jpg.

8.2. Paul Cézanne (French, Aix-en-Provence, 1839–1906), *View of the Domaine Saint-Joseph* [*La Colinne des Pauvres*], late 1880s. Oil on canvas, 25 5/8 × 32 in. (65.1 × 81.3 cm), accession no. 13.66, Catharine Lorillard Wolfe Collection, Wolfe Fund, 1913, The Metropolitan Museum of Art. https://www.metmuseum.org/art/collection/search/435885.

8.3. J. H. McGregor, "Head of the 'Old Man of Cro-Magnon,'" in Henry Fairfield Osborn, *Men of the Old Stone Age*, 2nd ed. (New York: C. Scribner's Sons, 1916), 293. Courtesy of Stanford University Libraries, Stephen Jay Gould Collection.

8.4. Charles Robert Knight (1874-1953), "Neanderthal Flintworkers, Le Moustier Cavern, Dordogne, France," 1920 painting, 173 × 381 cm, asset ID: ptc-618, AMNH Archives, Art Survey No. 1148: Neanderthal Flintworkers, Charles R. Knight, 1920. American Museum of Natural History.

8.5. Arthur Keith, *The Antiquity of Man* (London: Williams and Norgate, 1915), frontispiece. Stanford University Libraries.

8.6. Charles Knight, "Mural Paintings of Prehistoric Men and Animals," *Scribner's Magazine* 71, no. 3 (Jan.–June 1922), 279. Stanford University Libraries.

8.7. Charles Robert Knight (1874–1953), "Cro-Magnon Artists of Southern France—post restoration," 1920 painting created, 1995 photographed, asset ID: ptc-5375, AMNH Archives, Art Survey No. 1147: Knight, Charles

Robert. 1920. Cro-Magnon Artists of Southern France, AMNH Archives, Photographic Transparency Collection, 4 × 5: 5375. Courtesy of American Museum of Natural History.

8.8. "Osborn's Present Theory of the Ascent and Phylogeny of Man," in Henry Fairfield Osborn, "The Discovery of Tertiary Man," *Science* 71, no. 1827 (Jan. 3, 1930), 3. Stanford University Libraries.

8.9. "The Earliest Man Tracked by a Tooth: An 'Astounding Discovery' of Human Remains in Pliocene Strata; A Reconstruction Drawing by A. Forestier." In G. Elliott Smith, "Hesperopithecus: The Ape-Man of the Western World," *Illustrated London News* 160, 4340 (June 24, 1922): 942–43. Stanford University Libraries.

8.10. © C. G. Jung. *The Red Book: Liber Novus*, edited and introduced by Sonu Shamdasani, translated by Mark Kyburz, John Peck, and Sonu Shamdasani (New York: W. W. Norton, 2009). © 2009 Foundation of the Works of C. G. Jung, Zürich.

8.11. Beatrice Hinkle, *The Re-Creating of the Individual* (New York: Harcourt, Brace, 1923), between pages 124 and 125. Stanford University Libraries.

8.12. Drawing by Phillipps Ward. From Lillian Eichler, *The Customs of Mankind* (Garden City, NY: Doubleday, Page, 1924), frontispiece. Stanford University Libraries.

8.13. Rock painting, Zimbabwe, FBA-D401619. Frobenius-Institut, Frankfurt am Main.

8.14. "Battle of Fishes," 1926. André Masson (1896–1987) © ARS, NY. Sand, gesso, oil, pencil, and charcoal on canvas, 14 1/4 × 28 3/4 in. (36.2 × 73 cm). Purchase. Digital Image © The Museum of Modern Art/Licensed by SCALA/Art Resource, NY © 2023 Artists Rights Society (ARS), New York/ADAGP, Paris.

9.1. Thomas Cole, *The Subsiding of the Waters of the Deluge* (1829). Oil on canvas, 35 3/4 × 47 3/4 in. (90.8 × 121.4 cm), Smithsonian American Art Museum. Gift of Mrs. Katie Dean in memory of Minnibel S. and James Wallace Dean and museum purchase through the Smithsonian Institution Collections Acquisition Program, 1983.40.

9.2. Edward Hicks (American, 1780–1849), *Noah's Ark* (1846). Oil on canvas, 26 5/16 × 30 3/8 in. (66.8 × 77.2 cm). Philadelphia Museum of Art: Bequest of Lisa Norris Elkins, 1950, 1950-92-7.

9.3. George E. McCready Price, *The New Geology* (Mountain View, CA: Pacific Press, 1923), 520. Stanford University Libraries.

E.1. Joseph Graham, William Newman, and John Stacy, "The Geologic Time Spiral—A Path to the Past" (ver. 1.1): U.S. Geological Survey General Information Product 58, poster, 1 sheet (2008). http://pubs.usgs.gov/gip/2008/58/.

INDEX

Page numbers in *italics* indicate figures.

Academy of Natural Sciences, 45, 97, 99, 109, 113; fossil collection, 26; Philadelphia, 192, 195
Acadia (steamship), 61
Adam and Eve (Dubufe painting), 50, *51*
Adams, Henry, 60
Adams, Henry Brooks, 165
Adams, John, 2
Adams, John Quincy, 19
Adams, Sebastian, *Synchronological Chart of Universal History*, 189, *190*
Agassiz, Elizabeth Cary: tombstone of, *160*; wife of Louis, 165
Agassiz, Louis: Age of Fish (Reign of Fish), 152, 165, *195*, *196*, *197*, 257; archetypes of life, 194–95; articles for *Atlantic Monthly*, 171; on black people and climate, 171; on dawn of Creation and of America, 194; on Earth's antiquity, 185; *Études sur les Glaciers*, 152, *153*; fossil hunting, 211; on geology, 60; "glacial phenomena," 160–61; glacial theory of, 150, 152–55, 157; Ice Age catastrophism, 177; *Lake Superior*, 165–67, *166*; Lake Superior trip, 164–67; Lowell Lectures, 159; Morton and, 285n11; on Morton's fossil marine research, 98; naturalist, 184; personal magnetism of, 161; plates of vertebra and skulls, 127; polygenesis, 162; portrait with chalkboard drawings of Radiata, 164; *Principles of Zoology*, 161, *163*; *Recherches sur les Poissons Fossiles* (Investigations of fossil fish), 45, 152; rocks in coastal Massachusetts, *172*; theory of "periodical refrigeration," 151; tombstone of, 159, *160*
Age of Coal: Jevons, 62; question of, 65–66
Age of Fish (Reign of Fish), 152, 165, *195*, *196*, *197*, 257
Age of Mammals, 121–23, *124*, *125*, *127*, *131*, *135*; horned Loxolophodon Cornutus Cope, *133*. *See also* Cenozoic era
Age of Man, 121, 151, 180, 205, 211–12, 214–15; history of American West, 139–44; prehistory as "prophetic," 186; primordial life forms culminating in, 13
Age of Reptiles, *124*, *126*, *127*, *130*, *131*, 139, 205. *See also* Mesozoic era
Age of Reptiles, The (Zallinger), 5, *6*
Alabama: American Indians of, 108–9; fossil hunting in, 108–18; sketch of strata by Conrad, *113*
Albany Evening Journal, 50
Alberti, Friedrich August von, 199
Alvarez, Luis, 127
America, defining, 19–20
American bison, shrinking habitat (1876) map, *123*
American children's book, dinosaur images, 204, *204*
American Indians. *See* Native Americans
American Journal of Science and Arts, Silliman's, 71, 73

American Museum of Natural History (AMNH), 230–32, 253; cultural elites at, 244; Grant as trustee of, 233–34; Hall of the Age of Man, 234
American Naturalist, 209, 210
American Philosophical Society, 97, 111
American Revolution, 2, 32, 182–83
American Sociological Association, 222
American trilobites, Lesueur's drawings of, *31*
American West: history of, 122, 139–44; Western fossils, 186–87
Amherst Academy, 11
Amherst College, 46, *46*, 66, 69, 154, *157*; Hitchcock at, 75, 77
ammonite, 8, 97, 272
Ancient Monuments of the Mississippi Valley (Squier and Davis), 70
Ancient Society (Morgan), 222
Ancient West, creating, 122–39
Annales des sciences naturelles (journal), 105
Anning, Mary, 106, 127
Answers in Genesis, 269
Anthropocene, 147; idea of, 179; term, 13
Anthropoidal Ape, 229
Antiquity of Man (Keith), 238, *239*
ape, ape-like, 9, 132, 219, 221, 238, 241, 245
Apollo Belvedere, 184, 214
archaeology, 181
Aristotle, 12
Armory Show, 225, 248
Arp, Jean, 255, 257
Articulata, archetype of life, 195
Art Journal, 172
Art Students League, New York City, 235
Association of American Geologists, 154
Athapaskan peoples, 175–76
atheism, 188
Atlantic Monthly (journal), 171
Atwater, Caleb, 69
Auden, W. H.: on Freud, 245; psychoanalytic view of art, 253
Audubon, John James, 99
Aurignacian, 231

Badlands, 121, *124*
Barr, Alfred H., Jr., 255
Barton, Benjamin Smith: 20, 22
Battle of Fishes (Masson), 257; painting, *257*
Bayou Teche (Meeker), 86; painting, *88*
Beaux, Cecilia, 125
behaviorism, 273
belemnite, 93, 97, 104
Bergson, Henri, 249
Beringia, 29
Bible, 2; as historical text, 262–65; history of Hebrew people, 14–15
Bible Museum, 270
biblical flood. *See* Flood, biblical
Bibliotheca Sacra (journal), 76, 80
Bierstadt, Albert: *Valley of the Yosemite*, *178*; Yosemite paintings, 177
Big Bang, 33
biology, 181
black belt: black workers picking cotton in, *96*; Cretaceous, 95, 108, 112; term, 94, 95–96
Bleak House (Dickens), 205
Boltwood, Bertram: on "actual ages" for rocks, 3; on American rocks, 263
Book of Sand, The (Borges), 271
Borges, Jorge Luis, *The Book of Sand*, 271
Boule, Marcellin, 232
Boutelle, De Witt Clinton (*Trenton Falls Near Utica, New York*), 27
Brano, Joseph: 38, *39*, 45
Brazil, 92
Breuil, Henri: 232, 235
Bridgewater Treatise (Buckland), 47
Britain, coal deposits of, 73–74
British Museum, 45
Brongniart, Adolphe-Théodore: *Histoire des Végétaux Fossiles* (History of fossil vegetables), 70–71; study of fossil vegetables and coal formations, 70, 74–75
Brongniart, Alexandre: fossils of Paris Basin, 105; Green touring with, 42–43; *Histoire Naturelle des Crustacés Fossiles* (Natural history of fossil crustaceans),

34–35, 36, 37; plaster casts of Mosasaurus tooth, 104; study of fossil vegetables and coal formations, 70; on trilobites, 45; Vanuxem and, 98
Bronn, H. G., 81
Bronx Zoo, 234
Bronze Age, 187
Brush, George de Forest, 235
Buckland, William: 42, 45, 47
Buffalo Bill's Wild West show, 186
Buffalo Land (Webb), 126–27, 128–29
Buffon. *See* Leclerc, Georges-Louis (Comte de Buffon)
Bunker Hill Monument, 19
Bureau of American Ethnology, 140
Burnet, Thomas, 17
Burr, Aaron, 182

Calamites, 197, *198*
Calendar of the Dakota Nation, 141, *142*
Cambrian explosion, 33
Canadian Shield, 128, 29
Carboniferous period: 80, 84–86, *88*, 89, *90*, 91; Orra Hitchcock's drawings of, *65*, *80*, *81*, *82–83*, *85*; at Princeton museum, 197, *198*, 199; term, 64
Castelnau, François de, map of Silurian regions of North America, 55, *56*
catastrophes, catastrophism, 24, 73, 74, 84, 102–3, 117, 130, 139, 150, 152, 154, 162, 176, 177, 178, 265–70; term, 67
caveman, 9, 212, 213, 214, 219–21, 232–45, 247, 251, 253, 254, 257; term, 9
Caveman within Us, The (Fielding), 251
Cedarville University, 269
Cenozoic era, 121–23, *124*. *See also* Age of Mammals
Centennial Exhibition of 1876, Philadelphia's, 89, *90*
Central Park Zoo, 234
Century (magazine), Knight's illustration, 137, *138*
C'est ce que j'ai vu, motto, 98

Cézanne, Paul: *Colinne des Pauvres, La*, 225, 226, 248; on color sensibility, 225; deep time and Modern art, 255, 257; inspiration of, 225, 226
Chain of Life, The (Osborn), 235
Chaldeans, 14
change of climate, 146
Charles II (King), 158
Cheddar Man, 231
chemical revolution, 8
Child's Unconscious Mind, The (Lay), 251
Chinese, history of, 14
Christian eternity, deep time and, 10
Christian Science, 248
Church, Frederic Edwin: *The Icebergs* (painting), 169, *170*, *171*; *The North*, 169, *170*
Civil War, 60, 62, 73, 92, 118, 120, 147, 180, 218; climate determinism during, 169; Cretaceous "black belt" map during, 95; Lost Cause myth, 92; Reconstruction policies, 125
Claudius (Emperor), 52
climate change, modern idea of, 12, 179
climate oscillation, 146
Cluny, 235
coal: energy storage by God, 66; Hildreth proposing vegetable origin, 73
coal consumption, Jevons Curve, 62, *63*
coal deposits: subterranean, 68; United States, 66, 67
coalfields, as inexhaustible supply of cheap fuel, 68
coal power, manpower and, 66
Coal Question, The (Jevons), 62
coal sublime, 85–91; term, 85
cognition of deep time: 10–12; Lyell and, 68, 263–64; Young Earth Creationists and, 264–66
Cole, Thomas: American landscapes, 193; *The Subsiding of the Waters of the Deluge*, 260, *260*
Colinne des Pauvres, La (Cézanne), 225, 226, 248

College of New Jersey. *See* Princeton University
Columbus, Christopher, 14, 29
Comparative View of the Human and Animal Frame, A (Hawkins), 191
Comstock, John Lee, 43
Comte de Buffon. *See* Leclerc, Georges-Louis (Comte de Buffon)
Connecticut River, 77
Conrad, Solomon, 109, 111
Conrad, Timothy Abbott: Agassiz and, 151; fossil hunting in Alabama, 108, 111–14; on mysterious operations of nature, 103; paleontologist, 96; on physical condition of globe, 50; on Silurian strata, 54–55; sketch of strata in Alabama, *113*; son of Solomon Conrad, 109, 111; on trilobites, 46
Constitution, 96, 221
Cope, Edward Drinker: on boundaries between Age of Reptiles and Age of Mammals, 131; on carnivores, 208; illustrations of American dinosaurs, 209, *210*; on "monster of Mammoth Buttes," 132, *133*; paleontologist, 125; plates of vertebra and skulls, 127; on wealth in nature's creations, 126
Cornell Medical College, 248
Cotton Kingdom, 96, 101, 107, 112, 114
Crania Ægyptiaca (Morton), 99
Crania Americana (Morton), 70, 99, 109, *110*
craniology, craniometry: Age of Mammals, 133–34, 138–39; Guyot and, 214; "Hindoo," 210; Morton and, 6, 70, 99, *100*, 100–101, 109, *110*, 111, 118–20, *119*, 285n11; of Native Americans, 69–70, 109, *110*; Osborn and, 232–33; *233*
Creation: Agassiz on, 150–51, 162, *163*, 171; Genesis story of, 9, 14, *15*, 16, 20, 33, 39, 47, 50–52, 55–59, 104, 189, *190*; Green on, 44–45; Hitchcock on, 76, 84; Morton on, 102–4, 118; Nutt on, 115; at Princeton museum, 186, 193–95, 197–99, 204–11; Willard on 4004 BC date, 47, *48–49*

Creation; or, The Biblical Cosmogeny in the Light of Modern Science (Guyot), 194
Creation Museum, 259, 269; nonprofit status of, 270
Creation of the Earth (Hollar), 16
Creation Research Society, 269
Creation Research Society Quarterly (journal), 269
Creationism. *See* Young Earth Creationism
Creationists. *See* Young Earth Creationists
cretaceous, as chalky rock: 93
Cretaceous Life of New Jersey (Hawkins), 205–8, *208–9*
Cretaceous period: American reptiles of, 207–9; at Claiborne plantation, 112, *113*, 114; cotton plantations, 101; "Cretaceous mediterranean," 126; Deep South, 93–96; Jamaica, 106; primordial ocean of, 120; soils, 147; term, 93; as transatlantic formation, 6, 117–18
Crèvecoeur, Hector St. Jean de, 19
Croll, James, 175
Cro-Magnon, 231, 232, 234
Cro-Magnon Artists of Southern France (Knight), 238, 240, *240*, 242–43
Cro-Magnon Man, head of, *233*
Crystal Palace Exhibition, 190, *191*
Crystal Palace Park, 190, 199, 205
Customs of Mankind, The (Eichler), 253, 254
Cuvier, Georges: Adolphe-Théodore Brongniart and, 71; Agassiz and, 150, 152; Alexandre Brongniart and, 34; archetypes of life, 194–95; de Castelnau and, 55; fossils of Paris basin, 98; on idea of extinction, 22; plaster casts of Mosasaurus tooth, 104; reconstructing animals from bone fragments, 102, 150; view of fixed species, 102–3; on zoological categories, 162

Dadaists, 255–56
Dana, Charles, 248

Dana, James Dwight: on Agassiz, 161; on Earth "dragged slowly" in infancy, 197–98; glacialist, 172; *Manual of Geology*, 88, 140; on a time of disturbance, 130

Darwin, Charles, 161; Agassiz and, 167; deep time and cognition, 263; *The Descent of Man*, 212, 214, 219; evolution, 247; evolution of humans, 241; flattering Hitchcock, 77; *Origin of Species*, 59, 131, 183, 185, 205; origin of species as "mystery of mysteries," 101; on theory of evolution, 3, 81; theory of human ancestry, 264; theory of species transmutation by natural selection, 188; universal theory of development or evolution, 228; on unknowability of deep time, 266; Ward and, 223

David, Jacques-Louis, 50

Davis, Edwin, *Ancient Monuments of the Mississippi Valley*, 70

Dawson, John: "dawn animal of Canada" (*Eozoön canadense*), 57–58, 58; *The Story of the Earth and Man*, 57, 58

Declaration of Independence, 18

deep, meanings of, 2

Deep South: cotton economy of, 118–20; Cretaceous, 93–96; Native skulls from, 109; slavery in, 92, 94–96

deep time, 258, 262; antiquity and New World, 19–20; bolstering Manifest Destiny, 12; definition, 2; forging a national identity, 5; formulating of the idea of, 8–9; geological conceptions of, 70; immobile gallery of objects, 11–12; as inward journey, 10; making real to senses, 11; morality and, 13; new language of, 12–13; political triumph of, 22; revolution, 13, 29; short chronology and, 262–65; spreading idea of, 9; state and corporate power, 8; stratum as governing metaphor of, 24; term by McPhee, 2, 271; United States, 273; visualization of, 272, 273

Deerfield Academy, 77

Dekay, James, 38, 53

De la Beche, Henry, 105–8

DeLaski, John, 172–73

de Pauw, Cornelius, 18

Descent of Man, The (Darwin), 212, 214, 219

Desmarest, Anselme-Gaëtan, 34, 36

Devonian period, 8, 195, 196, 197, 198

Devonian Life of the Old Red Sandstone (Hawkins), 196, 198

Dickens, Charles, 205

Dickinson, Emily, 11

dinosaur(s), asteroid extinction theory, 127, 130; first US theories on beginning and end of Age of Reptiles, 126–27, 130–31; images directed at American children, 204, 204; term, 112

Dinosauria, category of, 205

Dinosaur National Monument, 1

Dippy the Diplodocus, 5

Dorset, England, 105, 199, 202–3

Draper, John William, 10

Du Bois, W.E.B., *The Souls of Black Folk*, 95

Dubufe, Claude-Marie, *Adam and Eve* painting, 50, 51

Dunbar, William, 20, 115

Durand, Asher, 193

Duria Antiquior (De la Beche), 106, 107

Durrie, George Henry, *Summer Landscape Near New Haven* painting, 157–58, 158

Dynamic Sociology (Ward), 222

Early Jurassic Marine Reptiles (Hawkins), 199, 202–3

Earth, age of, 1–2

Earth and Man, The (Guyot), 169, 184

Eaton, Amos: on chaos of geology, 24; diary entry, 45; Erie Canal field trip, 33; Gebhard and, 35; Rensselaer School founder, 31; sketched strata of Genesee Falls, 25; "swaggering European braggadocio," 38; trilobite hunting in Little Falls, 32; on trilobites, 46

Edward VII (King), 5

Egan, John, *Panorama of the Monumental Grandeur of the Mississippi Valley* painting, 115, *116*
Egyptians, history of, 14
Eichler, Lillian: *The Customs of Mankind*, 253; origin of handshake, 254
Eights, James, 46
Elasmosaurus, 127, *128*, 207, 209, *210*
Elementary Geology (Hitchcock), 80, *81*, 82–83, 155, *156*
Eliot, Charles, 183
E. M. Museum, 214; *George Washington at the Battle of Princeton* (Peale), 214, *217*; Hadrosaurus skeleton, *181*, 216; named for Elizabeth Marsh Libbey, 186; Princeton University, 215–16, *216*; Silurian creatures, 195, *196*; as temple of deep time, 194
energy, word, 66
Enlightenment, 221
Eocene epoch, 131, *132*, 288n14
Eohippus, 136, *137*, 229
Eozoön canadense, "dawn animal of Canada," 57–58, *58*
Erie Canal, 31, 60; Eaton group near Little Falls, 32; *Marquis de Lafayette* (boat), 30
erratic. *See* glacial erratic
Études sur les Glaciers (Studies of glaciers) (Agassiz), 152, *153*
eugenics, 258
Europe: America and, 5, 6; rock formations, 8
Evangeline (Longfellow), 61
Evidence as to Man's Place in Nature (Huxley), 212
evolution, denials of Darwin's theory, 188–89; 261–62
Expulsion from Paradise, The (Dubufe painting), 50, *51*
"Extermination of the American Bison, The" (Hornaday), 138, *139*
extinction: new idea of, 22
extinction event, term, 130

Falconer, Hugh, 211
Falkland Islands, 46

"Family of Man," Osborn's diagram, 241, 244
Featherstonhaugh, George, 109, 114
feminism, 9, 228–30; 248–50; 258
Feriday, William, 115
Fielding, William J., *The Caveman within Us*, 251
Flood, biblical, 2, 14–17, *17*, 32, 151, 260, 261; Green on, 40, 46; Ice Age theory and, 154, 167; origins of coal, 65, 73–74; Young Earth Creationists and, 265–66, 269
Forum (journal), 228
fossil collecting, 6–7, 96–108
fossil fuel: Hitchcock on, in United States, 75–76; term, 62
Foulke, William Parker, 192
Franco-Prussian War (1870–1871), 231
Franklin, Benjamin, 39, 40, 111
Frémiet, Emmanuel, 235
Freud, Sigmund: "Aetiology of Hysteria," 246; archaeological metaphors, 246–47; *The Interpretation of Dreams*, 246, *247*; Jung and, 245, 248; on unconscious mind, 246
Frobenius, Leo, 255

Ganoids, 152, 165
Garden of Eden, 9, 33, 50, *51*, 56, 59, 189, *190*, 259
Gebhard, John, 35
Geikie, James, 175, 185
Genesee Falls, Eaton's sketch of strata of, 25
Genesis, 2; account of the world, 14–15; aligning with geology, 47; animals in, 32–33; Christians and, 188; Creation account in, 265; creation of animals, 32; illustration (seventeenth-century) of story, *16*; story of Creation, 14, 39, 104; traditional reading of, 76; Young Earth Creationists, 268
Genesis Flood, The (Whitcomb and Morris), 266
Geological Evidences of the Antiquity of Man (Lyell), 212

geology: Agassiz lectures on at Harvard, 60; Emily Dickinson's study of, 11; ethnology and, 101, 119–20; Genesis and, 47, 184–85; Guyot lectures on, 186, 216; E. Hitchcock and, 75–76; Hodge rejection of, 189; human cognition and, 68, 263–66; of Paris basin, 34; progress in America, 63; purpose of, 21–22; Silliman public lectures on, 70; strata as governing metaphor of, 24, 26, 26, 27, 28, 29, 50; Young Earth Creationist version of, 259, 262, 264, 269

Geology of Pennsylvania, The (Rogers), 86

Gérôme, Jean-Léon, 235

Gilded Age, 39, 132, 134, 222, 234

Gilman, Charlotte Perkins: aspirations, 9; Hinkle following, 247–48; "Similar Cases" (poem), 228–29; Ward's vision of women, 228; *Women and Economics*, 230

glacial age/glacial era. *See* Ice Age

glacial erratic: Agassiz tombstone, 159, *160*; Church, *The Icebergs*, 169, *170*; in Europe, 148, 150; Guyot collection for Princeton, 185; Hitchcock on Agassiz's glacial theory, 155; Lake Superior, 165, *166*; in Massachusetts, *156*; in New Haven, *158*; Plymouth Rock as, 145

glacial theory, Agassiz, 150, 152–55, 157

glaciers of the North, 167–69, 171–73

God: Darwin's theory and, 188–89; existence outside of space and time, 10; Ice Age and, 177, 179; as "this great agent," 159–67; trilobites as first creation, 44–45

Goodrich, Samuel Griswold, *Peter Parley's Wonders of the Earth, Sea and Sky*, 204

Gould, Augustus, *Principles of Zoology*, 163

gradualism. *See* uniformitarianism

Grand Canyon, 1, 268; Holmes's "Panorama from Point Sublime," 143, *144*

Granger, Ebenezer, 69

Grant, Madison: American Museum of Natural History, 233–34; *The Passing of the Great Race*, 233

"Graywacke without petrifactions," 34, *35*

Great Chain of Being, 22, 162

Great Chalk Formation, 105

Great Ice Age, 238

Great Ice Age, The (Geikie), 175

Great Plains: fertility of modern, 126; Native groups in, 122–23

Great Refrigeration, 154

Great Salt Lake, 131

Green, Ashbel, 40

Green, Jacob: on antiquity of rocks and fossils, 11; depression after Robert's death, 40–41; journey and trilobites, 41–42; *A Monograph of the Trilobites of North America*, 38, 44–45; plate, *43*, *44*; portrait of, *41*; on Silurian deposits, 53; on time span between then and now, 10; touring with Alexandre Brongniart, 42–43; touring British Museum with William Buckland, 42; trilobite casts by Brano and, 38, *39*; on trilobite eyes, 50; trilobite models, 38–39; world of science and religion, 39–40

Green, Robert, 40

Green, Steve, 269–70

Green, William, 263

Grotto of Plants, Ohio, *73*, *74*

Gunning, Mary, 131

Gunning, William D., *Life-History of Our Planet*, 131, 132

Guyot, Arnold: deep time and, 189; *Creation; or, The Biblical Cosmogeny in the Light of Modern Science*, 194; discovery of prehistoric humans, 211, 214–15; E. M. Museum, 187–88, 194–215; *The Earth and Man*, 169, 184; geography, 184–85; on meaning of cool North, 168–69; naturalist, 184; Pestalozzian ideas, 185; Princeton and, 185–87; wall hanging of, *187*

gynaecoentric theory, 228

Hadrosaurus, 180, *181*, 192, 206–7, *208*, 215, 216

Halse Hall, 105–6

Ham, Ken, 269

Harlan, Richard, 96

Harper's (magazine), 249
Harvard, 59, 183, 247; Agassiz at, 60, 161, 184, 195
Harvard Museum of Comparative Zoology, 195
Haseltine, William, painting *Rocks at Nahant*, 171–72, *172*
Hatcher, John Bell, 125
Haüy, René Just, 98
Hawkins, Benjamin Waterhouse: biographical information and artistic style, 189–94; *Carboniferous Coal Swamp*, 197, *198*; *A Comparative View of the Human and Animal Frame*, 191; *Cretaceous Life of New Jersey*, 205–8, *208–9*; *Devonian Life of the Old Red Sandstone*, 195, *196*, 197; *Early Jurassic Marine Reptiles*, 199, 202–3, *204*; *Irish Elk and Palaeolithic Hunter*, 212, *213*; *Jurassic Life of Europe*, 205, *206–7*; *Pleistocene Fauna of Asia*, 210–11, *211*; *Silurian Shore at Low Tide*, 194 *195*; *Triassic Life of Germany*, 199, *200–201*; "On Visual Education as Applied to Geology," 191
Hayden, Ferdinand Vandeveer, 124
Henry, Joseph, 183
Hesperopithecus, 241, 243, 245
Heterodoxy Club, 248
Hibernia (steamship), 159
Hicks, Edward, *Noah's Ark*, 260, *261*
Hildreth, Samuel: biography of, 69; on coal deposits of Britain, 73–74; coal origins as vegetables, 73; Grotto of Plants, 74; Native American skulls, 70; as naturalist, 69
Hinkle, Beatrice: aspirations, 9; Jung and, 250; popularization of psychoanalysis, 247–50; *Psychology of the Unconscious*, 248; *The Re-Creating of the Individual*, 252
Hippopotamus, prehistoric, 155, *187*
Histoire des Végétaux Fossiles (History of fossil vegetables) (Adolphe-Théodore Brongniart), 70–71

Histoire Naturelle des Crustacés Fossiles (Natural history of fossil crustaceans) (Alexandre Brongniart), 34–35, *36*, 37
history, term (as opposed to prehistory), 181
History of Land Mammals in the Western Hemisphere, A (Scott), 134
Hitchcock, Charles, 172
Hitchcock, Edward: on Agassiz's glacier theory, 154–55, *156*; ancient bird tracks (Ichnolithology), 77, *78*; on bituminous specimens from Hildreth, 69; on Carboniferous, 84–85; on deep time, 76–77; education and early life, 75–76; *Elementary Geology*, 80, *81*, *82–83*, 155, *156*; on God's creative energy, 66; Lyell and, 75–77; on miracles, 80; Paleontological Chart, 80–81, *82–83*, 84; on order of strata, 24; son Charles, 172; on uniformitarianism, 76; visiting botanical garden, 89; wife Orra White Hitchcock welcoming, 79
Hitchcock, Orra White: Deerfield Academy, 77; greeting husband Edward returning from journey, 79; illustration of fossil plants, *65*; illustration of fossil forest, *85*; illustrations of glacial landscapes, 155; *156–57*; illustration of Lepidodendron, *81*; Paleontological Chart, *82–83*, 84; poster of coal strata, *80*; poster of Massachusetts strata, *26*; poster of trilobites, *46*
Hobby Lobby, 269
Hodge, Charles, 215; *What is Darwinism?* 188–89
Hodgkin, Thomas, 6
Hollar, Wenceslaus, 15, *16*
Holmes, William Henry, "Panorama from Point Sublime," 143, *144*
Holy Land, 15
Homo sapiens, 111, 221, 233
Hornaday, William, "The Extermination of the American Bison," 138, *139*
horses, ancient, Knight on, 135–36, *137*

Horticultural Hall, 89, *90*
Hosack, David, 37
Howe, Samuel Gridley, 171
Howells, William Dean, 229
Hudson River school, 193
Humboldt, Alexander von, 168
Hutton, James, 20, 34
Huxley, Thomas, 135; *Evidence as to Man's Place in Nature*, 212
Hyaelosaurus, 205, 207

Ice Age: as catastrophism, 177, 179; as Northern political agenda, 167–75; origin and American reception of theory, 145–67; in Sierra Nevada, 175–77, *178*
Icebergs, The (Church), painting, 169, *170*, 171
Ichthyosaur, ichthyosaurus, 106, 199, 202–3
Idealism, Agassiz's thesis of, 162
Iguanodon, 191, 205, 206, 225
Illustrated London News (journal), 241, 245
imperialism, coal and, 68
independence: American, 22–23; national self-definition, 18; political, 18–19
Indians. *See* Native Americans
Industrial Revolution, 8, 24, 42, 62, 64, 85, 93
Inness, George, *The Lackawanna Valley*, 86, *87*
Institut de Paléontologie Humaine, 235
Institute for Creation Research, 269
Internal Revenue Service, 270
Internal Structure of Fossil Vegetables, The (Witham), 71, 72
International Conference on Creationism, 269
Interpretation of Dreams, The (Freud), 246, 247
Irish Elk and Palaeolithic Hunter (Hawkins), 212, *213*
isothermal chart (Woodbridge), 168, *168*

Jackson, J. Hughlings, 247
Jamaica, 105–7
James, William, *The Principles of Psychology*, 247
Jardin des Plantes, 235
Jefferson, Thomas, 20, 111
Jefferson Medical College, 41
Jevons, William Stanley, *The Coal Question*, 62
Jevons curve, 62, *63*
Jim Crow South, 231
Jung, Carl: American popularizers of, 9; concept of the archetypes, 249; Freud and, 245, 248; Hinkle and, 250; psychoanalytic theories, 264; symbolizing primordial human myth, *250*, 251; *Wandlungen und Symbole der Libido*, 248
Jurassic, 107, 199–206, 207, 266; term, 9
Jurassic Life of Europe (Hawkins), 205, 206–7
Jurassic Park (film), 5

Kansas Pacific Railway, 126, 127
Keaton, Buster, *Three Ages* (film), 253
Keith, Arthur, *The Antiquity of Man*, 238, *239*
Keller, Ferdinand, 291n14
Keller, Gerta, 130
King, Clarence, 177
King of Siluria (Murchison), 53, *54*, 55
Klee, Paul: painter, 255, *257*; *Small Experimental Machine*, 257
Knight, Charles Robert: aspirations, 9; *Cro-Magnon Artists of Southern France*, 238, 240, *240*, 242–43; deep time and evolution, 266; landscape images of ancient horses, 135–36, *137*; low-browed Neanderthal hunters, 236–37; murals for Hall of the Age of Man, 235; nativity scene (1896) in *Century* magazine, 137, *138*; *Neanderthal Flintworkers*, 235, 236–37; paintings of prehistoric reptiles and mammals, 234–35; realism in art, 235; Titanotherium, 137–39

Labyrinthodons, 199, *200*
Lackawanna Valley, The (Inness), 86, *87*
Ladies' Repository (magazine), 55
Laelaps, 206, 208, *208*, 209
Lake Bonneville, 131

Lake Superior, Agassiz collecting data on trip to, 164–67
Lake Superior (Agassiz), 165–67, *166*
Lamarck, Jean-Baptiste, 102, 103
Lamarckianism. *See* transmutation of species
landscape (term), 84
Lartet, Louis, 232
law of superposition, 3–4
Lawrence Scientific School, 161
Lay, Wilfrid, *The Child's Unconscious Mind*, 251
Lea, Isaac, 96, 111
Leclerc, Georges-Louis (Comte de Buffon), 20, 22; on animals and plants in New World, 17; climate thesis, 18
LeConte, Joseph, 59, 60, 176–77
Le Corbusier, Swiss lake dwellings, 291–92n14
Leidy, Joseph: discovery of Hadrosaurus bones, 192; on Elasmosaurus, 127, 208; paleontologist, 125
Leopard's Spots, The (Stanton), 285n16
Lepidodendron, 197; Orra Hitchcock's illustration, *81*
Lesueur, Charles Alexandre: drawings of American trilobites, *31*; stratigraphic images of Mississippi, 114
Lewis, Samuel G., map of United States, *21*
Libbey, Elizabeth Marsh, E. M. Museum named for, 186
Liberal Club, 248
Life-History of Our Planet (Gunning), 131, *132*
Linnaeus, 35, 71
Little Falls, New York, 30, *32*
Little Ice Age (c. 1400–1850), 176
Livingstone, David, 53
Locke, John (philosopher), 17–18
Locke, John (Ohio naturalist), 69
London, Jack, 250
Longfellow, Henry Wadsworth, *Evangeline*, 61
Lost Cause myth, Old South, 92–93
Louvre, 235
Lovejoy, Arthur, 103
Loxolophodon Cornutus Cope, *133*

Lubbock, John, *Pre-historic Times*, 212
Lyell, Charles: Brongniart and, 74–75; cotton steamers as "paradise of geologists," 112; on creation of drift, 155; deep time and cognition, 26, 29, 68, 263–64; on Earth's antiquity, 185; *Geological Evidences of the Antiquity of Man*, 212; geological tour of North America, 61–62, 66–68, 75–79; Hitchcock and, 75–77; on long chronology, 265; on Morton, 118; at Morton's Mulberry Street house, 99; paleontologists adopting subdivisions of, 288n14; persona of, 161; receiving plaster casts of Mosasaurus tooth from Morton, 104; *Principles of Geology*, 26, 61, 66, 75, 76, 148; Silliman and, 70; *Travels in North America*, 87; uniformitarianism, 67–68, 75–76, 80; on unknowability of deep time, 266; on visit to United States, 64
Lyell, Mary Horner, 61–62

Maclure, William, *21*, 21–22, 37, 103
Madison, James, 182
magic lantern slide, 56, *57*
Magdalenian, 231
Mallery, Garrick, 140–42
Mallet, John, 118–19
mammoth, 151, 229, 238, 253
Mammerickx Microplate (Matthews, Müller, and Sandwell), 317
Manifest Destiny, 12, 177, 222
Mantell, Gideon, 104, 117–18, 152, 154, 285n11
Manual of Geology (Dana), 88, 140, 141
Marion, Antoine-Fortuné, 224, 225
marl pits, term, 97
Marquis de Lafayette (boat), 30
Marsh, Othniel: on fossil yields, 125; on horse evolution, 135; on Indians, 136; on mammalian life, 134; paleontologist, 125; Yale classroom, *136*
Massachusetts Institute of Technology, 183
mass extinctions, term, 130
Masson, André, *Battle of Fishes* (painting), 257, *257*

mastodon, 22, 23, 154
Matheron, Philippe, 224–25
mauvaises terres, 121, 124
Mayflower (ship), 29, 145
McCosh, James: deep time and, 189; Hawkins and, 193; non-Darwinian, 183–84, 186, 194; Princeton and, 183, 188; *Typical Forms and Special Ends of Creation*, 184
McEntee, Jervis, *Mount Desert Island, Maine* (painting), 173, *174*
McPhee, John, 2, 271
Meek, Fielding Bradford, 125
Meeker, Joseph: painting of Louisiana bayou, 86; *Bayou Teche*, *88*
Megalosaurus, 205, *206*
Men of the Old Stone Age (Osborn), 232
Mental Evolution in Man (Romanes), 247
Mesozoic era, 121, 124, 126, 131, 199. *See also* Age of Reptiles
Metropolitan Museum of Art, 225
Miller, Hugh, *The Old Red Sandstone*, 86, 197, *198*
Miocene epoch, 214, 288n14
Miró, Joan, 255
miracles: Hitchcock allows, 80; Lyell scorns, 67–68
Mississippian period, term, 9
Missouri Compromise of 1820, 167
Modern Age, Modern Period, 151, 219, 220, 221, 253
Modernism, 292n14
Modernist painters, 255, 257
modernity, caveman and, 9–10, 219–21
Mollusca, archetype of life, 195
Monograph of the Trilobites of North America (Green), 38, 44–45
Monongahela River, Brownville coal seam along, 86
Montesquieu, *Spirit of Laws*, 147
Moran, Thomas, 144
Morgan, J. P., 234
Morgan, Lewis Henry, *Ancient Society*, 222
Morrill Land Grant Act (1862), 175

Morris, Henry M., *The Genesis Flood*, 266
Morse, Samuel, 148, *149*, 158
Morton, Samuel George: Agassiz and, 285n11; Alabama fossils, 112–15, 117–18; as central bank of fossil specimens, 98–99; *Crania Americana*, 109, *110*; craniological studies, 100–101, 285n16; as "geologist," 98; Hildreth sending skulls to, 70; human anatomy and hierarchy, 119–20; as intermediary, 75; Mantell and, 117–18, 152; on Mosasaurus, 117; Natchez Indian skull, 110; New Jersey fossil history, 103–5; paleontologist, 96, 285n11; Philadelphia fossil group, 98–100; polygenesis, 120; portrait, *100*; racial science, 6, 120; ranking of human races, *119*; Sowerby agreement with, 38; species as "a primordial organic form," 102–3
Mosasaurus, 104, 117
Mount Desert Island, Maine (McEntee), painting, 173, *174*
Mt. Auburn Cemetery, Agassiz's tombstone, 159, *160*
Muir, John: on "glacial manuscripts of God," 146; landscape vision of, 176
Murchison, Roderick Impey: on Earth's antiquity, 185; portrait, *54*; Silurian System, 53, 54–55; *Siluria*, 55; *The Silurian System*, 52–53
Museum of Modern Art (MoMA), New York, 255

Napoleon, 103
Nation (magazine), 249
Nationalist (journal), 228
National Land Company, 126
Native American(s): Alabama, 108–9; at Badlands, *124*; calendars, 140–42, *142*, 144; deep time and, 7, 136, 140–42, 144; Lake Superior, 165; Pacific Northwest, 175–76; Plains Indians, 122, 125, 140; removal and genocide, 5; skulls of, 69–70, 109, *110*, 119; term, 144
Native peoples. *See* Native American(s)

natural resource(s), term, 8, 62
natural selection, Darwin's theory of evolution, 3, 10, 59, 222; and Christian certainty, 220; Hitchcock opposition to, 81; Hodge opposition to, 188; Huxley support of, 135; and Young Earth Creationists, 261–62
Neanderthal Flintworkers (Knight), painting, 235, 236–37
Nebraska Man, Osborn, 242–43, 245
Netherlands, 105
New Geology, The (Price), 266, 267
New Jersey: Cretaceous fossils in scene, 205–8, 208–9; marl pit workers, 97; paleontologists excavate, 97–98, 103–4; synchronous with Europe, 105–6, 112, 114, 117–18
New Philosophical Journal, 150
New World, 4; contrasted with Old World, 5–6; inferior to Old World, 16–18; meanings of term, 14–15
Niagara Falls, 195, 197
Noah's ark: Creation Museum, 259; story of, 32, 260
Noah's Ark (Hicks), 260, 261
Noah's flood. *See* Flood, biblical
North, The (Church), painting, 169, 170
North America: deepening of time in, 7; geological map of, 28, 29; Ice Age of, 175–77, 179
North American continent, glaciers of the North, 167–69, 171–73
Nova Scotia, 61
Nutt, Rush, 73, 108–9, 115

Oedipus complex, 246, 249
Ohio, 64; coal and, 68–75; Grotto of Plants, 73, 74; naturalists in, 69
Old Pennsylvania State House, 40
Old Red Sandstone, 52
Old Red Sandstone, The (Miller), 86, 197, 198
Old World: cultural difference with new, 5–6; 244; inferior to New, 43; superior to New, 16–18, 19

omnipotence, 12
origin, shifting meaning of, 59
Origin of Species, The (Darwin), 59, 131, 183, 185, 205
orthogenesis, 133
Osborn, Henry Fairfield: American Museum of Natural History (AMNH), 230–32, 253; aspirations, 9; causes and duration of extinctions, 130; "Family of Man," 241, 244; family tree of man, 241; head of Cro-Magnon Man, 233; *Hesperopithecus*, 241, 245; *Men of the Old Stone Age*, 232; Nebraska Man, 242–43, 245; "Old World culture," 244; paleontologist, 125
Osborn, Lucretia Perry: on aboriginal Americans, 243; *The Chain of Life*, 235; development of brain, 241
Ottoman Empire, 92
Outlines of Modern Christianity and Modern Science (Price), 264
Overland Monthly (journal), 175
Owen, David Dale: deep time and cognition, 10; on former existence of most remarkable races, 123–24, 124; geology as most "sublime" of natural sciences, 264; on lines of separation, 130; on trilobite discoveries, 55
Owen, Richard: on American fossil yields, 125; anti-transformist, 184, 190–91; on *Dinosauria* category, 205; on parallels between German Trias and British formations, 199

Pacific Northwest, Native peoples of, 175–76
paleobotany, term, 223, 227–28
Paleontological Chart: Carboniferous period, 84–85; Hitchcock's *Elementary Geology*, 81, 82–83, 84
paleontologist(s): international recognition of Morton as, 285n11; Philadelphia, 96–108; term, 96, 284n5

paleontology, 181
Paleozoic era, 266
"Panorama from Point Sublime" (Holmes), 143, *144*
Panorama of the Monumental Grandeur of the Mississippi Valley (Egan), painting, 115, *116*
Parkman, Francis, 165
Passing of the Great Race, The (Grant), 233
Peabody Museum (Yale), Lecture Room in, *136*
Peale, Charles Willson: on American nature, 19; *George Washington at the Battle of Princeton*, 217; museum, 97; natural history museum, 40; self-portrait, *22, 23*
Peking Man, 219
Pennsylvania coal seam, *Travels in North America* (Lyell), *87*
Pennsylvanian period, term, 9
Perkiomen lead mine, 98
Permian period, term, 9
Pestalozzi, Johann Heinrich, 185, 191
Peter Parley's Wonders of the Earth, Sea and Sky (Goodrich), *204*
Phelps, John Jay, 86
Philadelphia, paleontologists of, 96–108
Philadelphia's Centennial Exhibition, 89, *90*
Pilgrims, 145
Piltdown Man, 231
Pleistocene epoch, 209–10, 288n14; term, 146
Pleistocene climate, 232
Pleistocene Fauna of Asia (Hawkins), 210–11, *211*
Plesiosaurus, 199, 202, 208
Pliocene epoch, 209, 214, 241, 288n14
Plymouth Rock, 145
polygenesis: Agassiz's view of species fixity, 162; Morton on, 118, 120; Morton promoting doctrine of, 99–100
Popular Science Monthly (journal), 56, 175, 192, 235
Powell, John Wesley, 223
prehistoric, term, 212

prehistoric art: *Cro-Magnon Artists of Southern France* (Knight), 238, 240, *240*, 242–43; low-browed Neanderthal hunters, 236–37; men without chairs, 230–35, 238, 240–44
Prehistoric Park, *191*
Prehistoric Rock Pictures in Europe and Africa, Museum of Modern Art (MoMA) exhibition, 255, *256*
Pre-historic Times (Lubbock), 212
prehistory, term, 181, 212
Price, George McCready: *The New Geology*, 266, *267*; *Outlines of Modern Christianity and Modern Science*, 264; on Young Earth Creation, 264, 265
Primitive Culture (Tylor), 212
Princeton Theological Seminary, 188, 263
Princeton University, 180–82; E. M. Museum, 215–16, *216*; Guyot's wall hanging, *187*; Hawkins coming to, 190–93; reform at, 182–89
Principles of Geology (Lyell), 26, 61, 66, 75, 76, 148
Principles of Psychology, The (James), 247
Principles of Psychology, The (Spencer), 140
Principles of Zoology (Agassiz and Gould), 161–62, *163*
Protestant Reformation, 15
Psychic Factors of Civilization (Ward), 228
Psychology of the Unconscious (Hinkle), 248
P. T. Barnum's circus, 186
Pterodactyl, 127, 128–29, 204, 205, 207, *207*, 208–9, 258, 259, 270
Pterosaur. *See* Pterodactyl

Radiata, archetype of life, 162, *163, 164*, 195
radiometric dating, 3, 5, 267
Recherches sur les Poissons Fossiles (Investigations of fossil fish) (Agassiz), 152
Re-Creating of the Individual, The (Hinkle), 248–49, 252
Regent's Park, London, 89
Reign of Fish, 165; Agassiz on, 152

Reign of Man, 164
relative dating, 3–4
Rensselaer School, 30, 31
Revolutions of 1848, 184
Rhabododon priscum ("ancient fluted-tooth"), 224
Ritter, Carl, 184
roches moutonnées (polished stones), 153, 289n16
rock portraits, Boutelle's painting, 27
Rocks at Nahant (Haseltine), painting, 171–72, *172*
Rocky Mountains, 20–21
Rogers, Henry: 85; *The Geology of Pennsylvania*, 86
Romanes, George, *Mental Evolution in Man*, 247
Ross, E. A., 238
Rousseau, Jean-Jacques, 185
Royal Academy of Arts, London, 42
Royal Geographical Society, 53
Rudwick, Martin, 84

Saber-toothed cat, 5, *132*
Saharanpur Botanical Garden, 211
St. Hilaire, Étienne Geoffroy, 102
Saporta, Gaston de: *Le Monde des plantes avant l'apparition de l'homme*, 227; Ward and, 223–27
Say, Thomas, 96
Science (journal), 127, 241
Scientific Revolution, 24
Scofield Reference Bible, 263
Scopes Monkey Trial (1925), 219
Scott, William Berryman, 125; *A History of Land Mammals in the Western Hemisphere*, 134; paleontologist, 125
Scribner's Magazine, 240
Second Treatise of Government (Locke), 18
sensorium, 11, 47
short chronology. *See* Young Earth Creationism
Sigillaria, 197, *198*

Silliman, Benjamin: on America's "Gigantic vegetables," 62; on antiquity and New World, 19; on Genesis and geology, 47; glacialist, 148, 172; as Hildreth's fossil plant agent, 70; Hitchcock and, 75; iceberg theory, 148, 150; Lowell Lectures, 153; Morse painting of, *149*, 158; on Mosasaurus fossil, 104; on Ohio coal formations, 69; persona of, 161; supplier of American coal samples, 71; touring with Brongniart, 42; on trilobite casts, 45; on trilobite samples, 37, 38
Siluria (Murchison), 55
Silurian, term, 52
Silurian system, trilobites, 50–55
Silurian System, The (Murchison), 52–53
Sivatherium giganteum (four-horned mammal), 210, *211*
Sketch of Paleobotany (Ward), 223
skulls. *See* craniology
slavery, American, 92–96
Small Experimental Machine (Klee), 257
smallpox, 40
Smith, William "Strata," 24
Smithsonian Institution, 140
Société Helvétique des Sciences Naturelles, 150
sociology, 227–28
Solutrean, 231
Souls of Black Folk, The (Du Bois), 95
Southern Rhodesia (Zimbabwe), photograph of rock painting, 255, *256*
Sowerby, James de Carle, 38
Spencer, Herbert: *Principles of Psychology*, 140; survival of fittest, 222
Spirit of Laws, The (Montesquieu), 147
Spiritual Gifts (White), 264
Squier, Ephraim George, *Ancient Monuments of the Mississippi Valley*, 70
Statistics of Coal (Taylor), 67
Steno, Nicolaus, on law of superposition, 3
Stone Age, 187, 253

Story of the Earth and Man, The (Dawson), 57, 58
strata/stratum: Conrad's sketch of Alabama River's, *113*; Eaton's sketch of Genesee Falls, *25*; fossils buried in, 76; governing metaphor of deep time, 24, 26; Hitchcock's sketch of Massachusetts, *26*; rocks labeled "Graywacke without petrifaction," 34, *35*
Subsiding of the Waters of the Deluge, The (Cole), 260
Summer Landscape Near New Haven (Durrie), painting, 157–58, *158*
Sumner, William Graham, 222
Sun King, 43
surrealism, 258
surrealists, 255
Swiss lake dwellings, 188, 291–92n14
Synchronological Chart of Universal History (Adams), 189, *190*
synchronous, term, 8, 117
Synopsis of the Organic Remains of the Cretaceous Group of the United States (Morton), 117–18

Tait, Charles, 108–9, 11–12
Taylor, Richard, *Statistics of Coal*, 67
Teleosaurs, 205
telesis, 222
Theory of the Earth (Burnet), 17
Three Ages (film), 253
Ticknor, George, 212
Titanotherium (Titanothere), 137; as antiquity's bison, 137–39; Knight's posing of, 139; North American, *138*; rhinoceros-like, 121
Tlingit peoples, 175–76
tracks, dinosaur, 1, 77, *78*
transmutation of species: Darwinian, 188; Lamarckian, 102–3, 154, 191, 205; opposition to, 102–3, 150, 154, 188, 190–91, *191*, 197, 205, 206–7

Trenton Falls Near Utica, New York (Boutelle), 27
Triassic period, 198–99, *200–201*
Triassic Life of Germany (Hawkins), 199, *200–201*
trilobites, 30–33, 221; Alexandre Brongniart's image of, *36*; American, 33–38; divine message of, 47; drawings of American, *31*; as fossil messengers, 38–47, 50; plaster casts of, 38, *39*; sensorium of, 11; Silurian system, 50–55
Tylor, E. B., *Primitive Culture*, 212
Typical Forms and Special Ends in Creation (McCosh), 184

Uinta Mountains, 132
Uintatherium, 132, *134*, 221
uniformitarianism: ending Age of Reptiles, 130–31; Hildreth and, 75; Hitchcock and, 76, 80; iceberg theory and, 148; idea of multiple ice ages and 177; Lyell and 67–68
Union Pacific Rail Road, 122
United States: Canadian Shield, 29; Declaration of Independence, 18; deep time in, 273; deep time revolution in, 5; dinosaur diplomacy, 5; first geological map of, *21*; geological map of North America, *28*, 29; independence, 22–23; Lyell on visit to, 64; nationalism, 19; organizing natural history, 7–8; political independence, 18–19; political maps of, 118; rock formations, 8
United States Geological Survey, 8, 175, 223, 266
University of California, 59
University of Georgia, colored ceiling fresco at, 291n1
University of Michigan, 86, 207
University of Pennsylvania, 111
University of Wooster, 56, 57
Up from Slavery (Washington), 94
Ussher, James, 15, 47, 189, 212, 263

Valley of the Yosemite (Bierstadt), painting, 177, *178*
Vanuxem, Lardner, 97–98, 99
Vespucci, Amerigo, 14
virtual reality, 273

Wandlungen und Symbole der Libido (Jung), 248
Ward, Henry Augustus, 186
Ward, Lester Frank: aspirations of, 9; *Dynamic Sociology*, 222; on fossil plants, 222–23, 224, 225–30; on Gilman, 229–30; Hinkle citing, 250; prehistoric plants, 224; Saporta and, 223–27; *Sketch of Paleobotany*, 223; sociological theories of, 248
Washington, Booker T., *Up from Slavery*, 94
Washington, George, portrait of, 215, *216*, 217
Webb, William E., *Buffalo Land*, 126, 128–29
Webster, Daniel, 19
Wenlock Edge, 60
Western Civilization, 220
"We the People," 221
What Is Darwinism? (Hodge), 188–89
Whitcomb, John C., *The Genesis Flood*, 266
White, Ellen, *Spiritual Gifts*, 264

Whitney, Josiah, 176–77
Willard, Emma, 47, 48–49
Wilson, Woodrow, 19, 182, 216, 218
Winchell, Alexander, 86, 207
Wissler, Clark, 232
Witham, Henry, *The Internal Structure of Fossil Vegetables*, 71, *72*
women, as fossil collectors, 6–7
Women and Economics (Gilman), 230
Woodbridge, William, isothermal chart, 168, *168*
Worrall, Henry, artist, 126–27
Wright, Chauncey, 59

Yellowstone, 1, 223
Yosemite, 5, 146, 176–77
Youmans, E. L., 191–92
Young Earth Creationism, 266–69
Young Earth Creationists, 258–60, 264–65, 270

Zallinger, Rudolph: mural *The Age of Reptiles*, 5, *6*, 130
Zimbabwe, photograph of rock painting, 255, *256*

A NOTE ON THE TYPE

This book has been composed in Arno, an Old-style serif typeface in the classic Venetian tradition, designed by Robert Slimbach at Adobe.